Natural Computing Series

Series Editors: G. Rozenberg
Th. Bäck A.E. Eiben J.N. Kok H.P. Spaink
Leiden Center for Natural Computing

Philip F. Hingston · Luigi C. Barone ·
Zbigniew Michalewicz (Eds.)

Design by Evolution

Advances in Evolutionary Design

Foreword by David B. Fogel

 Springer

Philip F. Hingston
School of Computer and Information
Sciences
Edith Cowan University
2 Bradford St
Mt. Lawley, WA 6020, Australia
p.hingston@ecu.edu.au

Zbigniew Michalewicz
School of Computer Science
University of Adelaide
Adelaide, SA 5005, Australia
zbyszek@cs.adelaide.edu.au

Luigi C. Barone
School of Computer Science and
Software Engineering
The University of Western Australia
35 Stirling Highway
Crawley, WA 6009, Australia
luigi@csse.uwa.edu.au

ISBN: 978-3-540-74109-1   e-ISBN: 978-3-540-74111-4

Library of Congress Control Number: 2008932894

ACM Computing Classification (1998): I.2, F.1, J.2, J.3, J.5

© 2008 Springer-Verlag Berlin Heidelberg

Cover design: KuenkelLopka GmbH

Printed on acid-free paper

9 8 7 6 5 4 3 2 1

springer.com

Foreword

When we think of design, it is usually in the context of solving some sort of problem, such as the design of a bridge, a city, or a song. To be effective, the design must address a purpose to be achieved. In the case of a bridge or a city, the parameters of interest can be quantified. In the case of a bridge, they might include facets such as structural integrity, maximum carrying capacity, and cost, or in the case of a city, traffic flow, resource utilization, safety, and also again cost. In the case of a song, things are not as clear. The time-honored saying of beauty residing in the eye (or in this case, the ear) of the beholder remains true. Yet, a person can at least offer guidance about whether or not a song is to his or her liking, or perhaps more enjoyable than some other song. Thus, effective design requires some feedback mechanism to the designer.

As has been argued many times for more than 150 years, evolution, a design process that is ancient to the extreme, serves as a designer – the blind watchmaker – by using the process of random variation and natural selection, iterated over generations. The concept of design is usually connected with intelligence, and yet evolution – a simple process resulting from reproducing organisms with variation competing for finite resources – can itself efficiently design wondrous creatures, each of which finds different solutions to its primary problem: the problem of survival. It is natural to look to evolution for inspiration on how to create algorithms that can design solutions to our own challenging problems.

It should be no surprise then that computer scientists, engineers, mathematicians, roboticists, and biologists alike have made considerable efforts to program evolution into algorithms. Such efforts now span at least 55 years, dating back to Nils Barricelli's pioneering work on artificial life algorithms at von Neumann's laboratory at Princeton University in 1953. Even in the initial phases of this body of research into what is now called evolutionary computation there were examples of evolutionary algorithms used to create artificial intelligence, design mechanical devices, discover optimal strategies in games, and many other tasks that require the act of designing a solution to a problem.

Today, with computers that are over 10 billion times faster than the computers of the early 1950s, the range of design problems that can be explored with this approach is much broader. *Design by Evolution – Advances in Evolutionary Design* highlights some recent contributions in this field in the areas of biology, art, embryogeny, and engineering. You will find the chapters to be timely, sometimes controversial, and always thought-provoking.

Design and creativity are strongly connected concepts, but this is not a novel recognition. Over 40 years ago, Fogel et al. [2] argued that the process of evolution is analogous to the process of the scientific method, a process that can be simulated as an evolutionary program, in which "the process of induction [is] reduced to a routine procedure. If 'creativity' and 'imagination' are requisite attributes of this process, then these too have been realized." Even now, this concept may seem provocative to some. It was more than provocative in the 1960s, eliciting dismissive comments that such ideas were nothing more than "fustian [that] may unfortunately alienate many ... from the important work being done in artificial intelligence" [3]. Yet, what words of attribution then would be appropriate for an antenna designed for a space application by an evolutionary algorithm [4] that featured a brand new configuration and comments from traditional antenna designers of "we were afraid this was going to happen"? What words of attribution would be appropriate for an evolutionary program – *Blondie24* – that taught itself to play checkers at a level commensurate with human experts without using human expertise about how to play checkers [1], when human opponents routinely praised the program's "good moves" and even observed that the program was "clamping down on mobility"? *Blondie24* had no preprogrammed concept of mobility. Evidently, it created the concept of mobility.

Computers can become creative designers by using evolutionary processes. Random variation is a key component of this creativity, for random variation is what provides the element of surprise that can generate something truly novel, something beyond just shifting and sifting through combinations of pieces of existing solutions. It is ironic, perhaps, that so much of engineering is aimed at removing noise and yet noise is an integral part of the ingenuity that engineering requires!

Evolution is a creative process in every sense of the word creative. I expect that you will find the contributions in this book to provide a current accounting of the potential for using an evolutionary paradigm for creating novel designs in the four main topic areas discussed. I also expect that you will find these contributions creative in their own right, and illustrative of what we may expect in terms of automating the process of design in the future.

San Diego,
August 2007

David B. Fogel

References

1. Fogel, D.: Blondie24: Playing at the Edge of AI. Morgan Kaufmann, San Francisco, CA (2001)
2. Fogel, L., Owens, A., Walsh, M.: Artificial Intelligence through Simulated Evolution. John Wiley (1966)
3. Lindsay, R.: Artificial evolution of intelligence. Contemporary Psychology **13**(3), 113–116 (1968)
4. Lohn, J., Linden, D., Hornby, G., Kraus, B., Rodriguez, A., Seufert, S.: Evolutionary design of an X-band antenna for NASA's Space Technology 5 Mission. In: Proc. 2004 IEEE Antenna and Propagation Society International Symposium and USNC/URSI National Radio Science Meeting, vol. 3, pp. 2313–2316 (2004)

Preface

It is often stated that the 19th century was the century of chemistry, the 20th century was the century of physics and that the 21st century will be the century of biology. We wonder just who first said this. Perhaps Polykarp Kusch should be credited with the phrase "century of physics", from his 1955 acceptance speech for the Nobel Prize in Physics (though the quote was actually "We live, I think, in the century of science and, perhaps, even in the century of physics" [2]. It may have been US President Bill Clinton who first made the observation and coined the phrase "century of biology" [1], or maybe Kenneth Shine, President of the Institute of Medicine, who actually said "... the 20th century will be known as the century of physics and astronomy ... the 21st century will be the century of the life sciences in all their ramifications" [3]. Many people seem to have appropriated the idea and offer it as their own. But, whoever said it first, it is widely accepted now that advances in biology will lead to many great marvels in our lifetimes. One of the keys to this progress is our new and ever improving understanding of genetics and the processes of natural evolution. And much of this new understanding is driven by computation, which is where we, as computer scientists, come in.

Our own interests are in Artificial Intelligence (AI) – the attempt to model and understand intelligence through computational models. Evolutionary Computation (EC) is one way to do this using models based on the principles of natural evolution. We feel very fortunate to be working in EC, which allows us to indulge our fascination with both AI and evolution. This work has taken us into evolutionary design: two of us on evolving designs for ore processing plants, and the other on a wide range of real world applications. This is what motivated us to organise a special session on evolutionary design at the IEEE World Congress on Evolutionary Computation in Edinburgh in 2005 (CEC'05), and, partly as a response to the interesting ideas presented at that session, to embark on collecting together the contributions that make up this book. Along the way, we have realised that there are deep and beautiful connections between work done by scientists, engineers and artists, each with their own specific knowledge and culture, all striving to understand evolution as a design process or as a tool.

We would like to thank all our authors, our reviewers, Lucas Bradstreet and Mark Wittkamp, for their help in proof-reading the book, and, most especially, our area leaders for biology (Kay Wiese), art (Jon McCormack), embryogeny (Daniel Ashlock), and engineering (Kalyanmoy Deb), whose creative suggestions and diligence have made an immense contribution to the quality of the final result.

In this book, the reader will find chapters drawing on many diverse, yet interconnected topics. There are chapters on creating evolved art using EC along with L-systems, cellular automata and Mandelbrot sets, making images of biological life or beguiling abstract patterns. Chapters about simulations of biological processes that produce images so beautiful they must be called art. Chapters describing our progress from understanding how morphogenesis works in biology, to using such processes to create art or other physical artefacts, towards designing systems to control the process of morphogenesis itself. Chapters about using what we have learned to solve practical engineering design problems and even some more philosophical discussions about what this all means and how it relates to our place in the universe.

Whatever the reader's penchant, we hope there is something new here for everyone, that will educate, inspire, and delight.

Philip Hingston
Luigi Barone
Zbigniew Michalewicz

References

1. Clinton, W.: Remarks at the Morgan State University commencement ceremony (1997). http://avac.org/pdf/hvad_clinton_speech.pdf
2. Kusch, P.: Nobel banquet speech (1955). http://nobelprize.org/nobel_prizes/physics/laureates/1955/kusch-speech.html
3. Shine, K.: Welcome. In: Serving Science and Society Into the New Millennium. The National Academies Press (1998)

Contents

Part I

Biology

Evolutionary Design in Biology

Kay C. Wiese

Simon Fraser University, BC, Canada wiese@cs.sfu.ca

Much progress has been achieved in recent years in molecular biology and genetics. The sheer volume of data in the form of biological sequences has been enormous and efficient methods for dealing with these huge amounts of data are needed. In addition, the data alone does not provide information on the workings of biological systems; hence much research effort has focused on designing mathematical and computational models to address problems from molecular biology. Often, the terms bioinformatics and computational biology are used to refer to the research fields concerning themselves with designing solutions to molecular problems in biology. However, there is a slight distinction between bioinformatics and computational biology: the former is concerned with managing the enormous amounts of biological data and extracting information from it, while the latter is more concerned with the design and development of new algorithms to address problems such as protein or RNA folding. However, the boundary is blurry, and there is no consistent usage of the terms. We will use the term bioinformatics to encompass both fields. To cover all areas of research in bioinformatics is beyond the scope of this section and we refer the interested reader to [2] for a general introduction. A large part of what bioinformatics is concerned about is evolution and function of biological systems on a molecular level. Evolutionary computation and evolutionary design are concerned with developing computational systems that "mimic" certain aspects of natural evolution (mutation, crossover, selection, fitness). Much of the inner workings of natural evolutionary systems have been copied, sometimes in modified format into evolutionary computation systems. Artificial neural networks mimic the functioning of simple brain cell clusters. Fuzzy systems are concerned with the "fuzzyness" in decision making, similar to a human expert. These three computational paradigms fall into the category of computational intelligence (CI). While biological systems have helped to develop many of the computational paradigms in CI, CI is now returning the favor to help solve some of the most challenging biological mysteries itself. In many cases these probabilistic methods can produce biologically relevant results where exact deterministic methods fail. For an extensive overview of successful applications of CI algorithms to problems in bioinformatics please refer to [1].

The work presented in this section covers four chapters.

The first chapter by Tom English and Garrison Greenwood covers a discussion of intelligent design (ID) and evolutionary computation. The proponents of intelligent design try to establish that the complexities inherent in many biological systems are such that there is no chance they may have "evolved". Rather, it is proposed that elements of design are evident, and hence these biological systems were designed, not evolved. The chapter investigates the general claims made by ID and shows the logical flaws in this reasoning on a number of examples. Also, it highlights how some of the mathematics and statistical reasoning is flawed. In addition, it provides examples of how evolutionary algorithms can evolve complex systems that according to ID would exhibit elements of design, but obviously this is not the case since the system evolved from a few very basic rules without human intelligence.

The second chapter by Jennifer Hallinan is a review chapter on the topic of synthetic biology. The differences and commonalities between genetic engineering and synthetic biology are explored. While genetic engineering is a reality today and involves the modification of genetic or biological systems, synthetic biology aims to construct an organism from first principles. A noteworthy example of a successful synthetic biology experiment is that of the creation of a poliovirus from the knowledge of its DNA sequence alone. After discussing some minimal requirements for the successful practice of synthetic biology the chapter explores the topic of biological networks and motifs. Understanding gene networks and gene expression is essential towards understanding how cellular systems work. The chapter advises us that a more data-driven approach to synthetic biology will prove useful and it explores the potential of evolutionary computation to inform the field of synthetic biology.

Chapter 3 by Shuhei Kimura discusses the inference of genetic networks with an evolutionary algorithm. Two inference methods, the problem decomposition approach and the cooperative co-evolution approach, are presented. Both of these methods are evaluated and it is demonstrated that they can infer genetic networks of dozens of genes. The analysis of actual DNA microarray data usually requires genetic networks of hundreds of genes. To address this Kimura proposes a method to combine the cooperative co-evolution approach with a clustering technique.

Chapter 4 by Christian Jacob discusses examples of massively parallel, decentralized information processing systems and explores how swarm intelligence can be used to build and investigate complex systems. The examples studied include gene regulation inside a bacterial cell, bacteria and cell population simulations including chemotaxis, cellular orchestration and army ant raiding, herd behavior (predator prey), and human behavior (sales table rush). All of these examples are beautifully illustrated with graphics that show the evolution of the systems or the swarm behavior. This is very useful for understanding these complex, dynamic systems. Some insightful comments on swarm intelligence, evolutionary computation, and the re-use of software libraries are provided as well.

These chapters provide an interesting mix of problems and methods to show the diversity of biological applications that can be solved with evolutionary techniques. While biological systems have inspired new computational paradigms such as evolutionary computation or swarm intelligence, these techniques can now be used to further our understanding of biological systems themselves. We hope that you will enjoy these chapters.

Kay C. Wiese
Biology Area Leader

References

1. Fogel, G., Corne, D., Pan, Y. (eds.): Computational Intelligence in Bioinformatics. IEEE Press, Piscataway, NJ (2007)
2. Jones, N., Pevzner, P.: An Introduction to Bioinformatics Algorithms. The MIT Press, Cambridge, Massachusetts (2004)

Intelligent Design and Evolutionary Computation

Thomas English[1] and Garrison W. Greenwood[2]

[1] Bounded Theoretics, Lubbock, Texas, USA Thom.English@gmail.com
[2] Portland State University, Portland, Oregon, USA greenwd@ece.pdx.edu

> We ought, it seems to me, to consider it likely that the formation of elementary living organisms, and the evolution of those organisms, are also governed by elementary properties of matter that we do not understand perfectly but whose existence we ought nevertheless admit.
>
> Émile Borel, *Probability and Certainty*

1.1 Introduction

In the United States, a succession of lost legal battles forced opponents of public education in evolution to downgrade their goals repeatedly. By the 1980s, evolution was ensconced in the biology curricula of public schools, and references to the creator of life were illegal. The question of the day was whether instruction in creation, without reference to the creator, as an alternative explanation of life violated the constitutional separation of church and state. In 1987, the U.S. Supreme Court decided that it did, and intelligent design (ID) rose from the ashes of creation science. ID may be seen as a downgraded form of creation. While the creation science movement sought to have biology students introduced to the notion that *creation* is evident in the complexity of living things, the ID movement sought to have students introduced to the notion that *design, intelligence*, and *purpose* are evident.[3] ID preserves everything in the notion of *creation* but the *making*.

Although intellectual endeavor is secondary to sociopolitical action in the ID movement, the objective here is to assess the intellectual component. Separating the two is not always possible. Sometimes ID advocates formulate their

---

[3] The ID movement, led by the Discovery Institute, has downgraded its goals, and presently does not advocate teaching ID in public schools. The Discovery Institute continues, however, to advocate teaching the shortcomings of evolutionary theory. It bears mention also that the ID movement now distinguishes biological ID and cosmological ID. Here we focus on biological ID.

ideas in ways that make sense only in light of their sociopolitical objectives. The main intellectual offering of ID is the *design inference*, an ostensibly scientific adaptation of the classical argument from design. While the classical argument might indicate that a natural entity is too complex to have arisen unless created by an intelligent and purposive agent, and that the agent could only be God, a design inference eschews creation and declines to identify the agent, concluding that a non-natural and purposive intelligence designed the natural entity.

The sociopolitical ingenuity of the design inference is that, if taught as an alternative to evolution in public-school science classes, it would leave identification of the designer to schoolchildren. The faithful would conclude that science supports belief in God the Designer and disbelief in evolutionary theory. Whether ID's evasion of direct reference to God eventually will pass judicial scrutiny in the U.S. is unknown. The design inference has a legal vulnerability arising from the fact that *non-natural* intelligence is *supernatural*, and the supernatural is clearly linked with the religious in case law [24, p. 67]. In recent years, the ID movement has shifted to saying that intelligence is natural, but not material. (For an example of earlier usage, see [23].) Given that scientists conventionally regard nature to be material, the ID movement has changed the meaning of *natural* to suit itself, and for no apparent reason but to gain better legal footing. Similarly, many ID advocates call themselves evolutionists, falling back on a dictionary meaning of the term (a process of change in a given direction), rather than scientists' conventional interpretation (an undirected process of change deriving from random variation of offspring and natural selection). This chapter will use conventional scientific terminology.

There are two basic approaches to design inference: the argument from *irreducible complexity*, and the argument from *specified complexity*. The argument from irreducible complexity is a demonstration that a biological system with several or more parts could not serve any useful function if any of its parts were removed. This in effect shows that the system cannot have evolved directly. The argument from specified complexity demonstrates that an entity matches some pattern recognized by an intelligent agent, and that it is improbable that any match of the pattern would arise by natural (materialistic) causes in the history of the universe.

The ID movement has for some years considered evolutionary computation (EC) a threat. EC is the proving ground of ideas in evolutionary theory, according to William A. Dembski, billed by the ID movement as the "Isaac Newton of Information Theory" [13]. Positive results in EC apparently contradict Dembski's claims about "conservation of information" in chance-and-necessity processes. He and other ID advocates acknowledge this, and are at pains to show that the contradiction is only apparent [13]. Thus EC investigators have a potent means of challenging ID. In fact, the main biological claims of ID advocates such as Michael Behe [3] and Steven Meyer [28] are that evolution cannot account for certain biological innovations that occurred hundreds of millions of years ago, and it is easier to challenge the information-theoretic claims of Dembski with simulation and analysis than it is to challenge the specific biological claims of other ID advocates with new data on ancient events.

Research in artificial life is closely related to that in EC, and in the present context the artificial life program Avida will be considered an example of EC. Research with Avida has been reported in *Nature* [25], and it apparently contradicts the central claim Behe makes in *Darwin's Black Box* [3], namely that gradual evolutionary processes cannot generate irreducible complexity (defined below). In a brief intended for submission to a federal judge, Dembski does not deny that Avida generated irreducible complexity, but instead argues that it lacks biological relevance [16, p. 19].

1.2 Historical Background

The current arguments in favor of ID follow a common theme. In one way or another they all state that the universe we observe is so complex that it simply could not have developed by mere chance; some intelligent agent had to have been responsible. Today ID is the primary opponent of Darwin's theories of evolution and disputes often become heated.

Is this controversy – some would call it a war – of competing ideas something new? Surprisingly, no. In fact, this war began centuries ago and the battles conducted today are not all that different from those waged in the past. It is therefore instructive to see exactly what the arguments were in favor of creationism then, and how they compare to the arguments being made by the ID community today. That is the purpose of this section.[4]

Darwin (1809–1882) was among the first to offer a plausible explanation of nature's development [9]. His theory was accompanied by a large amount of empirical evidence, which strongly contributed to its initial acceptance. In spite of its critics, Darwin's theory was largely responsible for the demise of creationism in the late 19th century. It therefore seems somewhat ironic that the tables have turned – today ID is the major threat to the teaching of evolution!

The earliest ID arguments were overtly religious with no attempts to hide the intelligent designer's identity. Archbishop James Ussher (1581–1656) is best known for declaring the world was created in late October 4004 BC. [43].

> In the beginning God created Heaven and Earth, Gen. I. V. I. Which beginning of time, according to our Chronologie, fell upon the entrance of the night proceeding the twenty third day of October, in the year of the Julian Calendar, 710 [4004 BC].

---

[4] Some quotes appearing in this section have unusual grammar, spelling and punctuation. They were purposely kept that way to preserve the original wording of papers written in the 17th through 19th centuries.

Others interjected anatomy into the debate while still acknowledging God as the designer. John Ray (1627–1705) used the eye [37]:

> For first, Seeing, for instance, That the Eye is employed by Man and Animals for the sue of Vision, which, as they are framed, is so necessary for them, that they could not live without it; and God Almighty knew that it would be so; and seeing it is so admirably fitted and adapted this use, that all the Wit and Art of men and Angels could not have contrived it better is so well; it must needs be highly absurd and unreasonable to affirm, either that it was not designed at all for this use, or that it is impossible for man to know whether it was or not.

Of course early arguments were not restricted to just human anatomy. William Paley (1743–1805) compared the eye of a fish, which must process light refracted by water, with the eye of land animals, which must process light passing through air [35]. He believed such a subtle physical difference, while still performing the same function, was conclusive proof of a designer.

> Accordingly we find that the eye of a fish... is much rounder than the eye of terrestrial animals. What plainer manifestation of design can there be than this difference?

Remarkably the eye example is still used today, although the arguments are now more sophisticated with decades of biochemical research for support. Michael Behe attempts to discredit Darwin in the following way [4]:

> Neither of Darwin's black boxes – the origin of life or the origin of vision (or other complex biochemical systems) – has been accounted for by his theory.

Many early ID arguments were philosophical and had no real scientific content. This is not unexpected because science had made little progress by today's standards and the general populace had no background in it anyway. Perhaps the most famous argument made along this line is William Paley's story about finding a watch [35].

> In crossing a heath,[5] suppose I pitched my foot against a stone, and were asked how the stone came to be there; I might possibly answer that, for anything I knew to the contrary, it might have been there for ever: nor would it perhaps be very easy to show the absurdity of this answer. But suppose I had found a watch on the ground, and it should be inquired how the watch happened to be in that place; I should hardly think of the answer I had given before, that for anything I knew, the watch might have always been there. Yet why should not this answer serve for the watch as well as for the stone? ... when

---

[5] A tract of open wasteland.

we come to inspect the watch, we perceive ... that the watch must
have had a maker: that there must have existed, at some time, and at
some place or other, and artificer or artificers, who formed it for the
purpose which we find it actually to answer; who comprehended its
construction, and designed its use.

It should come as no surprise that such a widely referenced anecdote from the
past would be the basis for a philosophical counter-argument of today [10].

All appearances to the contrary, the only watchmaker in nature is the
blind forces of physics, albeit deployed in a very special way. A true
watchmaker has foresight: he designs his cogs and springs, and plans
their interconnections, with a future purpose in mind. Natural selection, the blind, unconscious automatic process Darwin discovered...
has no purpose in mind. It has no vision, no foresight, no sight at all.
If it can be said to play the role of watchmaker in nature, it is the
blind watchmaker.

Prior to the 20th century science had not progressed to a point where it could
provide much support to the proponents of evolution. Nevertheless, that did
not prevent some from disparaging scientific arguments anyway. For instance,
John Ray even went so far as to say scientific principles are meaningless without belief in a designer [37].

In particular I am difficult to believe, that the Bodies of Animals can
be formed by Matter divided and moved by that Laws you will or can
imagine, without the immediate Presidency, Direction and Regulation
of some Intelligent Being.

The last quote from Ray also demonstrates a new and significant change
in the creationist's arguments. Up to this point there were no misgivings
about identifying who the intelligent designer was: it was God from the Old
Testament of the Bible. Notice in Ray's quote there was no specific reference
to a Judeo-Christian God. Instead, the designer's identity was purposely kept
vague. The ID community today is very careful to avoid any references to a
Judeo-Christian deity. William Dembski, one of the leaders in the present day
ID movement, recently put it this way [12]:

Intelligent design is theologically minimalist. It detects intelligence
without speculating about the nature of the intelligence.

Another change in tactics along this same line was to make secular arguments
based on what a reasonable man would (or at least should) believe. What gave
some of these new ID arguments credibility was highly respected scientists
were making them and not theologians. For instance, the renowned naturalist
Louis Agassiz (1807–1873) took believers of evolution to task [1].

> The most advanced Darwinians seem reluctant to acknowledge the intervention of an intellectual power in the diversity which obtains in nature, under that plea that such an admission implies distinct creative acts for every species. What of it, if it were true?

The goal here was to put the Darwinist on the defensive. The argument went something like this: since there was no irrefutable proof that evolution can explain how the world developed, then any "reasonable person" should be open to alternative ideas – and of course creationism just happens to be one of those alternative ideas. These arguments were meant to intimidate, if not outright mock, those who wouldn't seriously consider creationist ideas. For example, in [1] Agassiz also said

> Have those who object to repeated acts of creation ever considered that no progress can be made in knowledge without repeated acts of thinking? And what are thoughts but specific acts of the mind? Why should it then be unscientific to infer that the facts of nature are the result of a similar process, since there is no evidence of any other cause? The world has arisen in some way or other.

This tactic of mocking unbelievers is still used today. For example, in 1990 Phillip Johnson (who was a lawyer and not a scientist) wrote the following [23]:

> What the science educators propose to teach as evolution and label as fact, is based not upon any incontrovertible empirical evidence, but upon a highly controversial philosophical presupposition. The controversy over evolution is therefore not going to go away as people become better educated on the subject.

Until the late 19th century science and religion were strongly intertwined. It is therefore not unexpected that creationist thought permeated early discussions about the origin of life. Even Darwin was not immune and recognized the role of a Creator. In his treatise *On the Origin of the Species* he stated

> To my mind, it accords better with what we know of the laws impressed upon matter by the Creator, that the production and extinction of the past inhabitants of the world should have been due to secondary causes, like those determining the birth and death of the individual.

Nowadays science and religion are at odds with each other over the evolution issue. Without exaggeration, ID advocates seek to return science to the days of natural philosophy, when scientific beliefs were tested against higher religious truths. ID supporters talk about "design theory" as science while mainstream scientists counter that it is really no science at all. Unfortunately, this debate has moved to the public arena where emotions often trump objective dialog.

1.3 What Is Intelligent Design?

Advocates of ID use the term *intelligent design* to name both their field of inquiry and a putative cause of certain natural phenomena. They refer to their body of beliefs as *intelligent design theory*. Note that the sense of *theory* here is not *scientific theory*.

In ID theory, information is a physical primitive, like matter and energy, which may enter the natural universe from without.[6] An *intelligence* is a non-natural source of information – i.e., it changes probabilities of events in the natural universe. When an intelligence increases the probability of an event that is in some sense meaningful or functional, it is *goal-directed* or *telic*. The central thesis of ID is that some natural entities exhibit such complex organization that the processes giving rise to them cannot have been entirely natural, but instead must have been directed (informed) to some degree by telic intelligence. The type of organization of interest to ID theorists is known as *specified complexity* (or *complex specified information*). An entity with specified complexity higher than ID theory says could have arisen by purely natural processes is said to be *intelligently designed*. ID theorists consider irreducible complexity as an indicator of high specified complexity.

ID theory says something outside of nature may cause an event within nature. In contrast, mainstream scientists embrace *methodological naturalism*, the working assumption that all natural phenomena have natural causes. ID theory allows theists to associate the intelligent causation of humans with that of one or more deities. In particular, the Biblical notion that humans, as intelligent entities with free will, are created in the image of God, and thus stand apart from nature in some aspects, is supported by the philosophy of intelligent design.

Many ID sites on the Internet (e.g., [21]) offer the definition, attributed to Dembski, "Intelligent Design is the study of patterns in nature that are best explained as the result of intelligence." Although this is a casual definition, its shortcomings are worth examining. First, ID theory permits an event to be explained in terms of both non-natural intelligence and natural antecedents. Design is not all-or-nothing. Second, ID does not study the patterns per se, but *which* patterns indicate that intelligence has contributed information.

The following sections discuss the two main approaches to design inference: *argument from irreducible complexity* and *argument from specified complexity*.

1.4 Irreducible Complexity

The champion of irreducible complexity in the ID community is Michael Behe, a biochemist. He has given two definitions of the term. William Dembski, a

---

[6] Recall that ID advocates say that intelligence is natural but not material, and that mainstream science holds that anything natural is material.

mathematician and philosopher, followed with a related definition. The artificial life program Avida has posed a challenge to the claim that evolution cannot give rise to irreducible complexity.

1.4.1 Behe's Definitions

In *The Origin of Species*, Charles Darwin wrote, "If it could be demonstrated that any complex organ existed, which could not possibly have been formed by numerous, successive, slight modifications, my theory would absolutely break down" [9]. In *Darwin's Black Box*, biochemist and intelligent design advocate Michael Behe responds with the claim that some biological systems are irreducibly complex:

> By irreducibly complex I mean a single system which is composed of several interacting parts that contribute to the basic function, and where the removal of any one of the parts causes the system to effectively cease functioning. An irreducibly complex system cannot be produced directly (that is, by continuously improving the initial function, which continues to work by the same mechanism) by slight, successive modifications of a precursor system, because any precursor to an irreducibly complex system that is missing a part is by definition nonfunctional. [...] Even if a system is irreducibly complex (and thus cannot have been produced directly), however, one can not definitely rule out the possibility of an indirect, circuitous route. As the complexity of an interacting system increases, though, the likelihood of such an indirect route drops precipitously. [3, pp. 39–40].

Behe holds that irreducible complexity is evidence for intelligent design. There are two logical flaws here, however, both of which Behe has acknowledged [24]. First, to treat evidence against Darwinian gradualism as evidence for intelligent design is to set up a false dichotomy. For instance, there could be some biological structure for which both explanations are wrong. Second, a biological system that is irreducibly complex may have precursors that were not irreducibly complex.

To help understand this idea, think of an irreducibly complex biological structure as a stone arch. The argument from irreducible complexity is that all the stones could not have been put in place simultaneously by evolutionary processes, and that the arch must be the product of intelligent design. But this ignores the possibility that preexisting structures were used opportunistically as "scaffolding" in gradual assembly of the arch. If the scaffolding is removed by evolution after the arch is complete, then the arch is irreducibly complex, but arguments that it could not have emerged gradually are wrong.

Interestingly, H. J. Muller, a geneticist who went on to win a Nobel prize, held in 1918 that *interlocking complexity* (identical to irreducible complexity) arose through evolution:

Most present-day animals are the result of a long process of evolution, in which at least thousands of mutations must have taken place. Each new mutant in turn must have derived its survival value from the effect which it produced upon the "reaction system" that had been brought into being by the many previously formed factors in cooperation; thus a complicated machine was gradually built up whose effective working was dependent upon the interlocking action of very numerous different elementary parts or factors, and many of the characters and factors which, when new, were originally merely an asset finally became necessary because other necessary characters and factors had subsequently become changed so as to be dependent on the former. It must result, in consequence, that a dropping out of, or even a slight change in any one of these parts is very likely to disturb fatally the whole machinery... [32]

Muller worked out interlocking complexity in more detail in a 1939 paper [33]. According to Orr [34], "Muller gives reasons for thinking that genes which at first improved function will routinely become essential parts of a pathway. So the gradual evolution of irreducibly complex systems is not only possible, it's expected."

Parts of the pathway may also arise from neutral mutations and gene duplication. A duplicate gene is available for mutation into a gene that serves a different function from the original. The mutated duplicate may serve a function similar to the original, and may come to be required by the organism. For instance, the genes for myoglobin, which stores oxygen in muscles, and hemoglobin, which stores oxygen in blood, are closely related, and there is strong evidence that one or both arose through duplication. Both are necessary to humans [34]. Another way in which parts of the pathway may arise is through *co-optation* (also known as cooption), the adaptation of an existing biological system to serve a new function. The role of neutral mutations, gene duplication, and co-optation in the evolution of systems deemed irreducibly complex by Behe will be discussed in the following.

Behe's examples of irreducible complexity have generally not stood up to scrutiny. He gives considerable attention to biochemical cascades – in particular, the blood-clotting cascade and the complementary cascade of the immune system. Here we focus on blood clotting. Behe says of cascades [3, p. 87]:

Because of the nature of a cascade, a new protein would immediately have to be regulated. From the beginning, a new step in the cascade would require both a proenzyme and also an activating enzyme to switch on the proenzyme at the correct time and place. Since each step necessarily requires several parts, not only is the entire blood-clotting system irreducibly complex, but so is each step in the pathway.

To this Orr responds [34]:

> [Behe] even admits that some genes in his favorite pathway – blood clotting – are similar. But he refuses to draw the obvious conclusion: some genes are copies of others. [...] But this implies that such systems can arise step by step. Behe avoids this conclusion only by sheer evasion: he brands gene duplication a "hypothesis," leaves the similarity of his favorite genes unexplained...

Miller [30] argues that the gene for fibrinogen has as its ancestor a duplicate of a gene that had nothing to do with blood clotting. A genetic sequence similar to that of the fibrinogen gene has been identified in the sea cucumber, an echinoderm [45]. Furthermore, the blood clotting cascade is not irreducibly complex, because a major component, the Hageman factor, is missing in whales and dolphins [42], and three major components are missing in puffer fish [22].

Behe also introduced what has since become the most widely cited example of irreducible complexity, the bacterial flagellum. However, the base of the flagellum is structurally similar to the type-3 secretory system (TTSS) of some bacteria [6]. Furthermore, with 42 distinct proteins in the flagellum, and 25 in the TTSS, there are 10 homologous proteins in the two structures. This constitutes evidence that the TTSS was co-opted in evolution of the flagellum. But ID proponents contend "the other thirty proteins in the flagellar motor (that are not present in the TTSS) are unique to the motor and are not found in any other living system. From whence, then, were these protein parts co-opted?" [31]

A simple response to this challenge is that non-TTSS homologs have been identified for 17 of the 42 flagellar proteins, leaving only $42 - 10 - 17 = 15$ proteins without known homologs. A more subtle response is that only 20 proteins appear to be structurally indispensable to modern flagella (i.e., 22 are not), and only two of them have no known homologs [36]. Thus most proteins of the flagellum are not unique to the flagellum, and the notion that the structure arose through co-optation is at least plausible. That half of the flagellar proteins are not structurally necessary suggests the flagellum is not irreducibly complex, but this ignores issues in the evolution of regulation [36].

ID advocates insist that such indirect evidence of co-optation is insufficient. As Miller [29] has pointed out, demanding direct evidence of the evolution of biochemical systems has advantages for the ID movement:

> Behe demands that evolutionary biologists should tell us exactly "how" evolution can produce a complex biochemical system. This is a good strategic choice on his part, because the systems he cites, being common to most eukaryotic cells, are literally hundreds of millions of years old. And, being biochemical, they leave no fossils.

In contrast, ID advocates might emphasize that the system of ossicles (small bones transmitting sound from the tympanic membrane to the cochlea) in the middle ear is irreducibly complex if not for direct fossil evidence of its evolution from a reptilian jawbone [39].

Responding to critics of *Darwin's Black Box*, Behe [5] points out limitations in his original definition of irreducible complexity:

It focuses on already-completed systems, rather than on the process of trying to build a system, as natural selection would have to do. [...] What's more, the definition doesn't allow for degree of irreducible complexity [...] irreducible complexity could be better formulated in evolutionary terms by focusing on a proposed *pathway*, and on whether each step that would be necessary to build a certain system using that pathway was selected or unselected.

Here he acknowledges that neutral mutations (which he refers to as "unselected") can give rise to irreducible complexity. He observes "if a mutation is not selected, the probability of its being fixed in a population is independent of the probability of the next mutation." And this motivates his "evolutionary" definition of irreducible complexity: "An irreducibly complex evolutionary pathway is one that contains one or more unselected steps (that is, one or more necessary-but-unselected mutations). The degree of irreducible complexity is the number of unselected steps in the pathway" [5].

Behe relates the degree of irreducible complexity directly to the improbability that evolution followed the pathway. "If the improbability of the pathway exceeds the available probabilistic resources (roughly the number of organisms over the relevant time in the relevant phylogenetic branch) then Darwinism is deemed an unlikely explanation and intelligent design a likely one." There are two serious errors in logic here. First, there is the fallacy of the false dichotomy, with a forced choice between Darwinism and ID when a third alternative might explain the pathway. The improbability of one explanation in terms of natural causation does not lend credence to an explanation in terms of non-natural causation. Second, there is a mathematical fallacy long exploited by creationists. When an evolutionist specifies a particular evolutionary, Behe proceeds as though evolution could have taken no other path, and computes an absurdly low probability that the system arose by evolution [5]:

To get a flavor of the difficulties [my adversary's] scenario faces, note that standard population genetics says that the rate at which neutral mutations become fixed in the population is equal to the mutation rate. Although the neutral mutation rate is usually stated as about 10^{-6} per gene per generation, that is for any random mutation in the gene. When one is looking at particular mutations such as the duplication of a certain gene or the mutation of one certain amino acid residue in the duplicated gene, the mutation rate is likely about 10^{-10}. Thus the fixation of just one step in the population for the scenario would be expected to occur only once every ten billion generations. Yet [my adversary's] scenario postulates multiple such events.

A quantity more relevant to falsifying evolutionary theory is the probability that *no* evolutionary pathway arrives at the system. (However, even this is ad hoc.) Behe and others in the ID movement essentially take a divide-and-conquer approach, dispensing with evolutionary pathways individually rather than collectively to discredit evolutionary theory.

1.4.2 Dembski's Definition

William Dembski, better known in ID circles for his notion of specified complexity, claims to have generalized Behe's notion of irreducible complexity, but in fact has greatly restricted the class of irreducibly complex systems [15]. The salient point of his modification of Behe's definition of irreducible complexity is that "we need to establish that no simpler system achieves the same basic function." For instance, a three-legged stool is irreducibly complex for Behe, but not for Dembski, because a block serves the same function as the stool. The import of the "no simpler system" requirement is that evolution cannot obtain an irreducibly complex biological system through successive improvements of simpler precursors performing the "same basic function." That is, Dembski rules out *direct* evolution of irreducibly complex systems by definition. If a putatively irreducibly complex system turns out to have emerged by a direct evolutionary pathway, his ready response is that the system was not irreducibly complex in the first place.

Turning to *indirect* evolution of irreducibly complex systems, Dembski falls back on argument from ignorance [15]:

> Here the point at issue is no longer logical but empirical. The fact is that for irreducibly complex biochemical systems, no indirect Darwinian pathways are known. [...] What's needed is a seamless Darwinian account that's both detailed and testable of how subsystems undergoing coevolution could gradually transform into an irreducibly complex system. No such accounts are available or have so far been forthcoming.

Thus Dembski adopts Behe's tactic of limiting the domain of investigation to that of maximum biological ignorance, suggesting that evolutionary findings do not generalize to biochemical systems. Given that the biochemical systems of interest to ID advocates originated hundreds of millions of years ago and left no fossil traces, Dembski does not risk much in demanding seamless and detailed evolutionary accounts.

Why should our relative ignorance of the evolution of irreducibly complex biochemical systems lead us to believe something other than that we are ignorant? "[W]ithout the bias of speculative Darwinism coloring our conclusions, we are naturally inclined to see such irreducibly complex systems as the products of intelligent design." Dembski claims, in other words, that evolutionists have had their native perception of the truth educated out of them.

1.4.3 Evolution of Complexity in Avida

The artificial life program Avida has provided evidence that irreducible complexity can evolve [25]. In the Avida environment, a digital organism is a virtual computer with an assembly language program as its genotype. An organism must code to replicate itself in order to generate offspring. The only logical operation provided by the assembly language is NAND ("not and"). Multiple instructions are required to compute other logical functions. In [25], the fitness of an organism is the sum of fitness values for distinct logical functions it computes in its lifetime. Nine logical functions are associated with positive fitness – the greater the inherent "complexity" of computing the function, the greater the contribution to fitness. The logical function contributing the most to fitness is EQU ("equals"). In 50 runs with a population of 3600 organisms, 23 gave rise to EQU.

The focus of [25] is on the dominant genotype in the final population of a particular run giving rise to EQU. A step in the evolution of the genotype comes when an ancestor has a genotype different from that of its parent. The final dominant genotype, which computes all nine logical functions, has 83 instructions, and is 344 steps removed from its first ancestor, which had 50 instructions.

> The EQU function first appeared at step 111 (update 27,450). There were 103 single mutations, six double mutations, and two triple mutations among these steps. Forty-five of the steps increased overall fitness, 48 were neutral and 18 were deleterious relative to the immediate parent. [25]

The step giving rise to EQU was highly deleterious. Thus the "evolution of a complex feature, such as EQU, is not always an inexorably upward climb toward a fitness peak, but instead may involve sideways and even backward steps, some of which are important."

The evolved code includes a component that is irreducibly complex in the sense of Behe [3].

> The genome of the first EQU-performing organism had 60 instructions; eliminating any of 35 of them destroyed that function. Although the mutation of only one instruction produced this innovation when it originated, the EQU function evidently depends on many interacting components. [25]

The code is not irreducibly complex in the sense of Dembski [15], because it has been established that 19 Avida instructions suffice to compute EQU. However, in another run there was a genotype that computed EQU unless any of 17 instructions were eliminated. The researchers determined by inspection that there was redundant computation of some critical operations. Thus Dembski's stringent definition of irreducible complexity, which requires a near-minimalist implementation, appears to have been satisfied by one of the 50 runs.

Lenski et al. [25] realize that critics of the study will complain that they "'stacked the deck' by studying the evolution of a complex feature that could be built on simpler functions that were also useful." In fact, EQU did not emerge when all simpler logical functions were assigned zero fitness. They contend that this "is precisely what evolutionary theory requires." That is, evolutionary theory holds that complex features emerge through successive steps, not by saltation, and that intermediate forms persist in a population only if they imbue the individuals that possess them with some advantage.

1.5 Specified Complexity

For centuries, design advocates, though not the present-day ID advocates, have advanced the *argument from improbability* [40]. The approach is to show some event in nature is very unlikely to have occurred by chance, and to therefore conclude God caused it. But this argument is fallacious. For one thing, low probability does not necessarily justify rejection of chance. When numbers are drawn in a lottery, for instance, it is certain that the chance outcome will be one that was highly improbable *a priori* [40]. Another problem with the argument is that assigning an identity to the cause of the event is unwarranted [14].

William Dembski has developed an analogous *argument from specified complexity*, which concludes that a natural event reflects the intervention of intelligence [11, 13, 18]. That is, a natural event contains information that was introduced purposely by an unidentified source outside of nature. This statement is not quite the same thing as saying intelligence caused the event, because the event may have resulted from a combination of natural causes and intelligent intervention. For instance, one may argue that natural evolutionary mechanisms, while operating as claimed by mainstream scientists, do not account fully for life on earth, and that intelligence has guided (added information to) evolutionary processes. Note that information may have entered continually, and that there may have never have been a discrete design event.

To conclude that a natural event reflects intelligent design, one must demonstrate that the event is improbable. Dembski describes improbable events as *complex*. One must also demonstrate that the event is *specified* in the sense that it exhibits a pattern that exists independently of itself [11, 13]. To be more precise, there must exist some "semiotic agent" that describes the event with a sequence of signs [18]. Specified complexity, or *complex specified information* (CSI), is a quantity that "factors in" the improbability of the event and the cost of describing it. Dembski claims that when the CSI of an event exceeds a threshold value, inference that intelligence contributed to the event is warranted.

1.5.1 Design Inference as Statistical Hypothesis Testing

In unpublished work [18], Dembski has reduced the argument from specified complexity [11, 13] to statistical hypothesis testing. The approach is derived from that of Fisher, with a null (or *chance*) hypothesis possibly rejected in favor of an alternative hypothesis. The chance hypothesis says natural causes account entirely for an event in nature, and the alternative hypothesis says the event reflects design (contains information that could only have come from without nature). The Fisherian approach requires specification of the rejection region prior to sampling. But the argument from specified complexity entails selection of a past event and subsequent definition of a rejection region in terms of a pattern found in the event. Dembski claims to have corrected for "data dredging" and *a posteriori* fitting of the rejection region to the event by including *replicational resources* and *specificational resources* as penalty factors in a test statistic [18].

Dembski's argument goes something like this. Suppose $\mathbf{H}$ is the chance hypothesis, and let E be an event in the sample space of $\mathbf{H}$. For any pattern describable by semiotic agent S, there is a corresponding event T containing all matches of the pattern. Dembski uses T to denote both the event and the pattern [18]. The probability (under the chance hypothesis) of matching the pattern is

$$\mathbf{P}(T \mid \mathbf{H}).$$

T serves as a rejection region, and it is possible to make the probability low enough to ensure rejection by choosing a very specific pattern that matches few events, or perhaps no event but E. A penalty factor counters such a "rigged" selection of the pattern.

The *specificational resources* used by S in identifying pattern T are

$$\varphi_S(T),$$

which gives the rank-complexity of the semiotic agent's description of the pattern. In essence, the agent enumerates its pattern descriptions from less complex (e.g., shorter) to more complex (longer), looking for matches of E. The rank-complexity is the least index of a description of pattern T in the enumeration. It is a count of how many descriptions the agent processed to obtain the description of matching pattern T. Dembski [18] considers

$$\varphi_S(T) \cdot \mathbf{P}(T \mid \mathbf{H})$$

"an upper bound on the probability (with respect to the chance hypothesis H) for the chance occurrence of an event that matches any pattern whose descriptive complexity is no more than T and whose probability is no more than $\mathbf{P}(T \mid \mathbf{H})$" for a fixed agent S and a fixed event E. The negative logarithm of this quantity,

$$\sigma = -\log_2[\varphi_S(T) \cdot \mathbf{P}(T \mid \mathbf{H})] \text{ bits,}$$

is *specificity*, a type of information [18]. As the probability of matching the pattern goes down, specificity goes up. As the number of patterns "dredged" by the semiotic agent goes up, specificity goes down. Maximizing specificity – and ultimately inferring design in event E – is a matter of finding in the event a simple pattern that is matched with low probability under the chance hypothesis.

Not only is the pattern chosen to obtain high specificity, but the event E and the semiotic agent S, and another penalty is required. The number of *replicational resources* is bounded above by the product of the number of semiotic agents available and the number of events that might have been considered. In applications to biology, Dembski uses as an upper bound Seth Lloyd's estimate of the number of elementary logical operations in the history of the universe, 10^{120} [27]. Dembski claims that if

$$10^{120} \cdot \varphi_S(T) \cdot \mathbf{P}(T \mid \mathbf{H}) < 0.5,$$

then the chance hypothesis is less likely to account for event E than the alternative hypothesis of intelligent design [18]. The CSI of event E is the penalized specificity,

$$\chi = -\log_2[10^{120} \cdot \varphi_S(T) \cdot \mathbf{P}(T \mid \mathbf{H})] \approx \sigma - 399 \text{ bits.}$$

Intelligent design is inferred for event $\mathbf{H}$ if $\chi > 1$ or, equivalently, specificity in excess of 400 bits [18]. Note that Dembski sometimes invokes the argument from specified complexity to reject chance in favor of human intelligence, and in these cases he sets the number of replicational resources smaller [18].[7]

1.5.2 Some Criticisms of Specified Complexity

It seems that Dembski, as a mathematician and philosopher, thinks more analytically than algorithmically. Most of the following addresses aspects of computation of CSI. It is important to keep in mind the adversarial aspect of the argument from specified complexity. Chance hypotheses should come from mainstream scientists, not ID advocates. They often will be analytically intractable, and design inferences will require direct computation of CSI. Below are listed some major criticisms of Dembski's arguments.

A Model of Nature Is Conflated with Nature Itself

Recall that the chance hypothesis is essentially that natural causes account entirely for a natural event. The design inference is a claim that natural causes

---

[7] ID advocates hold that human intelligence is not natural (materialistic). Thus humans can cause events with high levels of CSI.

alone do not suffice to explain the event. But in practice the chance hypothesis that is likely to be derived from a scientific model, and what is subject to rejection is not natural causation itself, but the model of natural causation. The distinction is of vital importance. If scientists do not understand some class of events, a chance hypothesis derived from their best model may be rejected in favor of design. The inability of the model to account for the event is treated as the inability of natural causation to account for the event. This constitutes a logically fallacious argument from ignorance. And as described above, ID advocates indeed focus on biological entities with histories that are very difficult to determine.

Key Aspects of CSI Are Not Explicit in Dembski's Treatment

In conventional mathematical terms, a "pattern" described by a semiotic agent is a *property*. A property T is a subset of some set U, and saying that $x \in U$ *has property* T is equivalent to saying that $x \in T$. Let D_S denote the set of all descriptions that semiotic agent S may emit. For all descriptions d in D_S, let $\varphi_S(d)$ be the rank-complexity of d described above. Let $D_S(E) \subseteq D_S$ be the set of all descriptions associated with event E by S. Finally, let

$$T_S(d) = \{\omega \in \Omega \mid \omega \text{ has the property } S \text{ describes with } d\},$$

where Ω is the sample space. This glosses over semantic interpretation of the descriptions in D_S. Nonetheless, it should convey that there is no way to determine the rejection region without knowing both its description and the semantics of the semiotic agent that generated the description. Then for all semiotic agents S and for all descriptions d in $D_S(E)$ the CSI is

$$\chi_S(d) = -\log_2[10^{120} \cdot \varphi_S(d) \cdot \mathbf{P}(T_S(d) \mid \mathbf{H})].$$

This appropriately indicates that CSI is associated with descriptions of event E. For completeness, one may define $\chi(E)$ as the maximum of $\chi_S(d)$ over all S and d, but maximization is infeasible in practice, and design inference requires only $\chi_S(d) > 1$ for some S and d.

"Divide-and-Conquer" Rejection of Disjunctive Hypotheses Is Permitted

When there are multiple chance hypotheses $\{\mathbf{H}_i\}$, they must be rejected jointly to infer intelligent design. Dembski fails to point out that the semiotic agent S and the description d in $D_S(E)$ must be held constant while rejecting all hypotheses [18]. This requirement is captured by generalizing the definition of χ_S to

$$\chi_S(d) = -\log_2[10^{120} \cdot \varphi_S(d) \cdot \max_i \mathbf{P}(T_S(d) \mid \mathbf{H}_i)].$$

CSI Is Not Computable

For Dembski, the physical (material) universe is discrete and finite, and so is Ω [11, 13, 18]. This would seem to bode well for computation of CSI, but problems arise from the fact that a semiotic agent may associate with event E a description of a property defined on an infinite set. Many finitely describable properties are not algorithmically decidable [26], irrespective of the nonexistence of infinite sets in the physical universe.

The value of $\mathbf{P}(T_S(d) \mid \mathbf{H})$ is the sum of $\mathbf{P}(\omega \mid \mathbf{H})$ over all points ω in rejection region $T_S(d)$. Its computation generally requires conversion of description d into an algorithm that decides which points in Ω have the described property. But if the described property is not decidable, $\mathbf{P}(T_S(d) \mid \mathbf{H})$ is computable only under special circumstances. This holds even if the initial "translation" of d into an algorithm is non-algorithmic.

Incomputable properties are especially likely to arise in the important case that Ω is a set of entities that describe or compute partial (not always total) recursive functions. An example is the set of all LISP programs of length not exceeding some large bound. A semiotic agent's description of program E will commonly refer to a nontrivial property of the function computed by E. But a key result in the theory of computation, Rice's theorem, implies that no algorithm decides whether other LISP programs compute functions with that property [26]. In other words, there is generally no algorithm to say whether programs in Ω belong to the rejection region. This indicates that for a wide range of computational entities CSI may be computed only for the form (e.g., the source code), and not the function. Note that some philosophers and scientists believe that brains compute partial (not total) recursive functions [19].

Some Design Hypotheses Call for Nonexistent Chance Hypotheses

In conventional statistical hypothesis testing, one begins with an alternative hypothesis and then selects a chance hypothesis. This does not carry over to the argument from specified complexity. An ID advocate may believe an event is designed, but mainstream scientists may not have provided an appropriate chance hypothesis to reject. The non-existence of the hypothesis (scientific model) may be due to scientific indifference or scientific ignorance.

As an example of scientific indifference, consider what is required to compute the CSI of the bacterial flagellum, which Dembski *qua* semiotic agent describes as a "bidirectional rotary motor-driven propeller" [18]. The sample space contains biological structures detached from whole phenotypes, and the chance hypothesis must associate probabilities of evolution with them. But nothing in evolutionary theory leads to such a hypothesis, and it is absurd to insist that scientists to supply one.

Ignorance is ubiquitous in science, and some phenomena (e.g., gravity) have resisted explanation for centuries. The inability of science to explain a class of events does not constitute the least evidence for ID. To suggest otherwise is to engage in a logical fallacy known as argument from ignorance.

Computation of CSI May Be Infeasible When Theoretically Possible

If Ω is the set of all biological structures (begging the question of how to define "biological structure") that have existed (begging the question of how to determine all structures of entities that have ever lived) or might have existed (begging the question of how to determine what might have lived), how will an algorithm efficiently locate the points in the sample space with the property "bidirectional rotary motor-driven propeller"? No approach other than exhaustive exploration of the sample space for points with the property is evident. The time required for such a computation makes it infeasible. Furthermore, the practicality of defining the sample space for an algorithm to operate upon is highly dubious.

Another feasibility issue is the cost of computing $\mathbf{P}(\omega \mid \mathbf{H})$ for a single ω in Ω. Suppose $\mathbf{P}(\omega \mid \mathbf{H})$ is, loosely speaking, the probability of evolution of ω, and that $\mathbf{H}$ is derived from a simulation model supplied by a scientist. The results of a simulation run usually depend upon initial conditions and parameter settings. There will virtually always be uncertainty as to how to set these values, and the consequence is that many runs of the simulation model (with various settings) will be required to obtain $\mathbf{P}(\omega \mid \mathbf{H})$.

Putative Innovations in Statistical Hypothesis Testing Have Not Passed Peer Review

Dembski's approach to design inference [18] is correct only if he has made monumental contributions to statistical hypothesis testing. There is nothing precluding publication of his statistical work in a peer-reviewed journal of mathematics or statistics. At the time of this writing, Dembski has not published any of his work. Consequently, one must regard his statistical reasoning with skepticism.

1.5.3 The Law of Conservation of Information

In earlier work [13], Dembski argues informally for a *law of conservation of information*, which does not specify that complex specified information is strictly conserved in natural processes, but that gain of CSI is bounded above by 500 bits. That is, a closed physical system may go from a state of lower CSI to a state of higher CSI, but the increase cannot exceed 500 bits. The bound corresponds to a putative limit on the improbability of events in the physical universe, as described below. Dembski regards evolutionary computations (ECs) as closed systems, and if an EC produces an apparent gain of more than

500 bits of CSI in its population, he argues that humans have surreptitiously (perhaps haplessly) added CSI to the process [13].

The 500-bit bound on CSI gain is the negative logarithm of the *universal probability bound* Dembski advocates in earlier work, 10^{-150} [11, 13]. He considers events with probability below the bound to be effectively impossible. Dembski [11] cites Émile Borel, who is quoted in the epigraph of this chapter, as a famous proponent of a universal probability bound. In fact Borel selects different bounds for different applications – they are hardly "universal" [7]. Some are much smaller, and some much larger, than Dembski's bound. In the work detailed above, Dembski indicates that "instead of a static universal probability bound of 10^{-150} we now have a dynamic one of $10^{-120}/\varphi_S(d)$" [18]. That is, the bound is adapted to the observer of an event and the observer's description of the event. This is in marked contrast with Borel's approach.

Dembski does not indicate in [18] how to rescue the law of "conservation" of information. He states, however, that $\varphi_S(d)$ should not exceed 10^{30} in practice, and observes that his old static bound of 10^{-150} is a lower bound on the dynamic bound. This suggests that Dembski may renew his claim that CSI gain cannot exceed 500 bits in a natural process. With the dependence of CSI upon observers and their descriptions of events, what it means to gain CSI is hardly obvious.

1.6 ID and Evolutionary Computation

Dembski has been at pains to argue, particularly in Chapter 4 of *No Free Lunch* [13], that the results of evolutionary computation violate his law of conservation of information, and that human investigators must be injecting their own intelligence into the EC programs under investigation. In particular, he has attacked Chellapilla and Fogel's study of co-evolution of checkers players [8], Ray's Tierra program for artificial life [38], Schneider's demonstration of gain of Shannon information in an evolutionary program [41], and Altshuler and Linden's evolutionary optimization of bent-wire antenna designs [2].

Dembski often cites the main "no free lunch" (NFL) theorem for optimization, which says in essence that if all objective functions are equally likely, then all optimizers that do not revisit points have identically distributed performance [44]. He takes this as an indication that performance is generally bad. Ironically, English [20] showed six years prior to the publication of Dembski's book that NFL arises as a consequence of (absolute) conservation of Shannon information in optimization, and that average performance is very good when test functions are uniformly distributed. In other words, NFL does not bode as poorly for EC as Dembski has thought.

Dembski has since responded [17] by analyzing search of "needles-in-a-haystack" functions, in which a few points in the domain are categorically good and the remainder are categorically bad [20]. He motivates the analysis by alluding to proteins as needles in the haystack of all sequences of amino acids. For each function, it is highly improbable that an arbitrarily selected search algorithm locates a good solution in feasible time. Dembski holds that successful search requires a prior "search for a search" [17]. This amounts to *displacement* of the search problem from the original solution space to the space of search algorithms. He argues that a search problem cannot be solved more rapidly with displacement than in the original space. Thus from his perspective, if an EC finds a good solution in feasible time, the choice of the EC was necessarily informed by intelligence [17].

There is nothing novel in the notion that it is sometimes necessary to "align" the search algorithm with the problem [44], but there is in the idea that alignment requires search [17]. How does one search for a search algorithm? Dembski is very vague about this. All one can possibly do, in black-box optimization, is to examine the value of at least one point in the search space (domain) and use the information to select an algorithm. But then one has initiated a search of the function. It follows that any search for a search may be embedded in an algorithm for search of the original solution space.

"Displacement" is a construct that makes it appear that intelligence creates information by selecting an effective search algorithm to locate a solution. In reality, humans are able to tune an EC to a fitness function only when the fitness function is not a black box. Only when one knows some property or properties of the fitness function can one select an EC that is expected to outperform random sampling. And how does one recognize properties of a function? Does one's intelligence create information? No, it seems much more reasonable to say that one has learned (acquired information) about functions and algorithms in the past, and that one uses this repository of information to match algorithms to functions. There is a great deal of empirical research aimed at learning which forms of EC handle which classes of functions well.

1.7 Conclusion

We have criticized ID theory for its intrinsic faults. But in the end the only way to understand the theory is as a veiled apologetic. Jews, Christians, and Muslims agree that the God of Abraham created the diverse forms of life on earth, imbuing only humans with a capacity to create ex nihilo. Although some of the faithful accept that religion and science are different belief systems leading to different beliefs, others insist that science must never contradict religion. ID theorists begin with religious beliefs about life and humanity, and attempt to show that contradictory beliefs held by almost all mainstream scientists are wrong. They hide their religious motivation because they hope their theory will find its way into science classes of public schools.

Irreducible complexity is the weakest part of the apologetics. Behe has had to concede what Muller pointed out decades before he was born, namely that indirect evolutionary pathways may give rise to irreducible complexity.

And there is good fossil evidence that the interconnected bones of the mammalian middle ear evolved from a reptilian jawbone. The Avida simulation is reasonably interpreted as generating irreducibly complex programs. ID advocates continue, however, to focus on irreducibly complex biosystems for which there are few historical data (e.g., the flagellum). They argue that evolutionary theory fails to account for the emergence of these systems when in fact there are few hard data.

The argument from specified complexity rests on an approach to statistical hypothesis testing that has not passed peer review. Even if the statistical foundation is sound, the argument is logically flawed. When it claims to reject purely natural causation in favor of design, it actually rejects a model. That is, if there is no good model of a phenomenon, then the argument from specified complexity reduces to argument from ignorance. Even with an excellent model, specified complexity is in some cases impractical to compute, or even incomputable.

Dembski's claim that all entities with high specified complexity are intelligently designed seems to have been falsified by various evolutionary computations. But Dembski argues constantly that experimenters have smuggled intelligence into the computations. Accumulating further computational evidence should be valuable, but in the end formal mathematical analysis may be required to settle the dispute.

References

1. Agassiz, L.: Evolution and permanence of type. Atlantic Monthly (1874)
2. Altshuler, E., Linden, D.: Wire-antenna designs using genetic algorithms. IEEE Antennas and Propagation Magazine **39**, 33–43 (1997)
3. Behe, M.: Darwin's Black Box: The Biochemical Challenge to Evolution. Free Press, New York (1996)
4. Behe, M.: Molecular machines: experimental support for the design inference. Cosmic Pursuit **1**(2), 27–35 (1998)
5. Behe, M.: A response to critics of Darwin's Black Box. Progress in Complexity, Information, and Design **1** (2002)
6. Blocker, A., Komoriya, K., Aizawa, S.: Type III secretion systems and bacterial flagella: insights into their function from structural similarities. Proceedings of the National Academy of Science USA **100**, 3027–3030 (2003)
7. Borel, E.: Probability and Life. Dover, New York (1962)
8. Chellapilla, K., Fogel, D.: Evolving an expert checkers playing program without using human expertise. IEEE Transactions on Evolutionary Computation **5**, 422–428 (2001)
9. Darwin, C.: On the Origin of Species by Means of Natural Selection – the Preservation of Favoured Races in the Struggle for Life. John Murray, London (1859)
10. Dawkins, R.: The Blind Watchmaker. W. W. Norton and Co., London (1986)
11. Dembski, W.: The Design Inference: Eliminating Chance Through Small Probabilities. Cambridge University Press, Cambridge (1998)

12. Dembski, W.: The intelligent design movement. In: J. Miller (ed.) An Evolving Dialogue: Theological and Scientific Perspectives on Evolution, pp. 439–443. Trinity Press International, Harrisburg, PA (2001)
13. Dembski, W.: No Free Lunch: Why Specified Complexity Cannot be Purchased Without Intelligence. Rowman and Littlefield, Lanham, MD (2002)
14. Dembski, W.: The Design Revolution. InterVarsity, Downers Grove, IL (2004)
15. Dembski, W.: Irreducible complexity revisited. Tech. rep. (2004). http://www.designinference.com/documents/2004.01.Irred_Compl_Revisited.pdf
16. Dembski, W.: Rebuttal to reports by opposing expert witnesses (2005). http://www.designinference.com/documents/2005.09.Expert_Rebuttal_Dembski.pdf
17. Dembski, W.: Searching large spaces. Tech. rep. (2005). http://www.designinference.com/documents/2005.03.Searching_Large_Spaces.pdf
18. Dembski, W.: Specification: the pattern that signifies intelligence. Tech. rep. (2005). http://www.designinference.com/documents/2005.06.Specification.pdf
19. Dietrich, E.: The ubiquity of computation. Think 2, 12–78 (1993)
20. English, T.: Evaluation of evolutionary and genetic optimizers: no free lunch. In: L. Fogel, P. Angeline, T. Bäck (eds.) Evolutionary Programming V: Proceedings of the Fifth Annual Conference on Evolutionary Programming, pp. 163–169. MIT Press, Cambridge, Mass. (1996)
21. Hartwig, M.: What is intelligent design? (2003). http://www.arn.org/idfaq/Whatisintelligentdesign.htm
22. Jiang, Y., Doolittle, R.: The evolution of vertebrate blood coagulation as viewed from a comparison of puffer fish and sea squirt genomes. Proceedings of the National Academy of Science USA 100(13), 7527–32 (2003)
23. Johnson, P.: Evolution as dogma: the establishment of naturalism. In: W. Dembski (ed.) Uncommon Descent: Intellectuals Who Find Darwinism Unconvincing. ISI Books, Wilmington, Delaware (2004)
24. Jones III, J.: Tammy Kitzmiller et al. v. Dover Area School District et al. Memorandum opinion in case no. 04cv2688, United States District Court for the Middle District of Pennsylvania (2005). http://www.pamd.uscourts.gov/kitzmiller/kitzmiller_342.pdf
25. Lenski, R., Ofria, C., Pennock, R., Adami, C.: The evolutionary origin of complex features. Nature 423, 129–144 (2003)
26. Lewis, H., Papadimitriou, C.: Elements of the Theory of Computation, 2nd edn. Prentice-Hall, Upper Saddle River, NJ (1998)
27. Lloyd, S.: Computational capacity of the universe. Physical Review Letters 88, 7901–7904 (2002)
28. Meyer, S.: The origin of biological information and the higher taxonomic categories. Proceedings of the Biological Society of Washington 117, 213–239 (2004)
29. Miller, K.: A review of Darwin's Black Box. Creation/Evolution pp. 36–40 (1996)
30. Miller, K.: The evolution of vertebrate blood clotting (no date). http://www.millerandlevine.com/km/evol/DI/clot/Clotting.html
31. Minnich, S., Meyer, S.: Genetic analysis of coordinate flagellar and type III regulatory circuits in pathogenic bacteria. In: Second International Conference on Design & Nature (2004)
32. Muller, H.: Genetic variability, twin hybrids and constant hybrids, in a case of balanced lethal factors. Genetics 3(5), 422–499 (1918)

33. Muller, H.: Reversibility in evolution considered from the standpoint of genetics. Biological Reviews of the Cambridge Philosophical Society **14**, 261–280 (1939)
34. Orr, H.: Darwin v. Intelligent Design (again). Boston Review (1996)
35. Paley, W.: Natural Theology – Evidences of the Existence and Attributes of the Deity: Collected from the Appearances of Nature (1802)
36. Pallen, M., Matzke, N.: From The Origin of Species to the origin of bacterial flagella. Nature Reviews Microbiology **4**, 784–790 (2006)
37. Ray, J.: The Wisdom of God Manifested in the Works of the Creation (1691)
38. Ray, T.: An approach to the synthesis of life. In: C. Langton, C. Taylor, J. Farmer, S. Rasmussen (eds.) Artificial Life II, pp. 371–408. Addison-Wesley, Reading, MA (1992)
39. Rich, T., Hopson, J., Musser, A., Flannery, T., Vickers-Rich, P.: Independent origins of middle ear bones in monotremes and therians. Science **307**, 910–914 (2005)
40. Rosenhouse, J.: How anti-evolutionists abuse mathematics. The Mathematical Intelligencer **23**, 3–8 (2001)
41. Schneider, T.: Evolution of biological information. Nucleic Acids Research **28**, 2794–2799 (2000)
42. Semba, U., Shibuya, Y., Okabe, H., Yamamoto, T.: Whale Hageman factor (factor XII): prevented production due to pseudogene conversion. Thrombosis Research **90**(1), 31–37 (1998)
43. Ussher, J.: The Annals of the Old Testament, From the Beginning of the World (1654)
44. Wolpert, D., Macready, W.: No free lunch theorems for optimization. IEEE Transactions on Evolutionary Computation **1**, 67–82 (1997)
45. Xu, X., Doolittle, R.: Presence of a vertebrate fibrinogen-like sequence in an echinoderm. Proceedings of the National Academy of Science USA **87**, 2097–2101 (1990)

Inference of Genetic Networks Using an Evolutionary Algorithm

Shuhei Kimura

Tottori University, 4-101, Koyama-Minami, Tottori, Japan
kimura@ike.tottori-u.ac.jp

2.1 Introduction

Genes control cellular behavior. Most genes play biological roles when they are translated into proteins via mRNA transcription. The process by which genes are converted into proteins is called *gene expression,* and the analysis of gene expression is one means by which to understand biological systems.

A DNA microarray is a collection of microscopic DNA spots attached to a solid surface forming an array. As each spot can measure an expression level of a different gene, this technology allows us to monitor expression patterns of multiple genes simultaneously. With recent advances in these technologies, it has become possible to measure gene expression patterns on a genomic scale. To exploit these technologies, however, we must find ways to extract useful information from massive amounts of data. One of the promising ways to extract useful information from voluminous data is to infer genetic networks. The inference of genetic networks is a problem in which mutual interactions among genes are deduced using gene expression patterns. The inferred model of the genetic network is conceived as an ideal tool to help biologists generate hypotheses and facilitate the design of their experiments. Many researchers have taken an interest in the inference of genetic networks, and the development of this methodology has become a major topic in the bioinformatics field.

Many models to describe genetic networks have been proposed. The Boolean network model is one of the more abstract models among them [1, 10, 14, 22]. This model classifies each gene into one of two states, ON (express) or OFF (not-express), then applies gene regulation rules as Boolean functions. The simplicity of inference methods based on the Boolean network model makes it possible to analyze genetic networks of thousands of genes. However, as gene expression levels are not binary but continuous in general, it is difficult for this approach to predict the effect of a change in the expression level of one gene upon the other genes [6].

The observed data on gene expression are normally polluted by noise arising from the measurement technologies, the experimental procedures, and the underlying stochastic biological processes. The Bayesian network model has often been used to handle these noisy data [9, 12, 26, 27, 40]. As the Bayesian network model treats the expression level of each gene as a random variable and represents interactions among genes as conditional probability distribution functions, this model can easily handle noisy data. Another advantage of this model is an ability to infer genetic networks consisting of hundreds of genes [9, 27]. These features have prompted the successful application of the Bayesian approach to the analysis of actual DNA microarray data [9,12,26,27].

Several groups have attempted to exploit the quantitative feature of gene expression data by models based on sets of differential equations. The following equations describe a genetic network in a genetic network inference problem based on a set of differential equations.

$$\frac{dX_i}{dt} = G_i(X_1, X_2, \cdots, X_N), \quad (i = 1, 2, \cdots, N), \tag{2.1}$$

where X_i is the expression level of the i-th gene, N is the number of genes in the network, and G_i is a function of an arbitrary form. The purpose of the genetic network inference problem based on the set of differential equations is to identify the functions G_i from the observed gene expression data.

In order to model a genetic network, Sakamoto and Iba [30] have proposed to use a set of differential equations of an arbitrary form. They tried to identify the functions G_i using genetic programming. The scope of their method has been limited, however, as the difficulty of identifying arbitrary functions restricts application to small-scale genetic network inference problems of less than five genes. It generally works better to approximate the functions G_i than to attempt to identify them. In many of the earlier studies, G_i has been approximated using models based on sets of differential equations of a fixed form [2, 3, 7, 34, 39]. When we use a model of fixed form to describe a genetic network, the genetic network inference problem becomes a model parameter estimation problem.

The S-system model [36], a set of differential equations of fixed form, is one of several well-studied models for describing biochemical networks. A number of researchers have proposed genetic network inference methods based on the S-system model [4,11,15–18,24,25,33–35,37]. Some of these inference methods, however, must estimate a large number of model parameters simultaneously if they are to be used to infer large-scale genetic networks [4, 11, 15, 33–35]. Because of the high-dimensionality of the genetic network inference problem based on the S-system model, these inference methods have been applied only to small-scale networks of a few genes. To resolve this high-dimensionality, a problem decomposition strategy, that divides the original problem into several subproblems, has been proposed [25]. This strategy enables us to infer S-system models of larger-scale genetic networks, and the author and his colleagues have proposed two inference methods based on this strategy, i.e., the problem decomposition approach [16, 17] and the cooperative coevolutionary approach [18]. As the strategy defines the inference of the S-system model of the genetic network as non-linear function optimization problems, their methods use evolutionary algorithms as the function optimizer. The rest of this chapter presents these two inference methods. Then, through the application of these methods to genetic network inference problems consisting of dozens of genes, some observations on and limitations of these methods are discussed.

2.2 S-system Model

The S-system model [36] is a set of non-linear differential equations of the form

$$\frac{dX_i}{dt} = \alpha_i \prod_{j=1}^{N} X_j^{g_{i,j}} - \beta_i \prod_{j=1}^{N} X_j^{h_{i,j}}, \quad (i = 1, 2, \cdots, N), \tag{2.2}$$

where α_i and β_i are multiplicative parameters called rate constants, and $g_{i,j}$ and $h_{i,j}$ are exponential parameters called kinetic orders. X_i is the state variable and N is the number of components in the network. In genetic network inference problems, X_i is the expression level of the i-th gene and N is the number of genes in the network. The first and second terms on the right-hand side of (2.2) represent processes that contribute to the increase and decrease, respectively, in X_i.

The term S of the S-system refers to the synergism and saturation of the investigated system. *Synergism* is the phenomenon in which two or more objects produce an effect greater than the total of their individual effects. *Saturation* is the condition in which, after a sufficient increase in a causal force, no further increase in the resultant effect is possible. Synergism and saturation, fundamental properties of biochemical systems, are both inherent in the S-system model. Moreover, steady-state evaluation, control analysis and sensitivity analysis of this model have been established mathematically [36]. These properties have led to a number of applications of the S-system model to the analysis of biochemical networks (e.g. [31, 38]).

The genetic network inference methods described in this chapter use this model to approximate the functions G_i given in Sect. 2.1. Thus, the inference of genetic networks is recast as the estimation problem for the model parameters α_i, β_i, $g_{i,j}$, and $h_{i,j}$. As the numbers of the rate constants (α_i and β_i) and kinetic orders ($g_{i,j}$ and $h_{i,j}$) are $2N$ and $2N^2$, respectively, the number of S-system parameters to be estimated is $2N(N + 1)$.

2.3 Genetic Network Inference Problem

2.3.1 Canonical Problem Definition

The purpose of the genetic network inference problem based on the S-system model is to estimate the model parameters from the time-series of observed DNA microarray data. This problem is generally formulated as a function optimization problem to minimize the following sum of the squared relative error:

$$f = \sum_{i=1}^{N} \sum_{t=1}^{T} \left(\frac{X_{i,cal,t} - X_{i,exp,t}}{X_{i,exp,t}} \right)^2, \tag{2.3}$$

where $X_{i,exp,t}$ is an experimentally observed gene expression level at time t of the i-th gene, $X_{i,cal,t}$ is a numerically computed gene expression level acquired by solving a system of differential equations (2.2), N is the number of components in the network, and T is the number of sampling points of observed data.

The number of S-system parameters to be determined in order to solve the set of differential equations (2.2) is $2N(N + 1)$, hence this function optimization problem is $2N(N+1)$ dimensional. Several inference methods based on this problem definition have been proposed [11, 15, 34]. If we try to infer the S-system models of large-scale genetic networks made up of many network components, however, this problem has too many dimensions for non-linear function optimizers [25].

2.3.2 Problem Decomposition

Given the high-dimensionality of the canonical problem definition, it can be difficult to use function optimizers for inferring S-system models of large-scale genetic networks. Maki and his colleagues [25] proposed a strategy for resolving the high-dimensionality by dividing the genetic network inference problem into several subproblems. In their strategy, each subproblem corresponds to a gene. The objective function of the subproblem corresponding to the i-th gene is

$$f_i = \sum_{t=1}^{T} \left(\frac{X_{i,cal,t} - X_{i,exp,t}}{X_{i,exp,t}} \right)^2, \tag{2.4}$$

where $X_{i,cal,t}$ is a numerically computed gene expression level at time t of the i-th gene, as described in Sect. 2.3.1. In contrast to Sect. 2.3.1, however, we obtain $X_{i,cal,t}$ by solving the following differential equation:

$$\frac{dX_i}{dt} = \alpha_i \prod_{j=1}^{N} Y_j^{g_{i,j}} - \beta_i \prod_{j=1}^{N} Y_j^{h_{i,j}}, \tag{2.5}$$

where

$$Y_j = \begin{cases} X_j, \text{ if } j = i, \\ \hat{X}_j, \text{ otherwise.} \end{cases} \tag{2.6}$$

$\hat{X}_j$ is an estimated time-course of the j-th gene's expression level acquired not by solving a differential equation, but by direct estimation from the observed time-series data. We can obtain $\hat{X}_j$'s using a spline interpolation [29], a local linear regression [5], or other interpolation techniques.

Equation (2.5) is solvable when $2(N+1)$ S-system parameters (i.e., $\alpha_i, \beta_i,$ $g_{i,1}, \cdots, g_{i,N}, h_{i,1}, \cdots, h_{i,N}$) are given. Thus, the problem decomposition strategy divides a $2N(N+1)$ dimensional network inference problem into N subproblems that are $2(N+1)$ dimensional.

2.3.3 Use of a Priori Knowledge

The genetic network inference problem based on the S-system model may have multiple optima, as the model has a high degree-of-freedom and the observed time-series data are usually polluted by noise. To increase the probability of inferring a reasonable S-system model, we introduced a priori knowledge of the genetic network into the objective function (2.4) [16].

Genetic networks are known to be sparsely connected [32]. When an interaction between two genes is clearly absent, the S-system parameter values corresponding to the interaction (i.e., kinetic orders; $g_{i,j}$ and $h_{i,j}$) are zero. We incorporated the knowledge into the objective function (2.4) by applying a penalty term, as shown below:

$$F_i = \sum_{t=1}^{T} \left(\frac{X_{i,cal,t} - X_{i,exp,t}}{X_{i,exp,t}} \right)^2 + c \sum_{j=1}^{N-I} (|G_{i,j}| + |H_{i,j}|), \tag{2.7}$$

where $G_{i,j}$ and $H_{i,j}$ are given by rearranging $g_{i,j}$ and $h_{i,j}$, respectively, in descending order of their absolute values (i.e., $|G_{i,1}| \leq |G_{i,2}| \leq \cdots \leq |G_{i,N}|$ and $|H_{i,1}| \leq |H_{i,2}| \leq \cdots \leq |H_{i,N}|$). The variable c is a penalty coefficient and I is a maximum indegree. The maximum indegree determines the maximum number of genes that directly affect the i-th gene.

The penalty term is the second term on the right-hand side of (2.7). This term forces most of the kinetic orders down to zero. When the penalty term is applied, most of the genes are thus disconnected from each other. The term does not penalize, however, when the number of genes that directly affect the i-th gene is lower than the maximum indegree I. Thus, the optimum solutions to the objective functions (2.4) and (2.7) are identical when the number of interactions that affect the focused (i-th) gene is lower than the maximum indegree. Our inference methods use (2.7) as an objective function to be minimized.

2.4 Problem Decomposition Approach

2.4.1 Estimation of S-system Parameters

In order to estimate the S-system parameters, we must minimize the objective function (2.7). While any type of function optimizer can be used for this purpose, we used GLSDC (a Genetic Local Search with distance independent Diversity Control) [19], a method based on the real-coded genetic algorithm [16–18]. GLSDC has two features important for the inference of large-scale genetic networks; it works well on multimodal function optimization problems with high-dimensionality and it can be suitably applied to parallel computers. The following is a pseudocode description of GLSDC.

1. *Initialization*
   As an initial population, create n_p individuals randomly. Set Generation $= 0$ and the iteration number of the converging operations $N_{iter} = N_0$.
2. *Local Search Phase*
   Apply a local search method to all individuals in the population. As a sophisticated local search algorithm, the modified Powell's method [29] was used.
3. *Adaptation of N_{iter}*
   If the best individual in the population shows no improvement over the previous generation, set $N_{iter} \leftarrow N_{iter} + N_0$. Otherwise, set $N_{iter} = N_0$.
4. *Converging Phase*
   Execute the exchange of individuals according to the genetic operators N_{iter} times.
5. *Termination*
   Stop if the halting criteria are satisfied. Otherwise, Generation $\leftarrow$ Generation $+ 1$ and go to step 2.

Readers can find more detailed information on GLSDC in the paper [19].

Recently, however, several groups have proposed useful techniques to solve the genetic network inference problem based on the S-system model. Voit and Almeida [37] reformulate the estimation of the S-system parameters as a set of algebraic equations. Similarly, Tsai and Wang [35] converted a set of differential equations (2.2) into a set of algebraic equations. The computational costs of these inference methods are lower since they do not require solving any differential equation. Imade and his colleagues [11], on the other hand, have developed an inference system running on a grid computing environment. Grid computing is the pooling of computational resources into a single set of shared services. As the grid computing environment provides us with powerful computational resources, this method has an ability to infer larger-scale genetic networks. The use of these new techniques should improve the inference ability of the method presented in this chapter.

2.4.2 Estimation of Initial Gene Expression Level

To solve the decomposed subproblem, we first need to solve the differential equation (2.5). This can only be accomplished by inputting the initial expression level of the gene (the initial state value for the differential equation) in addition to the S-system parameters. We can obtain the initial gene expression level from the observed time-series data, but only when the data are unpolluted by noise. Given that this condition generally cannot be met, the initial gene expression level must be estimated together with the S-system parameters.

Again, we need to estimate the initial gene expression level when applying an inference algorithm to a realistic genetic network inference problem. However, the simultaneous estimation of the initial gene expression level and the S-system parameters makes the function optimization problem higher dimensional, and this is inconvenient for function optimizers. To skirt this problem, we proposed the method of estimating the initial gene expression level and the set of S-system parameters alternately [17]. When we use this method to estimate the initial expression level of the i-th gene, we fix the S-system parameters to the values of some candidate solution for the i-th subproblem. Because the initial expression level of the i-th gene is a unique variable and the rest of the model parameters are fixed, the estimation of the initial expression level of the i-th gene is formulated as a one-dimensional function minimization problem. The objective function of this estimation problem is

$$F_i^{initial} = \sum_{t=1}^{T} \gamma^{t-1} \left(\frac{X_{i,cal,t} - X_{i,exp,t}}{X_{i,exp,t}} \right)^2 , \qquad (2.8)$$

where $X_{i,cal,t}$ is acquired by solving (2.5), and γ ($0 \leq \gamma \leq 1$) is a discount parameter. As the fixed S-system parameters are not always optimal, the calculated time-course of gene expression may differ greatly from the actual time-course. When the calculated time-course is incorrect, the algorithm should not fit the time-course, especially the latter half of it, into the observed data. The discount parameter γ was introduced for this reason.

A golden section search [29] was used to solve this one-dimensional function minimization problem. (Golden section search is a technique for finding an optimum of a single-dimensional objective function by successively narrowing the interval inside which the optimum seems to exist.) When multiple sets of time-series data are given as the observed data, the one-dimensional search was applied to all of the sets.

2.4.3 Algorithm

As mentioned above, we can only solve a decomposed subproblem by estimating both the S-system parameters and the initial gene expression level.

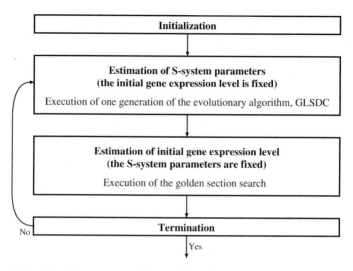

Fig. 2.1. Framework of the problem decomposition approach

The algorithm presented here handles these calculations alternately, estimating the initial gene expression level at the end of the every cycle (generation) of GLSDC (see Fig. 2.1) [17]. When the algorithm estimates the initial gene expression level, the S-system parameters are fixed to the values of the best candidate solution. In contrast, the initial gene expression level is fixed to the value obtained by optimizing the objective function (2.8) when the algorithm estimates the S-system parameters. This approach fails to estimate the initial expression level at the first generation, however, as the calculation only becomes possible after the estimation of the S-system parameters. We overcome this obstacle in the algorithm by using the value obtained from the observed time-series data as the initial gene expression level at the first generation. This inference method is referred to as the *problem decomposition approach.*

One run of this approach solves one subproblem corresponding to one gene. As N subproblems must be solved to infer a genetic network consisting of N genes, we need to perform N runs. A parallel computer can assist us in this task, however, as it can be used to execute multiple runs of the problem decomposition approach in parallel.

2.5 Cooperative Coevolutionary Approach

2.5.1 Concept

The problem decomposition approach described above can be used to infer large-scale genetic networks. When the time-series data are noisy, however,

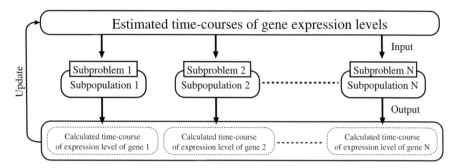

Fig. 2.2. The cooperative coevolutionary model for the inference of genetic networks

the model inferred by this approach fails in the computational simulation of genetic networks.

Although the computational simulation is performed by solving the set of differential equations (2.2), the problem decomposition approach estimates the model parameters without solving it. While the canonical (non-decomposed) approach uses the set of differential equations (2.2) in order to estimate the parameters, the problem decomposition approach uses the differential equations (2.5) for this purpose. As explained earlier, we cannot solve (2.5) without first obtaining the estimated time-courses of the gene expression levels, $\hat{X}_j$'s. The problem decomposition approach estimates the time-courses, $\hat{X}_j$'s, directly from the observed time-series data using some interpolation method. When $\hat{X}_j$'s are estimated correctly, optimum solutions obtained by the problem decomposition approach and the canonical approach will completely coincide with each other. The reason for this coincidence is that the optimum parameter values should make the solutions of the set of differential equations (2.2) and the differential equations (2.5) identical. If, on the other hand, the given time-series data are noisy, $\hat{X}_j$'s will often be difficult to estimate correctly. When incorrect $\hat{X}_j$'s are applied, the optimum solutions of the decomposed subproblems do not always coincide with the optimum solution of the non-decomposed problem. This means that the parameters obtained by solving the subproblems do not always provide a model [i.e., the set of differential equations (2.2)] that fits into the observed data. As such, the inferred model in the problem decomposition approach is not yet suitable for the computational simulation of genetic networks.

In the subproblem corresponding to the i-th gene, the time-course of the i-th gene's expression level is calculated by solving the differential equation (2.5). When optimizing the i-th subproblem, the function optimizer searches for the S-system parameters which lead to the best fit of the calculated expression time-course of the i-th gene into the observed data. Thus, the calculated time-courses obtained by solving the subproblems are the most suitable for $\hat{X}_j$'s. If we can always use the calculated time-courses of the gene

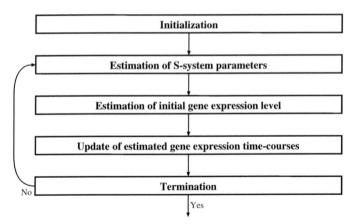

Fig. 2.3. Framework of the cooperative coevolutionary approach

expression levels as $\hat{X}_j$'s, optimizing the subproblems should provide a model that fits into the observed data.

A cooperative coevolutionary model (see, e.g. [23, 28]) can use the time-courses of the gene expression levels obtained by solving the subproblems as $\hat{X}_j$'s. The cooperative coevolutionary algorithm consists of several subpopulations, each of which contains competing individuals in each subproblem. The subpopulations are genetically isolated, i.e., the individuals within a subpopulation only mate with each other. The only interactions to take place between the subpopulations occur when the fitness values are calculated. When the coevolutionary model is applied to the genetic network inference method, the subpopulations only interact with each other through the gene expression time-courses. When the i-th subproblem is solved, the calculated expression time-courses of the other genes, that are obtained from the best individuals of the other subproblems at the previous generation, are used as $\hat{X}_j$'s (see Fig. 2.2).

2.5.2 Algorithm

Our group has proposed a cooperative coevolutionary algorithm for inferring genetic networks, what we call the *cooperative coevolutionary approach*, based on the foregoing concept [18]. The cooperative coevolutionary approach simultaneously solves all of the decomposed subproblems, which weakly interact with each other. The following is a description of this algorithm (see also Fig. 2.3).

1. *Initialization*
   N subpopulations, each corresponding to one subproblem, are generated. Each subpopulation consists of n_p randomly created individuals. The initial estimations of the time-courses of the gene expression levels, $\hat{X}_j$'s,

meanwhile, are directly calculated from the observed time-series data. Set Generation = 0.

2. *Estimation of S-system parameters*

   One generation of GLSDC [19] is performed to optimize the objective function (2.7) for each subpopulation. When the algorithm calculates the fitness value of each individual in each subpopulation, the differential equation (2.5) is solved using the estimated time-courses of the gene expression levels, $\hat{X}_j$'s. At this point, we require an initial level of gene expression (an initial state value for the differential equation) together with the S-system parameters. The initial expression level of the i-th gene is obtained from its estimated gene expression time-course, hence the value of $\hat{X}_i(0)$ is used for $X_{i,cal,0}$.

3. *Estimation of initial gene expression level*

   The objective function (2.8) is minimized on each subpopulation. The golden section search is used as the function optimizer. When the initial expression level of the i-th gene is estimated in this step, the S-system parameters are fixed to the best candidate solution of the i-th subproblem.

4. *Update of estimated gene expression time-courses*

   In this step, we calculate the time-courses of the gene expression levels obtained from the best individuals of the subpopulations, each of which is given as a solution of the differential equation (2.5). When the algorithm solves the differential equation, the new initial gene expression level estimated in the previous step is used as the initial state value. The old gene expression time-courses are then updated to the calculated time-courses. The updated gene expression time-courses are used as $\hat{X}_j$'s in the next generation.

5. *Termination*

   Stop if the halting criteria are satisfied. Otherwise, Generation ← Generation +1 and return to the step 2.

The cooperative coevolutionary approach is suitable for parallel implementation. Accordingly, we ran the calculations on a PC cluster.

2.6 Experiment on an Artificial Genetic Network

Next, we applied the problem decomposition approach and the cooperative coevolutionary approach to an artificial genetic network inference problem of 30 genes. We used the S-system model to describe the target network for the experiment. Figure 2.4 shows the network structure and model parameters of the target [24]. The purpose of this artificial problem is to estimate the S-system parameters solely from the time-series data obtained by solving the set of differential equations (2.2) on the target model.

Fifteen sets of noise-free time-series data, each covering all 30 genes, were given as the observed data in this case. Eleven sampling points for the time-

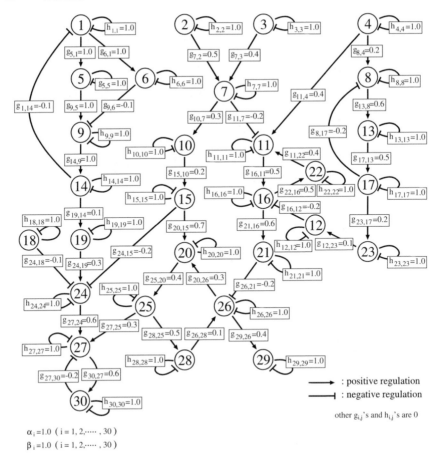

Fig. 2.4. The target network model. The S-system parameters and the network structure are shown

series data were assigned on each gene in each set. While these sets of time-series data could be obtained by actual biological experiments under different experimental conditions in a practical application, this experiment obtained them by solving the set of differential equations (2.2) on the target model. In this experiment, we estimated $2 \times 30 \times (30 + 1) = 1860$ S-system parameters and $30 \times 15 = 450$ levels of initial gene expression. The penalty coefficient c was 1.0, the maximum indegree I was 5 and the discount coefficient γ was 0.75. The following parameters were used in GLSDC applied here; the population size n_p is $3n$, where n is the dimension of the search space of each subproblem; and the number of children generated by the crossover per selection n_c is 10. Five runs were carried out by changing the seed for the pseudo random number generator. Each run was continued until the number of generations reached 75.

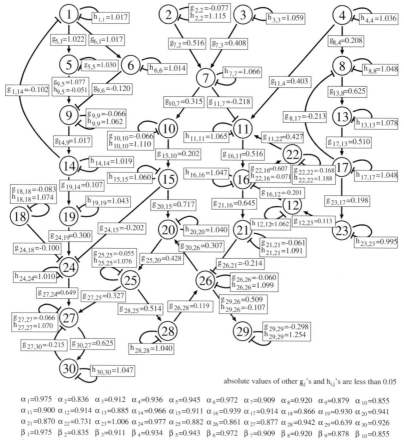

Fig. 2.5. A sample of the model inferred by the problem decomposition approach

When sufficient amounts of noise-free data were given as observed data, the models obtained by the two inference approaches presented in this chapter were almost identical. Figure 2.5 shows a sample model inferred by the problem decomposition approach. When the j-th gene positively regulates the i-th gene in the S-system model, $g_{i,j}$ is positive and/or $h_{i,j}$ is negative. Similarly, the negative value of $g_{i,j}$ and/or the positive value of $h_{i,j}$ reflect the negative regulation from the j-th gene to the i-th gene. The values of $g_{i,j}$ and $h_{i,j}$, on the contrary, are zero when the j-th gene does not regulate the i-th gene. In this experiment, we assume that, when the absolute values of $g_{i,j}$ and $h_{i,j}$ are less than 0.05, the i-th gene is unaffected by the j-th gene. Thus, although the inference methods were unable to estimate the parameter values with perfect precision, they succeeded in inferring the interactions between genes

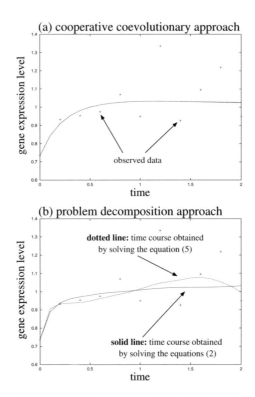

Fig. 2.6. Samples of calculated time-courses obtained from **(a)** the cooperative co-evolutionary approach, and **(b)** the problem decomposition approach, in the experiment with noisy data. Solid line: solution for the set of differential equations (2.2), where the estimated values are used as the model parameters. Dotted line: time course obtained at the end of the search, i.e., the solution of the differential equation (2.5). +: noisy time-series data given as the observed data

correctly (see Fig. 2.5). When we focus only on the structure of the inferred network, the averaged numbers of false-positive and false-negative regulations of the problem decomposition approach were 4.8 ± 5.1 and 0.4 ± 0.8, respectively, and those of the cooperative coevolutionary approach were 0.6 ± 1.2 and 0.0 ± 0.0, respectively. When used to solve the genetic network inference problem consisting of 30 genes, the problem decomposition approach required about 49.1×30 hours on a single-CPU personal computer (Pentium III 1GHz) and the cooperative coevolutionary approach required about 46.5 hours on a PC cluster (Pentium III 933MHz $\times$ 32 CPUs).

As mentioned above, when the given data were unpolluted by noise, the models obtained by the two inference methods were almost identical. In real-life application, however, DNA microarray data are usually noisy. When noisy data are given, the models inferred by the two methods slightly differ. To confirm the difference, we tested both inference approaches using noisy time-series data generated by adding 10% Gaussian noise to the time-series data computed by solving the differential equations on the target model. Figure 2.6 gives samples of the calculated gene expression time-courses obtained from the two approaches in the experiment with noisy data. Specifically, the figure shows the calculated time-courses obtained by solving the ˙set of differential equations (2.2) and the differential equation (2.5), respectively. When using the coevolutionary approach, the time-course obtained by solving the set of differential equations (2.2) was almost identical to that obtained by solving the differential equation (2.5) (see Fig. 2.6a). When using the problem decomposition approach, the time-courses calculated by these equations greatly differed (see Fig. 2.6b).

Both approaches use differential equation (2.5) to calculate time-courses of gene expression levels when inferring S-system models of genetic networks. In (2.5), however, the perturbation in the i-th gene does not affect the expression levels of the other genes. While (2.5) remains useful for inferring genetic networks, it is of no help to biologists who need a model to generate hypotheses or facilitate the design of experiments. When attempting to analyze an inferred genetic network, we thus must use the set of differential equations (2.2) as our model for computational simulation. We therefore conclude that the problem decomposition approach will not always produce a suitable model for computational simulation, as the model sometimes fits poorly into the observed data. Given that the time-courses obtained from the differential equation set (2.2) are almost identical to those obtained from (2.5), the cooperative coevolutionary approach provides us with a suitable model.

As just mentioned, the models inferred by the cooperative coevolutionary approach are suitable for the computational simulation. On the other hand, when the observed data are noisy, both inference methods erroneously infer many interactions absent from the target network (false-positive interactions) and fail to infer some interactions present in the target model (false-negative interactions). Table 2.1 and Table 2.2 show the averaged numbers of the false-positive and false-negative regulations of the two methods in the experiments using different numbers of sets of noisy time-series data. The tables show that, although the addition of more data decreased the numbers of the erroneous regulations, the inferred models still contained many false-positive regulations. The number of the false-positive regulations is, however, difficult to reduce because of the maximum indegree I introduced into the objective function (2.7). In order to make the obtained models more reasonable, we must find ways to reduce the number of the erroneous regulations. The use of further a priori knowledge about the genetic network should be one of promising means for this purpose.

Table 2.1. The numbers of the false-positive and false-negative regulations of the models inferred by the problem decomposition approach in the experiments with different sets of noisy time-series data given

The number of time-series sets	The number of false-positive regulations	The number of false-negative regulations
10	216.8 ± 1.5	17.0 ± 2.3
15	205.2 ± 3.3	9.4 ± 1.2
20	194.8 ± 2.3	4.2 ± 2.2
25	191.8 ± 3.6	2.6 ± 1.5

Table 2.2. The numbers of the false-positive and false-negative regulations of the models inferred by the cooperative coevolutionary approach

The number of time-series sets	The number of false-positive regulations	The number of false-negative regulations
10	218.4 ± 2.8	18.8 ± 1.3
15	205.8 ± 2.4	8.4 ± 1.4
20	201.2 ± 3.4	4.2 ± 1.5
25	195.6 ± 5.6	3.2 ± 1.0

2.7 Inference of an Actual Genetic Network

The inference methods described in this chapter can be used to infer large-scale genetic networks consisting of dozens of genes. The analysis of actual DNA microarray data, on the other hand, requires the handling of many hundreds or even thousands of genes. This task lies far beyond the powers of current inference methods based on the S-system model. One way to improve these inference capabilities may be to use a clustering technique to identify genes with similar expression patterns and group them together [6, 8]. By treating groups of similar genes as single network components, the inference approaches can analyze systems made up of many hundreds of genes. Based on this idea, the author and his colleagues proposed a method to combine the cooperative coevolutionary approach with the clustering technique [18]. In this section, we present the results of our attempt to apply the combined method for the analysis of cDNA microarray data on *Thermus thermophilus* HB8 strains.

Two sets of cDNA microarray time-series data, i.e., wild type and *UvrA* gene disruptant, were observed. Each data set was measured at 14 time points. Our clustering technique [13] grouped the 612 putative open reading frames (ORFs) included in the data into 24 clusters. An ORF is a sequence locating between the start-code sequence and the stop-code sequence of a gene. As we treated the disrupted gene, *UvrA*, as single network component, the target system consisted of $24 + 1 = 25$ network components. The time-series data

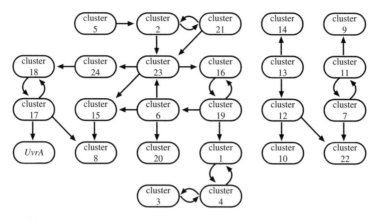

Fig. 2.7. The core network structure of the models inferred in the experiment using the actual DNA microarray data

of each cluster were given by averaging the expression patterns of the ORFs included in the cluster.

Figure 2.7 shows the core network structure when the interactions are inferred by the combined method more than nine times within 10 runs. The inferred network model seems to include many erroneous interactions due to a lack of sufficient data. The method also infers a number of reasonable interactions, however. Many of the ORFs in clusters 6, 7, 10, 15, 16, 19 and 22 are annotated to be concerned with "energy metabolism", and most of these clusters are located relatively near to one another in the inferred model. We also see from the figure that clusters 12 and 23 are located nearby the clusters of "energy metabolism". Only a few of the ORFs in the clusters 12 and 23, however, are annotated to be concerned with "energy metabolism". This suggests that some of the hypothetical and unknown ORFs included in the clusters 12 and 23 may work for "energy metabolism" or related functions. In this experiment, as the amount of the measured time-series data was insufficient, it is hard to extract many suggestions from the inferred network. We should obtain more meaningful results in either of two ways: by using more sets of time-series data obtained from additional biological experiments, or by using further a priori knowledge about the genetic network.

This section presented a method to combine the cooperative coevolutionary approach with a clustering technique in order to analyze a system consisting of hundreds of genes. The combined use of the inference method and clustering technique is a less than perfect solution for analyzing actual genetic networks, however, as few genes are likely to have the same expression pattern at any one time. A target of future research will therefore be to develop an inference method capable of handling hundreds of components.

2.8 Conclusion

In this chapter, we formulated the inference of the S-system model of the genetic network as several function optimization problems. Then, on the basis of this problem definition, two inference methods, i.e., the problem decomposition approach and the cooperative coevolution approach, were presented. Experimental results showed that, while the inference abilities of these methods are almost identical, the models inferred by the cooperative coevolutionary approach are suitable for computational simulation. These methods can infer S-system models of genetic networks made up of dozens of genes. However, the analysis of actual DNA microarray data generally requires the handling of hundreds of genes. In order to resolve this problem, this chapter also presented a method to combine the cooperative coevolution approach with a clustering technique. As the S-system model has an ability to exploit the quantitative feature of the gene expression data, it should be a promising model for the inference of the genetic network. However, as mentioned before, the methods described in this chapter are still insufficient for analyzing actual genetic networks since they have several drawbacks we must settle. A target of future research will be to develop an inference method based on the S-system model, that can handle hundreds of network components and can infer few erroneous regulations.

Sets of differential equations can capture the dynamics of target systems, and, with this capability, they show promise as models to describe genetic networks. The purpose of the genetic network inference problem based on the set of differential equations is to obtain a good approximation of the functions G_i given in (2.1), as described above. Models which accurately represent the dynamics of target systems may provide us with good approximations of G_i. In most cases, however, the parameters of these complicated models are difficult to estimate. To easily obtain a better approximation than the S-system model, the author and his colleagues proposed an approach to define the genetic network inference problem as a function approximation task [20]. Based on this new problem definition, they also proposed the inference methods derived from the neural network model [20] and the NGnet (Normalized Gaussian network) model [21], respectively. However, none of the proposed models yet seem sufficient. Another target for future research will be to develop an appropriate model for describing genetic networks and its inference method.

Acknowledgments

The problem decomposition approach and the cooperative coevolutionary approach were published in references [17] and [18], respectively. The method to combine the cooperative coevolutionary approach with the clustering technique and the analysis of *Thermus thermophilus* HB8 strains were also presented in reference [18].

The author thanks Dr. Mariko Hatakeyama of RIKEN Genomic Sciences Center for her useful suggestions and comments, and the reviewers for reviewing the draft of this chapter.

References

1. Akutsu, T., Miyano, S., Kuhara, S.: Inferring qualitative relations in genetic networks and metabolic pathways. Bioinformatics **16**, 727–734 (2000)
2. Bourque, G., Sankoff, D.: Improving gene network inference by comparing expression time-series across species, developmental stages or tissues. Journal of Bioinformatics and Computational Biology **2**, 765–783 (2004)
3. Chen, T., He, H., Church, G.: Modeling gene expression with differential equations. In: Proceedings of the Pacific Symposium on Biocomputing 4, pp. 29–40 (1999)
4. Cho, D., Cho, K., Zhang, B.: Identification of biochemical networks by S-tree based genetic programming. Bioinformatics **22**, 1631–1640 (2006)
5. Cleveland, W.: Robust locally weight regression and smoothing scatterplots. Journal of American Statistical Association **79**, 829–836 (1979)
6. D'haeseleer, P., Liang, S., Somogyi, R.: Genetic network inference: from co-expression clustering to reverse engineering. Bioinformatics **16**, 707–726 (2000)
7. D'haeseleer, P., Wen, X., Fuhrman, S., Somogyi, R.: Linear modeling of mRNA expression levels during CNS development and injury. In: Proceedings of the Pacific Symposium on Biocomputing 4, pp. 42–52 (1999)
8. Eisen, M., Spellman, P., Brown, P., Botstein, D.: Cluster analysis and display of genome-wide expression patterns. Proceedings of the National Academy of Sciences of the United States of America **95**, 14,863–14,868 (1998)
9. Friedman, N., Linial, M., Nachman, I., Pe'er, D.: Using Bayesian networks to analyze expression data. Journal of Computational Biology **7**, 601–620 (2000)
10. Ideker, T., Thorsson, V., Karp, R.: Discovery of regulatory interactions through perturbation: inference and experimental design. In: Proceedings of the Pacific Symposium on Biocomputing 5, pp. 302–313 (2000)
11. Imade, H., Mizuguchi, N., Ono, I., Ono, N., Okamoto, M.: Gridifying: An evolutionary algorithm for inference of genetic networks using the improved GOGA framework and its performance evaluation on OBI grid. Lecture Notes in Bioinformatics **3370**, 171–186 (2005)
12. Imoto, S., Goto, T., Miyano, S.: Estimation of genetic networks and functional structures between genes by using Bayesian network and nonparametric regression. In: Proceedings of the Pacific Symposium on Biocomputing 7, pp. 175–186 (2002)
13. Kano, M., Nishimura, K., Tsutsumi, S., Aburatani, H., Hirota, K., Hirose, M.: Cluster overlap distribution map: visualization for gene expression analysis using immersive projection technology. Presence: Teleoperators and Virtual Environments **12**, 96–109 (2003)
14. Kauffman, S.: Metabolic stability and epigenesis in randomly constructed genetic nets. Journal of Theoretical Biology **22**, 437–467 (1969)
15. Kikuchi, S., Tominaga, D., Arita, M., Takahashi, K., Tomita, M.: Dynamic modeling of genetic networks using genetic algorithm and S-system. Bioinformatics **19**, 643–650 (2003)

16. Kimura, S., Hatakeyama, M., Konagaya, A.: Inference of S-system models of genetic networks using a genetic local search. In: Proceedings of the 2003 Congress on Evolutionary Computation, pp. 631–638 (2003)

17. Kimura, S., Hatakeyama, M., Konagaya, A.: Inference of S-system models of genetic networks from noisy time-series data. Chem-Bio Informatics Journal **4**, 1–14 (2004)

18. Kimura, S., Ide, K., Kashihara, A., Kano, M., Hatakeyama, M., Masui, R., Nakagawa, N., Yokoyama, S., Kuramitsu, S., Konagaya, A.: Inference of S-system models of genetic networks using a cooperative coevolutionary algorithm. Bioinformatics **21**, 1154–1163 (2005)

19. Kimura, S., Konagaya, A.: High dimensional function optimization using a new genetic local search suitable for parallel computers. In: Proceedings of the 2003 Conference on Systems, Man & Cybernetics, pp. 335–342 (2003)

20. Kimura, S., Sonoda, K., Yamane, S., Matsumura, K., Hatakeyama, M.: Inference of genetic networks using neural network models. In: Proceedings of the 2005 Congress on Evolutionary Computation, pp. 1738–1745 (2005)

21. Kimura, S., Sonoda, K., Yamane, S., Matsumura, K., Hatakeyama, M.: Function approximation approach to the inference of normalized Gaussian network models of genetic networks. In: Proceedings of the 2006 International Joint Conference on Neural Networks, pp. 4525–4532 (2006)

22. Liang, R., Fuhrman, S., Somogyi, R.: Reveal, a general reverse engineering algorithm for inference of genetic network architectures. In: Proceedings of the Pacific Symposium on Biocomputing 3, pp. 18–29 (1998)

23. Liu, Y., Yao, X., Zhao, Q., Higuchi, T.: Scaling up fast evolutionary programming with cooperative coevolution. In: Proceedings of the 2001 Congress on Evolutionary Computation, pp. 1101–1108 (2001)

24. Maki, Y., Tominaga, D., Okamoto, M., Watanabe, S., Eguchi, Y.: Development of a system for the inference of large scale genetic networks. In: Proceedings of the Pacific Symposium on Biocomputing 6, pp. 446–458 (2001)

25. Maki, Y., Ueda, T., Okamoto, M., Uematsu, N., Inamura, Y., Eguchi, Y.: Inference of genetic network using the expression profile time course data of mouse P19 cells. Genome Informatics **13**, 382–383 (2002)

26. Ong, I., Glasner, J., Page, D.: Modelling regulatory pathways in Escherichia coli from time series expression profiles. Bioinformatics **18**, S241–S248 (2002)

27. Pe'er, D., Regev, A., Elidan, G., Friedman, N.: Inferring subnetworks from perturbed expression profiles. Bioinformatics **17**(S215–S224) (2001)

28. Potter, M., De Jong, K.: Cooperative coevolution: an architecture for evolving coadapted subcomponents. Evolutionary Computation **8**, 1–29 (2000)

29. Press, W., Teukolsky, S., Vetterling, W., Flannery, B.: Numerical Recipes in C, 2nd edn. Cambridge University Press, Cambridge (1995)

30. Sakamoto, E., Iba, H.: Inferring a system of differential equations for a gene regulatory network by using genetic programming. In: Proceedings of the 2001 Congress on Evolutionary Computation, pp. 720–726 (2001)

31. Shiraishi, F., Savageau, M.: The tricarboxylic acid cycle in Dictyostelium discoideum. Journal of Biological Chemistry **267**, 22,912–22,918 (1992)

32. Thieffry, D., Huerta, A., Pérez-Rueda, E., Collado-Vides, J.: From specific gene regulation to genomic networks: a global analysis of transcriptional regulation in Escherichia coli. BioEssays **20**, 433–440 (1998)

33. Tominaga, D., Horton, P.: Inference of scale-free networks from gene expression time series. Journal of Bioinformatics and Computational Biology **4**, 503–514 (2006)
34. Tominaga, D., Koga, N., Okamoto, M.: Efficient numerical optimization algorithm based on genetic algorithm for inverse problem. In: Proceedings of the Genetic and Evolutionary Computation Conference, pp. 251–258 (2000)
35. Tsai, K., Wang, F.: Evolutionary optimization with data collocation for reverse engineering of biological networks. Bioinformatics **21**, 1180–1188 (2005)
36. Voit, E.: Computational Analysis of Biochemical Systems. Cambridge University Press, Cambridge (2000)
37. Voit, E., Almeida, J.: Decoupling dynamical systems for pathway identification from metabolic profiles. Bioinformatics **20**, 1670–1681 (2004)
38. Voit, E., Radivoyevitch, T.: Biochemical systems analysis of genome-wide expression data. Bioinformatics **16**, 1023–1037 (2000)
39. Yeung, M., Tegnér, J., Collins, J.: Reverse engineering gene networks using singular value decomposition and robust regression. Proceedings of the National Academy of Sciences of the United States of America **99**, 6163–6168 (2002)
40. Yu, J., Smith, V., Wang, P., Hartemink, A., Jarvis, E.: Advances to Bayesian network inference for generating causal networks from observational biological data. Bioinformatics **20**, 3594–3603 (2004)

3

Synthetic Biology: Life, Jim, but Not As We Know It

Jennifer Hallinan

Newcastle University, Newcastle upon Tyne, UK `J.S.Hallinan@ncl.ac.uk`

3.1 Introduction

Frankenstein, Mary Shelley's classic tale of horror, warns of the perils of hubris; of the terrible fate that awaits when Man plays God and attempts to create life. Molecular biologists are clearly not listening. Not content with merely inserting the occasional gene into the genome of an existing organism, they are developing a whole new field, Synthetic Biology, which aims to engineer from first principles organisms with desirable, controllable qualities.

Synthetic biology is not a new idea. Ever since the nature of the genetic code was elucidated by Watson and Crick in 1953, the imagination of biologists has leapt ahead of their technology. By the late 1970s, that technology had advanced to the point where Szybalski and Skalka [57] could suggest "The work on restriction nucleases not only permits us easily to construct recombinant DNA molecules and to analyse individual genes, but also has led us into the new era of 'synthetic biology' where not only existing genes are described and analysed but also new gene arrangements can be constructed and evaluated." (cited in [14]).

The difference between the existing, well established field of genetic engineering and that of synthetic biology is largely one of attitude. Counterintuitively, molecular biologists who engage in genetic engineering are thinking as biologists, taking an existing complex system and tweaking it slightly by adding a gene or modifying a couple of regulatory processes, but essentially respecting the integrity of the existing system. Synthetic biologists, in comparison, think as engineers, with the aim of developing a toolbox of biological devices and modules which can be combined in a multitude of ways to produce a system with predictable properties. Synthetic biology takes the classical engineering strategies of standardization of components, decoupling of processes and abstraction of theory and implementation, and attempts to apply them to biological systems [1].

Synthetic biology is a field which has attracted considerable interest within the molecular biology community for the last decade or so, and there are a number of recent reviews addressing different aspects of the field, which together provide a broader perspective than is covered here [1,6,14,20,35,42,54]. In this review we address specifically the potential of computational intelligence approaches, in particular evolutionary computation (EC), to inform the field of synthetic biology, which is currently dominated by molecular biologists and engineers. Excellent overviews of EC are given in [19, 26, 45], and with specific reference to its use in bioinformatics in [18].

3.2 The Basics

All living organisms are constructed from a genome made up of DNA, which is a deceptively simple string of four molecules: adenine (A), guanine (G), thymine (T) and cytosine (C). Nucleotides are read by the cellular machinery in groups of three, known as codons. Each codon codes for a specific amino acid, so a length of DNA codes for a string of amino acids (via an intermediary called RNA, a similar, but chemically slightly different, molecule to DNA). A string of amino acids folds up to form a three-dimensional protein.

The "Central Dogma" of molecular biology, proposed in the late 1950s by one of the discoverers of the structure of DNA, Francis Crick, was that genetic information flows from DNA through RNA to protein, and that proteins do the work of building cells and bodies. This straightforward, mechanistic view revolutionized biology; from being a primarily observational discipline (natural history) it became an empirical science. Genetic engineering – the mechanistic manipulation of the genetic material – became feasible, and over the past half century it has been an extraordinarily successful and productive field of endeavour. Although genetically modified organisms (GMOs) have acquired an aura of menace in the public eye due to the large agricultural companies' engineering of crops and the subsequent media backlash (the "Frankenfoods" debate), genetic engineering has been hugely beneficial to society in somewhat less well-publicised applications, such as the cheap commercial production of large amounts of insulin for the treatment of diabetes [12].

Genetic technologies are improving all the time, and there is widespread belief that the goal of synthetic biology – the engineering of artificial organisms with desirable characteristics – is within reach. The basic technological requirements for engineering organisms were identified by Heinemann and Panke [32] as: standardized cloning; de novo DNA synthesis; and providing what they refer to as an "engineering chassis" for biology. The first two requirements are already standard in every biology laboratory around the world, and have well and truly demonstrated their utility, whilst the latter is the subject of considerable research.

Synthetic biology does not necessarily aim to construct organisms from scratch; much of the work currently dubbed "synthetic biology" is merely genetic engineering rebadged. However, the ultimate aim, in the eyes of many, would be to design and build an organism from first principles. And a minimal organism has, in fact, been constructed more-or-less from scratch: the poliovirus [9]. Poliovirus has a small RNA genome; it is only 7.5 kilobases in length and codes for five different macromolecules. It has been completely sequenced, and the sequence (the "recipe" for polio) is freely available from the large online databases which are the lifeblood of bioinformatics. Because RNA is chemically less stable than DNA, the researchers who re-engineered polio downloaded the RNA sequence information, translated it into its DNA equivalent, and then used a commercial DNA synthesis service to produce the required DNA. A naturally occurring enzyme known as reverse transcriptase was then used to convert the DNA back into RNA, which was added to a mixture of mouse cellular components. The RNA copied itself and produced new, infectious, disease-causing viruses.

Although the poliovirus synthesis is a major advance – the first time an organism has been built from scratch using only information about its genome sequence – there are a couple of caveats. Firstly, polio is a very small, simple organism. Active only when it is within a host cell, it has all the parasite's advantages of being able to outsource a lot of its cellular machinery to the host cell. Free-living organisms are generally much larger, with more complex genomes, and are consequently harder to construct. Further, it is now apparent that the Central Dogma is woefully incomplete; cellular metabolism relies not just upon the synthesis of proteins, but upon a bewildering variety of non-protein-coding RNAs [41]. There are also a whole suite of modifications to DNA, RNA and proteins, collectively known as epigenetic modifications, which are equally important to gene regulation and expression [34]. Moreover, the poliovirus was not constructed completely without the assistance of an existing biological system; the medium in which it was grown, although cell-free, consisted of components of mammalian host cells, without which replication would have been impossible. And, of course, the polio was not engineered to do anything except be a polio virus.

Some idea of the magnitude of the task of creating a viable organism from scratch has been provided by investigations into the minimal genome needed by an independently living cell. A number of workers have addressed this issue, and estimates for the minimal number of genes needed to sustain life tend to be around the 200 range (see, e.g. [46]), although Forster and Church [20] theorise that only 151 genes are essential. The organism[1] with the smallest known genome is *Mycoplasma genitalium*. Its genome contains 482 protein coding genes (along with 43 genes which produce RNA but not protein). Every genome appears to carry some redundancy, however, Glass et al. [25], using mutation to disrupt *M. genitalium* genes in a systematic

---

[1] I refer here only to free-living organisms. Parasites may have smaller genomes, because they can hijack some of the metabolic functions of their hosts. Viruses are, of course, the extreme example of this, to the point where some people query whether they're alive at all.

manner, estimate that only 382 of the protein-coding genes are essential to life. Interestingly, it appears that at least some of the RNA-coding genes are absolutely essential [56].

Synthetic biologists rely upon the assumption that a complex system such as an organism can be deconstructed into a manageable number of components, the behaviour of which can be understood, and which can be put together in a variety of ways to produce organisms with different behaviours, in the same way in which electronic components can be combined in different ways to make a microwave oven or a computer. Endy [14] provides the following list of minimal requirements for the successful practice of synthetic biology:

1. the existence of a limited set of predefined, refined materials that can be delivered on demand and that behave as expected;
2. a set of generally useful rules (that is, simple models) that describe how materials can be used in combination (or alone);
3. a community of skilled individuals with a working knowledge of and means to apply these rules.

Although this list provides a comforting illusion of engineering control over biological processes, Endy does, however, concede that so far this has not proven to be a straightforward process, commenting, somewhat drily, that "it is possible that the designs of natural biological systems are not optimized by evolution for the purposes of human understanding and engineering" [14]. The complex system of interactions within the cell must somehow be simplified to the point where biologists can understand it, comprehend the relationship between the network topology and its dynamics, and use this understanding to predict ways in which the networks can be engineered to produce reliably repeatable behaviours. The most popular approach to this task is currently that of network motifs.

3.3 Networks and Motifs

Biological systems consist of networks of interactions. At a cellular level, these interactions are between genes, gene products (protein and RNA), metabolites and biomolecules. Gene expression, and hence cellular phenotype, is controlled by a complex network of intracellular interactions. Gene networks are not only complex, they tend to be very large. Even the 482-gene genome of *Mycoplasma genitalium* has the potential to produce a network of over 200,000 interactions, even if we assume that each gene only produces one product (an assumption which is certainly false in most eukaryotes). Understanding gene networks is essential to understanding the workings of cells, but their size and complexity make this a challenging undertaking.

Networks are usually modelled computationally as graphs. A graph is a collection of nodes and links. In gene networks nodes usually represent genes (or proteins, or some similar biological agent). Links can be undirected, when the relationship between the agents is symmetrical, as with protein-protein interaction networks; in this case they are known as edges. Directed links, known as arcs, occur in networks such as genetic regulatory networks, where the fact that gene A regulates gene B does not necessarily mean that gene B regulates gene A. Graph theory is a field of mathematics with a long history (see, for example, [16]) and graph analysis has been applied to networks in fields as diverse as sociology, economics, computer science, physics and, of course, biology (see, e.g. [27]). There is a large body of literature on the computational modelling of genetic networks. For reviews of this broader literature, see [7, 10, 15, 23, 50, 52].

Endy's point 1 – the existence of a limited set of predefined, refined materials that can be delivered on demand – can be translated, in network terms, into the existence of a set of genetic components whose behaviour is understood and which can be combined at will. One approach to the identification of these components is the study of network motifs; small subsets of nodes with a given pattern of connectivity. The idea that networks might be decomposable into more-or-less distinct motifs with comprehensible functions was suggested by Milo et al. [44], and has since been the subject of considerable research (e.g. [4, 48]). This research is predicated upon the assumption that if enough motifs can be identified and characterized in terms of their dynamic behaviour, the behaviour of an entire network can be reconstructed. Indeed, several authors believe that the analysis of network motifs is not just the best but the only way in which large, complex networks can be analysed [31, 47, 58]. It has been asserted that "Identifying and characterizing the design principles which underlie such feedback paths is essential to understanding sub-cellular systems" [47].

Small network motifs have, indeed, been found to be over-represented in the gene networks of model organisms such as the gut bacterium *E. coli* [13,53] and the Bakers' yeast *Saccharomyces cerevisiae* [36, 60]. The assumption is that these motifs occur more often than would be expected by chance, which must imply that they are under positive selection pressure, and hence must be performing a useful, and hopefully comprehensible biological function. However, not all researchers accept that network motifs are the key to understanding the behaviour of complex networks, arguing that complex behaviour is likely to be an emergent network property rather than an additive function of individual modules [30].

If we assume that we can understand the behaviour of individual network motifs, it would appear to be a relatively straightforward matter to engineer them into genomes for the purpose of introducing new behaviour into organisms. However, naturally occurring modules are not optimized for operation within a cellular context that is not their own, and outside of their evolutionary context may not be functional at all. In addition, they are difficult to modify and it may be impossible to find an appropriate natural module that performs a desired task [1].

Although the practical significance of network motifs in general remains unproven, there is a class of network motifs likely, on theoretical grounds, to be of particular interest to those interested in network dynamics; feedback loops [39, 40]. A feedback loop may be defined as a path through a network which begins and ends at the same node; in gene network terms this is a set of regulatory interactions in which the expression of one gene ultimately feeds back to affect the activity of the gene itself. Feedback loops may be positive or negative, depending upon the parity of the number of negative interactions in the loop. Negative feedback loops tend to act within biological systems to maintain homeostasis, the maintenance of internal stability in the face of environmental fluctuations [5], whereas positive feedback loops tend to produce amplification of existing trends, or even runaway increases in dynamic behaviour [17].

Systems involving negative feedback loops tend to settle to a steady state, which may be stable or unstable. If it is stable, the result is generally a damped oscillation, tending towards the stable state; if it is unstable the result is a sustained oscillation. Oscillatory systems are crucial to all organisms, from cyanobacteria to elephants. The most obvious biological oscillatory system (or set of systems) is that underlying circadian rhythms in gene expression. The expression of many genes in almost all light-sensitive organisms varies in a periodic manner over the course of a day, affecting physical traits such as sleep–wake cycles, cardiovascular activity, endocrinology, body temperature, renal activity, physiology of the gastro-intestinal tract, and hepatic metabolism [22].

Negative feedback loops amongst the genes involved in circadian rhythms have been modelled in organisms as diverse as the fungus *Neurospora* [55], the plant *Arabidopsis thaliana* [38] and the fruit fly *Drosophila* [51]. The behaviour of a negative feedback loop depends partly upon its length. A one element loop generates a single, stable, steady state; a two element loop produces a single steady state which is approached to and departed from in a periodic way; while a loop with three or more elements can generate damped or stable oscillations depending upon parameter values. In addition, negative feedback tends to suppress the noise in biological systems which arises from the small number of molecules which may be available to take part in the processes of transcription and translation [47].

In contrast, positive feedback loops promote multistationarity; that is, the existence of a number of different stable states [17, 58]. Multistationarity is essential to development, since different cell types represent different stable states in the gene expression space of the organism. Multistationarity is frequently studied in the context of bistable switches in regulatory networks. In a bistable switch there are two stable states, between which the system can be moved by an external stimulus. Bistable switches are essentially a memory for the cell, since the state in which it finds itself is dependent upon the history of the system. Such switches have been the subject of both computational modelling (e.g. [43]) and in vivo investigation.

The theory-based approach described above is one way in which genetic circuits can be designed, but it has not so far been completely successful. As Sprinzak and Elowitz [54] point out, the study of existing gene circuits is complicated by our incomplete knowledge of the organisms in question. The circuits identified as important may be either incompletely understood or may contain extraneous, confusing interactions. Successful synthetic biology requires a more hands-on approach.

The first Intercollegiate Genetically Engineered Machine (iGEM) competition was held at the Massachusetts Institute of Technology in November 1995, with the stated aim of increasing interest and expertise in the synthetic biology field amongst students of biology and engineering. Although a wide variety of reasonably ambitious projects were proposed, the actual achievements were, although promising, more modest. Perhaps the most impressive outcome of the competition was a bacterial circuit that is switched between different states, producing differently coloured biochemicals, by red light. The research team grew a lawn of bacteria on an agar plate, and projected a pattern of light onto it to generate a high-definition "photograph" [37]. iGEM is an engaging way to recruit the synthetic biologists of the future, but the "engineer from scratch" approach will probably not be feasible in the immediate future. What, then, is the best way to do synthetic biology with current technology?

The most obvious way in which to engineer cells to produce a desired behaviour is to work with the already highly-optimized products of natural evolution. An existing system can be modified according to a careful design based upon an understanding of the system of interest. This is the most widely used approach to date, and has been successfully applied to a number of well-characterised biological circuits.

3.4 Model Biological Circuits

There are several biological control systems which are relatively simple and occur in easily-cultured bacteria, and hence have been studied in considerable detail for several decades. Circuits which respond to changes in their environment are obviously of interest to synthetic biologists. Fortunately, several such systems have been studied in detail, one of which is the lac operon in bacteria.

The lac operon is one of the all-time classics of molecular biology. An operon is a collection of genes whose functions are related, and which are usually located close together on a chromosome, under the control of a single transcriptional promoter. The lac operon was the first to be identified, and consists of a set of genes which can be switched on or off as needed, to deal with the sugar lactose as an energy source in the medium in which bacteria grow [33]. The lac operon is a natural target for synthetic biologists. For example, the engineering approach to synthetic biology was successfully applied to the lac operon in the gut bacterium *E. coli* by Atkinson et al. [2], who produced both bistable switch and oscillatory behaviour by engineering carefully designed modifications into the already well-understood genetic circuit. The major problem with this approach is that it relies upon a detailed understanding of the system to be modified. The lac operon has been intensively studied and manipulated for nearly half a century, and so is ideal for engineering, but very few other circuits are understood in anything like such detail.

Another type of circuit which has been the subject of much experimentation is the bistable switch. This is a system which can exist in either of two stable states, between which it can be flipped by a specific trigger. Such switches are of interest because they are widespread in genetic systems, are extremely important in determining the behaviour of the system, and are, in principle at least, relatively easy to understand. The simplest version of a bistable switch involve a system of two genes, each of which produces a protein which inhibits the transcription of the other. So if gene a is being actively transcribed, intracellular levels of its protein, A will be high, gene b will be inhibited and intracellular level of protein B will be low. If some event represses a or activates b, the switch will flip and the cell will enter the alternate state. A bistable switch is the simplest form of cellular memory, since once in a particular state the system will tend to stay there, even after the stimulus that flipped it into that state is gone. Understandably, bistable switches are ubiquitous in biological systems.

In real systems a simple two-repressor switch as described above is unlikely to work, given the complex genetic context in which it is embedded. Biological bistable switches tend to be somewhat more complex. A classic example is the lysis/lysogeny switch of bacteriophage (phage) λ.

Bacteriophages are viruses which infect bacteria. Being viruses, they have small genomes, and since they grow in bacterial cells they are easy to culture and collect. Phage λ has therefore been intensively studied, and much is known about its regulatory interactions [49]. Hard upon its entry into a bacterial cell, λ must decide which of two pathways it will take: lysis or lysogeny. Under the lytic pathway the viral DNA is replicated and packed into around 100 new virions, which are then released into the environment by rupture (lysis) of the host cell membrane. In contrast, if the lysogenic pathway is followed, the viral DNA becomes integrated into the host DNA and its lytic genes are repressed by a repressor known as CI. Although the lysogenic state is very stable, it can be switched to the lytic state. This occurs if an enzyme called RecA becomes activated, at which point it cleaves CI and hence inactivates it. The decision as to whether to enter the lytic or the lysogenic pathway depends upon levels of another viral protein, CII, which in turn is controlled by the physiological state of the cell. Once one of the states is entered it is maintained by a network of interactions: for the lysogenic state this network of interactions is dominated by CI and in the lytic state it is dominated by an enzyme called Cro, which also acts as a repressor.

The phage λ lysis/lysogeny switch is one of the most intensively studied of biological circuits, and hence is a tempting target for synthetic biologists. Because it is so well understood, it is possible to manipulate it in a wide variety of ways. Although more complex than the lac operon, it is entirely possible to engineer desired changes into the phage λ switch, but an intriguing alternative to designed engineering is to harness the power of chance.

3.5 Evolutionary Approaches to Synthetic Biology

The simplest way in which to incorporate stochasticity is to genetically engineer small networks of transcription factor encoding genes with varying topologies in an existing organism, and then to screen them for interesting behaviours. Guet et al. [28] did exactly this using three transcriptional regulators in the gut bacterium *Escherichia coli* and found that simple binary logical circuits, analogous to computational "and" and "or" gates could be produced by networks with a variety of different topologies.

A further refinement to the in vivo approach is the addition of selection. Applying artificial selection pressure to existing cells can lead to the evolution of solutions which would not necessarily be apparent to human ingenuity. This approach is particularly valuable for generating behaviours in which the nonlinearity and feedback in the system make its behaviour hard to predict in advance. Artificial evolution in vivo involves generating an initial pool of mutants which are then selected for their ability to exhibit a desired behaviour. This was proposed, in the context of genetic regulatory circuits, by Yokobayashi, Weiss and Arnold [61], who demonstrated that artificial evolution could rapidly convert a non-functional circuit in phage λ into a functional one, and that circuits consisting of linked logic gates could be evolved.

Atsumi and Little [3] extended this approach to modify the existing genetic circuitry controlling the λ lysis / lysogeny switch, replacing Cro with an alternative repressor with very different properties. They evolved several different types of mutants with different regulatory behaviour which they proceeded to analyse in excruciating biochemical detail, demonstrating that this approach is valuable not only from a pragmatic point of view, but also as a generator of interesting experimental subjects. Circuits generated by artificial evolution vary in the details of their organization and behaviour; studying them in detail increases our understanding of the different ways in which similar behaviour can be generated.

From a computational point of view artificial evolution in vivo is a fascinating but remote topic. But the same sort of approach can be taken in silico, generating potential network designs which can then be implemented in real cells. A computational evolutionary approach is potentially much more rapid, controllable and inexpensive than in vivo experiments. An EC approach could also help us deal with the problem of incomplete knowledge; Andrianantoandro, Basu, Karig and Weiss [1] suggest that many of the problems currently encountered in synthetic biology, arising from our lack of knowledge about the details of the systems we are trying to manipulate, can be overcome by using populations of cells, so that fluctuations in the behaviour of individual cells are buffered by the overall behaviour of the cell population as a whole. This approach clearly lends itself to modelling via a computational evolution approach such as the classic genetic algorithm, in which evolution involves a population of competing (or cooperating) individuals.

Several researchers have applied EC approaches to problems related to, although not directly relevant to, synthetic biology. For example, Deckard and Sauro [11] used an evolutionary strategy to evolve "signalling networks", modelled as systems of ODEs, which could perform mathematical operations such as multiplication by constants, square roots and natural logarithms. EC has also been used to study the regulation of flux in simple metabolic models [24], and to optimise parameter values in reaction rates in signal transduction models [8, 59]. While such models provide valuable insights into the topology and dynamics of abstract network models, they are not immediately relevant to the hands-on engineering of organisms.

EC has to date been underused in synthetic biology. Most of the work has involved deliberate design, or artificial evolution in vivo, as described above. But the potential utility of computational intelligence is starting to be appreciated. Simple networks exhibiting bistable switch behaviour were produced using computational evolution by Francois and Hakim [21]. They used a simple evolutionary algorithm operating upon a population of gene networks. As well as the promotion and repression of transcription, proteins in their model could interact to form complexes, and could be post-transcriptionally modified. Post-transcriptional modification is known to be an important component of genetic regulatory systems, but it is still relatively poorly understood, and hence is rarely included in gene network models. Networks were subjected to a range of "mutation operators" including modification of kinetic constants and addition and deletion of genes, and were scored according to how closely their behaviour matched the target behaviour. The researchers found that bistable switches could be evolved in less than 100 generations, while oscillatory circuits could be achieved in a few hundred generations.

Francois and Hakim's results are particularly interesting because they reveal the variety of network topologies which can produce bistable behaviour, few of which had the "classical" two-repressor design described above. This observation was also made by Deckard and Sauro with reference to their evolved mathematical networks. Some of the bistable switch network designs were very similar to known biological switches such as the lac operon [33]. The networks also tended to be relatively robust to variations in parameter values and to noise, a good indication that the network topology is the most important factor in producing the observed behaviour.

The relationship between network topology and dynamics is not well understood, and Francois and Hakim demonstrate how EC approaches can produce insights into this relationship. In order to advance synthetic biology, these insights must be translated into useful biological systems. Computational and laboratory approaches to the design of living systems are essential complements to each other.

Computational modelling was tightly integrated with laboratory experimentation in an elegant set of experiments carried out by Guido et al. [29] using phage λ promoters inserted into the lac operon in *E. coli*. They started by engineering the system in such a way that no regulation occurred, in order to examine the behaviour of the unregulated system. They used data on the transcriptional activity of this system to build a least-squares mathematical model of the system, parameterized using the best of 2000 random initial guesses for the model parameter values. The biological system was then modified to incorporate various combinations of gene activation and repression and the model parameters were fitted to these new data sets. The deterministic model was then modified to incorporate stochastic noise, and it was observed that fluctuations in the concentrations of transcription factors have very minor effects on the variability of the expression levels. The model was then used to generate testable predictions as to how changes in the system would affect expression levels. The researchers conclude that their simple model had captured many, but not all, of the sources of variability in the biological system.

The work of Guido et al. is interesting for several reasons. Their tightly integrated and iterative modelling and experimental processes maximize the information gained from both the computational and the laboratory experiments, and this is almost certainly a good model for future progress in synthetic biology. They also demonstrate the potential of using computational models of regulatory subsystems to predict the behaviour of larger, more complex networks, an approach which is considerably more sophisticated than the search for network motifs discussed above, which has been the focus of most research in this area.

3.6 Conclusions

Synthetic biology is the culmination of the essentially engineering approach to biology which started with the deciphering of the molecular structure of DNA in 1953. In little more than half a century we have gone from wondering whether the informational molecule of life was protein or DNA, to manipulating the characteristics of existing organisms, to a point where the construction of entirely new life forms is a realistic ambition. Advances in technology have played an essential role in this progression. While biologists tend to lay emphasis upon the development of automated systems for DNA analysis and synthesis, it is becoming increasingly apparent that the mathematical and simulation capacities of computers will be equally essential for the achievement of this goal.

Synthetic biology to date has been based upon what biologists think they know about the organisms with which they work. This approach has been moderately successful, but as Endy puts it "At present, we do not have a practical theory that supports the design of reproducing biological machines, despite great progress in understanding how natural biological systems couple and tune error detection and correction during machine replication to organism fitness" [14]. Part of the problem may be an over-reliance on theory, with a concomitant failure to extract full value from the large amounts of data and sophisticated computational algorithms currently available.

The need for data-driven, as opposed to theory-driven, approaches to synthetic biology is starting to be acknowledged. In a recent review, Heinemann and Panke [32] stress the need for quantitative analysis, which they interpret as mathematical models, for understanding and hence engineering complex biological systems. This is a step in the right direction; however, the very factors which make biological systems difficult to understand and predict also make them hard to model using classical mathematical approaches such as systems of ordinary differential equations, which describe the rates of change of continuously changing quantities modelled by functions. The more flexible, data-driven and heuristic approach of computational intelligence has enormous potential to contribute to synthetic biology.

For a start, EC algorithms mimic the processes by which life was originally created and shaped. Biological evolution is, after all, the only force which has produced life so far. From a more pragmatic point of view, EC is uniquely fitted to deal with complex, incompletely specified problems in which the goal may be clear, although the means by which to achieve it is not. Although our understanding of biological systems is constantly and rapidly improving, the field of synthetic biology is likely to be in this situation for many years yet. Until we really understand how living organisms work, the "engineering design" approach will only be successful for small, simple problems. And even when (or if!) we do have a full understanding of biological complexity, our designs will be limited by human imaginations. Evolution, real or computational, is not so limited, and may produce novel and effective solutions which are completely unintuitive to the human mind. EC algorithms, being stochastic, also provide multiple solutions to any given problem, many of which will be equally good. EC thus provides the opportunity to study the "whys" as well as the "hows" of complex biological behaviour.

The field of systems biology is in its very early stages, and new tools and techniques are under active investigation. Evolutionary computation has been employed, and found useful, but undoubtedly has more to offer, particularly if used in close conjunction with laboratory experiments. Although EC is never guaranteed to produce the optimum solution to any problem, it may well provide "good-enough" solutions to many of the problems of synthetic biology. And sometimes, as Victor Frankenstein would undoubtedly agree, good enough is the best you can hope for, particularly when trying to create life.

References

1. Andrianantoandro, A., Basu, S., Karig, D., Weiss, R.: Synthetic biology: New engineering rules for an emerging discipline. Molecular Systems Biology **2**(2006.0028) (2006)
2. Atkinson, M., Savageau, M., Myers, J., Ninfa, A.: Development of genetic circuitry exhibiting toggle switch or oscillatory behaviour in Escherichia coli. Cell **113**, 597–607 (2003)
3. Atsumi, S., Little, J.: Regulatory circuit design and evolution using phage lambda. Genes and Development **18**, 2086–2094 (2004)
4. Banzhaf, W., Kuo, P.: Network motifs in natural and artificial transcriptional regulatory networks. Journal of Biological Physics and Chemistry **4**(2), 85–92 (2004)
5. Becskei, A., Serrano, L.: Engineering stability in gene networks by autoregulation. Nature **405**, 590–593 (2000)
6. Benner, S., Sismour, A.: Synthetic biology. Nature Reviews Genetics **6**, 533–543 (2005)
7. Bolouri, H., Davidson, E.: Modelling transcriptional regulatory networks. BioEssays **24**, 1118–1129 (2002)
8. Bray, D., Lay, S.: Computer simulated evolution of a network of cell-signalling molecules. Biophysical Journal **66**(4), 972–977 (1994)
9. Cello, J., Paul, A., Wimmer, E.: Chemical synthesis of poliovirus cDNA: generation of infectious virus in the absence of natural template. Science **297**, 1016–1018 (2002)
10. de Jong, H.: Modelling and simulation of genetic regulatory systems: a literature review. Journal of Computational Biology **9**(1), 67–103 (2002)
11. Deckard, A., Sauro, H.: Preliminary studies on the in silico evolution of biochemical networks. ChemBioChem **5**, 1423–1431 (2004)
12. Dickman, S.: Production of recombinant insulin begins. Nature **329**(6136), 193 (1987)
13. Dobrin, R., Beqand, Q., Barabasi, A., Oltvai, S.: Aggregation of topological motifs in the Escherichia coli transcriptional regulatory network. BMC Bioinformatics **5**(10) (2004)
14. Endy, D.: Foundations for engineering biology. Nature **438**, 449–453 (2005)
15. Endy, D., Brent, R.: Modelling cellular behaviour. Nature **409**, 391–395 (2001)
16. Erdös, P., Rényi, A.: On random graphs. Publicationes Mathematicae **6**, 290 (1959)
17. Ferrell Jr, J.: Self-perpetuating states in signal transduction: positive feedback, double-negative feedback and bistability. Current Opinion in Cell Biology **14**, 140–148 (2002)
18. Fogel, G., Corne, D.: Evolutionary Computation in Bioinformatics. Morgan Kaufmann, Boston (2003)
19. Fogel, L.: Intelligence Through Simulated Evolution: Four Decades of Evolutionary Programming. Wiley (1999)
20. Forster, A., Church, G.: Toward synthesis of a minimal cell. Molecular Systems Biology **2**, 1–10 (2006)
21. Francois, P., Hakim, V.: Design of genetic networks with specified functions by evolution in silico. Proceedings of the National Academy of Sciences of the USA **101**(2), 580–585 (2004)

22. Gachon, F., Nagoshi, E., Brown, S., Ripperger, J., Schibler, U.: The mammalian circadian timing system: from gene expression to physiology. Chromosoma **113**(3), 103–112 (2004)

23. Gilman, A., Larkin, A.: Genetic "code": Representations and dynamical models of genetic components and networks. Annual Reviews of Genomics and Human Genetics **3**, 341–369 (2002)

24. Gilman, A., Ross, J.: Genetic-algorithm selection of a regulatory structure that directs flux in a simple metabolic model. Biophysical Journal **69**(4), 1321–1333 (1995)

25. Glass, J., Assad-Garcia, N., Alperovitch, N., Yooseph, S., Lewis, M., Maruf, M., Hutchinson, C., Smith, H., Venter, J.: Essential genes of a minimal bacterium. Proceedings of the National Academy of Sciences of the USA (2006)

26. Goldberg, D.: Genetic Algorithms in Search, Optimization and Machine Learning. Addison-Wesley, Boston (1989)

27. Gross, J.: Graph Theory and its Applications. Chapman and Hall/CRC, Boca Raton (2006)

28. Guet, C., Elowitz, M., Hsing, W., Leibler, S.: Combinatorial synthesis of gene networks. Science **296**(5572), 1466–1470 (2002)

29. Guido, N., Wang, X., Adalsteinsson, D., McMillen, D., Hasty, J., Cantor, C., Elston, T., Collins, J.: A bottom-up approach to gene regulation. Nature **439**, 856–860 (2006)

30. Hallinan, J., Jackway, P.: Network motifs, feedback loops and the dynamics of genetic regulatory networks. In: Proceedings of the 2005 IEEE Symposium on Computational Intelligence in Bioinformatics and Computational Biology, pp. 90–96. IEEE Press (2005)

31. Hasty, J., McMillen, D., Collins, J.: Engineered gene circuits. Nature **420**, 224–230 (2002)

32. Heinemann, M., Panke, S.: Synthetic biology – putting engineering into biology. Bioinformatics **22**(22), 2790–2799 (2006)

33. Jacob, F., Monod, J.: Genetic regulatory mechanisms in the synthesis of proteins. Journal of Molecular Biology **3**, 318–356 (1961)

34. Jaenisch, R., Bird, A.: Epigenetic regulation of gene expression: how the genome integrates intrinsic and environmental signals. Nature Genetics **33**, 245–254 (2003)

35. Kaern, M., Blake, W., Collins, J.: The engineering of gene regulatory networks. Annual Review of Biomedical Engineering **5**, 179–206 (2003)

36. Lee, T., Rinaldi, N., Robert, F., Odom, D., Bar-Joseph, Z., Gerber, G.: Transcriptional regulatory networks in Saccharomyces cerevisiae. Science **298**, 799–805 (2002)

37. Levskaya, A., Chevalier, A., Tabor, J., Simpson, Z., Lavery, L., Levy, M., Davidson, E., Scouras, A., Ellington, A., Marcotte, E., Voigt, C.: Synthetic biology: engineering Escherichia coli to see light. Nature **438**, 441–442 (2005)

38. Locke, J., Millar, A., Turner, M.: Modelling genetic networks with noisy and varied experimental data: the circadian clock in Arabidopsis thaliana. Journal of Theoretical Biology **234**, 383–393 (2005)

39. Mangan, S., Alon, U.: Structure and function of the feed-forward loop network motif. Proceedings of the National Academy of Sciences of the USA (2003)

40. Mangan, S., Zaslaver, A., Alon, U.: The coherent feedforward loop serves as a sign-sensitive delay element in transcription networks. Journal of Molecular Biology **334**(2), 197–204 (2003)

41. Mattick, J.: Non-coding RNAs: the architects of molecular complexity. EMBO Reports **2**(11), 986–991 (2001)
42. McDaniel, R., Weiss, R.: Advances in synthetic biology: on the path from prototypes to applications. Current Opinion in Biology **16**, 476–483 (2005)
43. Michael, D., Oren, M.: The p53-Mdm2 module and the ubiquitin system. Seminars in Cancer Biology **13**(1), 49–58 (2003)
44. Milo, R., Shen-Orr, S., Itzkovitz, S., Kashtan, N., Chklovskii, D., Alon, U.: Network motifs: simple building blocks of complex networks. Science **298**, 824–827 (2002)
45. Mitchell, M.: An Introduction to Genetic Algorithms. MIT Press, Cambridge, MA (1996)
46. Mushegian, A., Koonin, E.: A minimal gene set for cellular life derived by comparison of complete bacterial genomes. Proceedings of the National Academy of Sciences of the USA **93**(19), 10,268–10,273 (1996)
47. Orrell, D., Bolouri, H.: Control of internal and external noise in genetic regulatory networks. Journal of Theoretical Biology **230**, 301–312 (2004)
48. Przulj, N., Wigle, D., Jurisica, I.: Functional topology in a network of protein interactions. Bioinformatics **20**(3), 340–348 (2004)
49. Ptashne, M.: A Genetic Switch: Phage Lambda and Higher Organisms. Cell Press and Blackwell Scientific Publications, Cambridge MA (1992)
50. Reil, T.: Models of gene regulation – a review. In: C. Maley, E. Boudreau (eds.) Artificial Life 7 Workshop Proceedings, pp. 107–113. MIT Press, Cambridge, MA (2000)
51. Ruoff, P., Christensen, M., Sharma, V.: PER/TIM-mediated amplification, gene dosage effects and temperature compensation in an interlocking-feedback loop model of the Drosophila circadian clock. Journal of Theoretical Biology **237**, 41–57 (2005)
52. Schilling, C., Schuster, S., Palsson, B., Heinrich, R.: Metabolic pathway analysis: basic concepts and scientific applications. Biotechnology Progress **15**(3), 296–303 (1999)
53. Shen-Orr, S., Milo, R., Mangan, S., Alon, U.: Network motifs in the transcriptional network of Escherichia coli. Nature Genetics **31**, 64–68 (2002)
54. Sprinzak, D., Elowitz, M.: Reconstruction of genetic circuits. Nature **438**, 443–448 (2005)
55. Sriram, K., Gopinathan, M.: A two variable delay model for the circadian rhythm of Neurospora crassa. Journal of Theoretical Biology **231**, 23–38 (2004)
56. Szathmary, E.: Life: In search of the simplest cell. Nature **433**, 469–470 (2006)
57. Szybalski, W., Skalka, A.: Nobel prizes and restriction enzymes. Gene **4**, 181–182 (1978)
58. Thomas, R., Thieffry, D., Kaufman, M.: Dynamical behaviour of biological regulatory networks I – biological role of feedback loops and practical use of the concept of the loop-characteristic state. Bulletin of Mathematical Biology **57**(2), 247–276 (1995)
59. Tsuchiya, M., Ross, J.: Application of genetic algorithm to chemical kinetics: Systematic determination of reaction mechanism and rate coefficients for a complex reaction network. Journal of Physical Chemistry A **105**(16), 4052–4058 (2001)
60. Wuchty, S., Oltvai, Z., Barabasi, A.: Evolutionary conservation of motif constituents in the yeast protein interaction network. Nature Genetics **35**(2), 176–179 (2003)

61. Yokobayashi, Y., Weiss, R., Arnold, F.: Directed evolution of a genetic circuit. Proceedings of the National Academy of Sciences of the USA **99**(26), 16,587–16,591 (2002)

4

Dancing with Swarms: Utilizing Swarm Intelligence to Build, Investigate, and Control Complex Systems

Christian Jacob

Department of Computer Science, Faculty of Science and Department of Biochemistry & Molecular Biology, Faculty of Medicine, University of Calgary, Calgary, Alberta, Canada cjacob@ucalgary.ca

We are surrounded by a natural world of massively parallel, decentralized biological "information processing" systems, a world that exhibits fascinating emergent properties in many ways. In fact, our very own bodies are the result of emergent patterns, as the development of any multi-cellular organism is determined by localized interactions among an enormous number of cells, carefully orchestrated by enzymes, signalling proteins and other molecular "agents". What is particularly striking about these highly distributed developmental processes is that a centralized control agency is completely absent. This is also the case for many other biological systems, such as termites which build their nests – without an architect that draws a plan, or brain cells evolving into a complex 'mind machine' – without an explicit blueprint of a network layout.

Obviously, being able to understand, build and harness the emergent properties of such systems would be highly beneficial. Designers of complex systems could utilize their adaptability and robustness. Such systems would construct themselves through self-organization. However, system designers and programmers are facing an enormous challenge. How can we actually build highly distributed systems of which we have only limited understanding? Would we have to invent new ways of building, maintaining, and controlling such systems? It seems to be necessary to explore a completely new mindset for programming and system control:

> "It is no longer possible to use traditional, centralized, hierarchical command and control techniques to deal with systems that have thousands or even millions of dynamically changing, communicating, heterogeneous entities. [...] the type of solution **swarm intelligence** offers is the only way of moving forward, [we] have to rethink the way [we] control complex distributed systems."

Eric Bonabeau, 2003, co-author of *Swarm Intelligence* [7]

4.1 The Emergence of Complexity

Despite the fact that we have become so used to living with machines and devices, which seem to be relatively easy to handle from a user perspective, these mostly represent extremely complex systems, built by carefully engineered top-down design processes. Cybernetics and systems sciences are experiencing a revival, mostly in combination with the sciences of complexity. Studying designs from an integrated systems perspective has become increasingly important as we recognize that the understanding of emergent properties of complex systems that are built and designed from a bottom-up perspective prove to be more and more advantageous [3, 42, 68]. Artificial neural networks and artificial immune systems as adaptive units that learn from training data and do not rely on large-scale programming are just one aspect of the success story of the integrative systems paradigm [13, 39, 45, 55, 72]. Advances in small-scale technologies and manufacturing enable us to build systems in a self-organizing manner – with a large number of interacting entities [12, 54]. Engineers, physicists, computer scientists, biologists, and life scientists are also getting a better understanding of complex systems as a whole, and how observations and behaviors on higher levels are effects of system-specific, underlying emergent properties with bottom-up and top-down feedback loops [8, 23, 40, 51, 63, 65].

We also want to know increasingly more details about vastly complex systems that we can either currently build on our own – such as nano-devices for medication and minimally invasive diagnostics – or natural systems that we do have to understand – such as regulatory processes inside our human bodies, or complex inter-relationships within eco-, economic or social systems [27, 38, 50]. Medicine has made major advances. However, too many of all the intricate details to make our human bodies work are still mysteries to us. Will we ever get definite answers to questions such as: How does the immune system work? How does the human brain work? How can we re-program cells that had a 'program crash' and have become cancerous? What do we understand about gene expression and the complex regulatory mechanisms involved? And from an educational point of view: Do we adequately prepare our next generations of medical doctors, biologists, software developers, and engineers to equip them with these new mindsets necessary to cope with the challenges that come with the complexities of natural systems? How do we make them reveal their secrets and utilize these for our engineered systems, which we want to build, understand, and control?

4.2 Emergent and Agent-Based Computing

Agent-based computing and simulation approaches to study emergent phenomena are starting to become more and more prominent within computational and mathematical modeling [1, 20, 22, 73]. It seems necessary to explore a completely new mindset for programming, system control, and design. Engineers, physicists, computer scientists and biologists already benefit from simulation tools that are being developed to investigate agent interactions in virtual laboratories such as StarLogo [62], REPAST [2], Swarm [49], and Breve [67].

Furthermore, some of these programming, exploration and simulation tools are now entering the classrooms – from kindergarten to high school to university. The latest release of StarLogo TNG [44], for example, comes with a graphical programming interface, a terrain editor, an agent shape customizer and many other features to build complex scenes of interacting agents, observe the emergent patterns (such as the spreading of an epidemic triggered by person-to-person contact), and investigate the influences of system parameters. Using tools like these, our current generation of students acquires a better appreciation of how observable effects on a global level are the consequence of an intricate ensemble of rather simple but local interactions among a multitude of "agents". Hence, our next-generation of researchers and engineers will gain a more in-depth comprehension of the fine balances within ecosystems, how traffic control systems based on collective intelligence can help avoid gridlocks on our city roads, how a people-centered architectural perspective may change the way we design our buildings, or how nano devices and swarm-based medical simulations will make true personalized medication possible.

4.3 Chapter Overview: An Emergent Computing Tour

This chapter will take us on a virtual tour of emergent computing systems – from very small scale to large scale. We approach the question of design from a swarm intelligence perspective, in order to explore bottom-up designs of dynamic systems with emergence properties. We will discuss example implementations of agent-based systems that have been designed and investigated over the last few years in the *Evolutionary & Swarm Design Laboratory* at the University of Calgary.

We start with a demonstration of a gene regulatory system within a bacterial cell in Sect. 4.4, where we show how a seemingly simple on/off switch has a rather intriguing realization, which – despite (or rather because of) its swarm-based implementation – is functional, reliable, and highly robust.

From this example of a metabolic process inside a single bacterium, we go one level higher in Sect. 4.5, where we look at interaction patterns among a population of bacterial cells in a simulated petri dish. We demonstrate that even simple interaction rules can quickly lead to intricate modes of interaction.

The focus in Sect. 4.6 is on how human cells of different types as well as messenger molecules shape an orchestrated network to control blood clotting. How a blood clot is formed is understood to some extent, but all its details are still not completely revealed, which turns this complex interaction process into yet another challenge and useful testbed for swarm-based computer simulations.

From an orchestra of cells within the human body we proceed to the world of social insects, which have so far provided a rich reservoir of inspirations for novel computer algorithms in optimization, traffic engineering and swarm intelligence. Section 4.7 shows computer models of raiding patterns within different species of army ants. The actual ways of exploration, food attack and collection back to the nest depends on specific ant characteristics as well as on the food distribution.

Section 4.8 takes us from insect societies to predator–prey systems in a spatial, two-dimensional world. We use this example to demonstrate how seemingly 'intelligent' strategies might not result from conscious strategic planning at all, but could as well be effects from collective interactions driven by simple rules. This predator–prey scenario also shows the role that the environmental setup plays when one tries to understand and interpret agent behaviors.

Looking at interactions in human societies, it is interesting to see how observations made among social insects, such as army ants, have parallels in the interaction patterns of human crowds. In Sect. 4.9 we explore this aspect in a simulation of a sales event, with customers rushing to and from sales desks.

Finally, swarm intelligence systems have also entered the artistic scene. Interactive computer installations that promote emergent systems behaviors engage the user in a scientific exploration, but at the same time push the envelope for modern multi-media art. Section 4.10 gives a brief overview of the *SwarmArt.com* initiative, which resulted from a close collaboration between a Calgarian artist (Gerald Hushlak) and computer scientists from the University of Calgary.

4.4 Inside a Cell: Gene Regulation

What distinguishes a neuron cell from a skin cell, or a liver cell from a heart cell? As all cells in our body share the same genetic information encoded in the DNA (deoxyribonucleic acid) and all cells have an identical copy of this DNA in their nucleus, how is it that there are different cell types, which not only can have a wide variety of shapes and functional designs but also attend to different tasks within our bodies? Cell differentiation is all about a cell's genome and its proteome, the set of proteins that it manufactures depending on the set of genes that are currently switched on.[1] Different cell types have different sets of genes switched on and off. But how do these on/off switches actually work?

It turns out that even the simplest procaryotic (bacterial) cells without a nucleus have evolved intriguing mechanisms for regulating the expression of

---

[1] Biologists talk about the 'expression' of genes in the form of proteins.

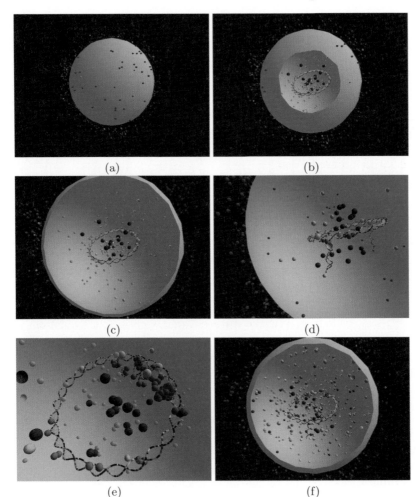

(a)   (b)   (c)   (d)   (e)   (f)

Fig. 4.1. A swarm-based model of gene regulation processes around the lactose operon in the bacterium *E. coli* **(a)** Looking at the outside of the cell immersed in lactose **(b)** Zooming inside the cell reveals the bio-molecular agents in the cytoplasm **(c)** Free polymerases (in brown) eventually dock onto the DNA and start scanning (in pink) one of the strands **(d)** The processes of transcription (from DNA to the single-strand copy of messenger RNA/mRNA) and translation (from mRNA to amino acid chains and folded proteins) are simulated. The yellow spheres represent lactose permeases, which – once close to the cell wall – enable lactose to enter the cytoplasm **(e)** The green repressors gather around the promoter site which marks the starting point of the β-galactosidase gene. The 'cloud' of these regulatory repressors prevents polymerases from docking onto the DNA at this location, hence keeping the subsequent genes switched off **(f)** More bio-molecular agents are added into the simulation, such as allolactose, acetylase, glucose, galactose, phosphate, ATP, and cAMP

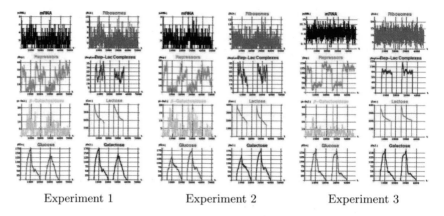

Experiment 1   Experiment 2   Experiment 3

Fig. 4.2. The concentrations of bio-molecular agents during a lactose operon simulation for three independent experiments as in Fig. 4.1. Note that there is inherent noise, which is a natural consequence of the agent-based spatial system. The switching behavior also shows a high degree of robustness. Each experiment plot shows the concentrations of mRNA, ribosomes, repressors, repressor-lactose complexes, β-galactosidase, lactose, glucose, and galactose (from left to right, top to bottom)

genes in the form of proteins as key building blocks of cells. The bacterium *Escherichia coli* (*E. coli*) has been studied for almost a century as a model organism to shed light on various designs of gene regulatory mechanisms [5, 37, 52, 59, 60].

Over the last few years, our lab has constructed 3D agent-based models of the so-called *Lactose Operon* [11, 31, 32, 70]. This system constitutes an active inhibiting switch as a result of a swarm of proteins interacting with a regulatory section (the promoter and operator regions) on the DNA. The bacterium *E. coli* is one of the main micro-organism species living in the lower intestines of mammals (including humans). One of the energy sources for these bacteria is lactose. That is, when you drink milk, these bacteria are getting prepared for a boost of energy. Figures 4.1(a) and (b) show a virtual bacterium immersed in lactose. Particles of lactose enter the cell by help of lactose-permease enzymes, which are represented as yellow spheres in Fig. 4.1(d).

Although the bacterium is enveloped by lactose, there is one crucial problem. *E. coli* can not use lactose in its original form, but has to break it down into two components: glucose and galactose. Glucose constitutes the actual energy-carrying components the bacterium is able to utilize in its metabolism. Now here comes a very neat design cue: In order to break down lactose a particular enzyme, β-galactosidase, has to be present inside the cell. However, it would be a waste of energy for the cell to produce this helpful enzyme all the time. The expression of β-galactosidase is therefore normally stopped, i.e., its gene is switched off. This inactivation is achieved by so-called repressors, which attach onto a region (the so-called operator) just before the β-galactosidase gene section on the DNA. Figure 4.1(e) shows a swarm of repressors (in green) that hover around the regulatory region for β-galactosidase and block the reading machinery – in the form of polymerases – from getting to the genome section encoding for β-galactosidase, lactose permease, and acetylase.[2]

At this point the cutting enzymes are almost completely inactivated. This actually is another intriguing aspect of the switch. As it turns out, repressors that have grabbed onto the DNA tend to fall off every now and then. This 'design flaw' is compensated for by the cloud of repressors that have high affinity for the same region. Once one repressor falls off the DNA, another one is quickly around to take its position. Therefore, by keeping a relatively low number (usually less than twenty) of repressors around, the cell manages to reliably keep the expression of the operon inactivated.

So when is the gene for β-galactosidase switched on? Once lactose enters the cytoplasm (the inside of the cell) and comes in contact with a repressor, they form a repressor-lactose complex, which completely changes the shape of the repressor. As a consequence, the repressor is no longer able to grab onto the DNA, hence looses its regulatory (suppressive) function. Eventually, with more lactose being distributed inside the cell, all repressors become disfunctional, do not block polymerases from accessing the operon and therefore start expressing β-galactosidase,[3] which, in turn, breaks down lactose into glucose and galactose.

Of course, this creates a negative feedback loop. As soon as all lactose is broken down and there is no more lactose influx into the cell, repressors can resume their task of deactivating β-galactosidase expression. The cell is then back to its initial state again. Figure 4.2 illustrates both the stochasticity and robustness of this simulated switch. Details regarding the actual interaction rules are described in [31, 32]. Our lab has also constructed a similar swarm-based model of the more complicated λ-switch involving a bacteriophage [30], which shows related properties and is also driven by emergent behaviors from local interactions of biomolecular agents.

4.5 Bacteria and Cell Populations: Chemotaxis

From examples of metabolic and regulatory processes inside a single bacterium, we go one level higher, where we look at interaction patterns among

---

[2] This is the actual *operon*, a set of genes that is switched on and off together as one module. The permease brings lactose into the cell. The acetylase is believed to detoxify thiogalactosides, which are transported into the cell along with lactose [60].

[3] At the same time, as part of the operon, lactose permeases and acetylases are also produced.

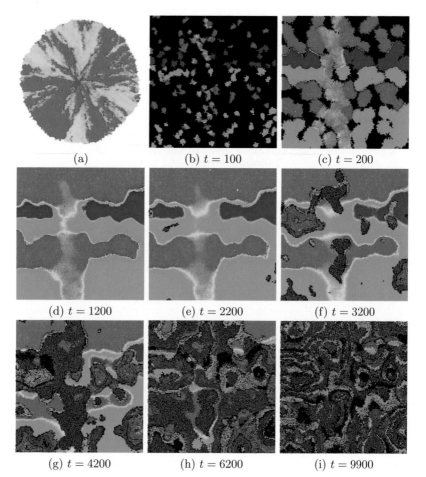

Fig. 4.3. Evolution of a bacterial culture with color genes. **(a)** 100 'color' bacteria (represented by yellow and blue color) are placed around the center of a simulated agar plate. The bacteria grow and multiply, and radial patterns emerge as food is depleted in the centre which drives bacteria outward. These patterns are similar to *in vivo* observations [48] **(b)**–**(i)** An analogous simulation, now with four types of bacteria distributed on an agar plate of size 1000×1000. The emergent patterns also resemble those found in cellular automaton simulations of excitable media [21]

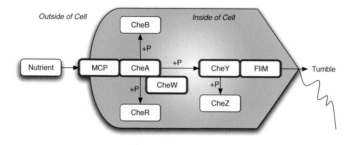

Fig. 4.4. The simplified *E. coli* chemotaxis signal transduction pathway (derived from [41]) used in our model [26,58]. A nutrient gradient is sensed by the bacterium that triggers an internal cascade of signals and molecular interactions, which result in an action such as tumbling by controlling the rotational movements of their flagella tails

a population of bacterial cells on a simulated petri dish. Bacteria are remarkably adaptive and can therefore survive in a wide variety of conditions, from extreme temperature environments of volcanoes to soil to the inside of bodies of higher organisms. Bacteria such as *E. coli* are relatively easy to handle in wet labs, thus provide perfect model organisms, and – due to their relative simplicity (which nevertheless poses enormous challenges for computational modeling) – are good testbeds for computer simulations. Apart from the interesting dynamics inside such single-cell organisms (as described in Sect. 4.4), bacteria display intriguing evolutionary patterns when they interact in large quantities (which is almost always the case).[4] The resulting emergent dynamics are not unlike the ones observed in social insects, such as ants and bees (see Sect. 4.7), where local interactions result in collective intelligence [24,28]. Such swarm- and agent-based paradigms have been used to model decentralized, massively-parallel systems – from vehicular traffic to ant colonies, bird flocks, and bacterial ecosystems [25,26,46,57,58,61].

Figure 4.3 shows the evolution among three types of simulated bacteria over time. This system (described in more detail in [26,58]) is an application of *in silico* simulation and modeling to the study of signal transduction pathways [56]. A *signal transduction pathway* (STP) describes the proteins, DNA, regulatory sites (operons) and other molecules involved in a series of intracellular chemical reactions, which eventually result in a physical action of the cell (Fig. 4.4). In this case, metabolism and chemotaxis in the bacterium *E. coli* is modeled. Through chemotaxis, bacteria navigate within their environ-

---

[4] As an illustrative example of a high-density effect, recent photographic techniques developed with *E. coli* bacteria result in a resolution of 100 megapixels per square inch [47].

ment and seek out food by following nutrient gradients. Signal transduction is modeled through an editable, evolvable rule-based artificial chemistry [16, 71] that defines all bindings and productions of proteins.

By encoding the genotype of each bacterium as a virtual DNA sequence we are also able to study how mutation and inheritance propagate through generations of bacteria. Intra-cellular complexes and molecules that match (pre-defined) behavioral patterns trigger observable behaviors, such as tumbling, swimming or cell-division. This approach allows for the construction and investigation of arbitrary STPs, which result in a large diversity of bacterial behavior patterns.

For the evolutionary studies described in this section (Fig. 4.3), each bacterium, n, carries characteristic color genes $g_n = (r_n, g_n, b_n)$, which are expressed as the red, green, and blue (RGB) color components. Each time a bacterium splits into two, the RGB genes are mutated within a radius r, resulting in new gene settings $g'_n = (r_n + \chi(r), g_n + \chi(r), b_n + \chi(r))$. Here $\chi(r)$ generates a uniformly distributed random number from the interval $[-r, r]$.

The bacteria circle (Fig. 4.3a) is an emergent phenomenon observed *in vivo* [48]. When a single bacterium is placed into agar and allowed to grow and multiply, a circle emerges as food is depleted in the centre and bacteria move outward. An interesting pattern forms in simulations when this initial setup is combined with bacteria which mutate only one of their three color components as they evolve. The culture in Fig. 4.3(a) begins with 100 bacteria in the centre, all with the same color gene, and a uniform food distribution. After 360, 000 iterations the culture has reached the edge of the square plate with a population peak of about 1, 000, 000 bacteria. There are distinct groupings of dark and light colored segments as well as mixed regions where bacteria with a variety of colors aggregate.

The triangular 'slices' radiating from the center of the circle show how an early subset of the population is able to propagate its genetic color over time. Some segments of the population exist within large pockets of similar individuals while others mix more thoroughly with non-similar bacteria. This visualization illustrates how locality and inheritance are related. Furthermore, it shows the non-uniformity of an evolving bacterial environment, an important fitness consideration.

Figure 4.3(b) shows 100 small, independent cultures distributed throughout the environment, with different initial color genes, mutation rates and swimming speeds. Figures 4.3(c) through (e) show a smoothing effect over the emitted colors, which merge into larger colored regions with bright boundaries, where bacteria with high mutation rates are located. The remaining images illustrate how the movement of bacteria and the decay of colors begin to form seemingly coordinated patterns. The bacteria continue to emit colors and more intricate patterns emerge, which we revisit from an art perspective in a later section (Fig. 4.12).

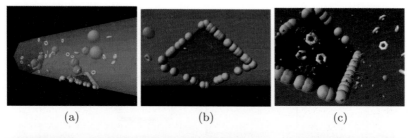

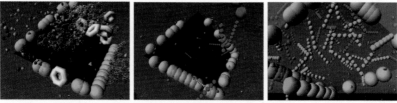

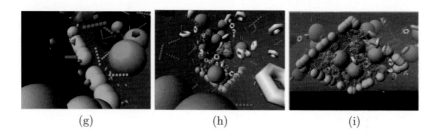

Fig. 4.5. Blood clotting: **(a)** Partial view of the simulated blood vessel with a trauma site at the bottom. Red blood cells, platelets (white) and fibrinogens (blue) are moving through the vessel section **(b)** Close-up of the damaged tissue site with collagen lining **(c)** Platelets (white) and activated platelets (transparent wire-frame) as well as activating factors **(d)** Agglomerations of activated platelets **(e)** Fibrinogens (blue) and fibrins (green) **(f)** An initial mesh formed by sticky fibrins **(g)** Red blood cells get caught in fibrin network **(h)** View from inside the blood vessel towards the wound site **(i)** Final clotting with red blood cells caught in a network of fibrins and activated platelets

4.6 The Swarming Body: Cellular Orchestration

In the previous section, we looked at cells of equal type and functionality, with clear, externally observable behaviors, such as swimming, tumbling or cell division. So, from a design and modeling point of view, these systems and their interaction dynamics seem to be less complicated than those biological systems in which an extensive cascade of orchestrated actions is necessary to achieve a final outcome – such as in blood clotting. The *coagulation* of blood is a highly complex process during which bio-molecular agents form solid clots and stop blood flow in a damaged vessel. The focus in this section

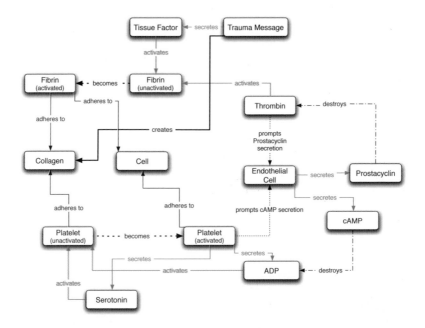

Fig. 4.6. Overview of bio-molecular agents and their interactions involved in the blood clotting cascade

is on how human blood-related cells of different types as well as messenger molecules form an orchestrated network to achieve blood clotting. How a blood clot is formed is understood to a large extent, but all its details are still not completely revealed [64].

A three-dimensional model of a section of a human blood vessel is shown in Fig. 4.5(a). Blood-related agents flow through the vessel section. At the bottom of the vessel is a trauma site, through which blood particles can escape. Figure 4.5(b) shows a close-up of the damaged endothelial blood vessel lining, around which the tips of collagen fibers are visible. Collagens are long, fibrous structural proteins and most abundant within connective tissue in mammals. When donut-shaped platelets come into contact with collagen they become activated (compare Fig. 4.6). Active platelets release messenger molecules in the form of several types of coagulation factors and platelet activating factors (Fig. 4.5c). After platelets have become activated, they tend to become sticky and adhere to collagen. Figure 4.5(d) shows several agglomerations of platelets. These platelets alone, however, are not enough to stop the escape of blood particles through the wound site. One of the proteins released by activated platelets is thrombin, which converts soluble fibrinogen into insoluble strands of fibrins which attach to collagen and form a weblike structure

(Fig. 4.5e,f). Red blood cells get caught in these fibrin networks (Fig. 4.5g), an effect that reinforces the actual clotting in combination with the activated platelets (Fig. 4.5i).

Again, what is interesting about these processes involved in coagulation is that all interactions among the bio-molecular and cellular agents can be described and implemented as activation rules based on local collisions among the agents involved (Fig. 4.6). The overall outcome leading to the formation of a web that manages to stop the escape of red blood cells is an emergent effect of this orchestration. Contrary to an orchestra, however, there is no central conductor. All actions are triggered locally; all system components work on a completely independent and distributed basis. This is what agent-based systems like this one can facilitate. Our lab has built similarly complicated swarm-based models of the processes within lymph nodes and at blood vessel-tissue interfaces involved in immune system responses to viral and bacterial infections [34, 36].

4.7 Colony Swarms: Army Ant Raids

From an 'orchestra' of cells within the human body we proceed to the world of social insects, which have so far provided a rich reservoir of inspirations for novel computer algorithms in optimization [13, 17, 43], traffic engineering [25, 57] and swarm intelligence [7]. We show computer models of raiding patterns within different species of army ants. The actual ways of exploration, food attack and collection back to the nest depends on specific ant characteristics as well as on the food distribution.

Army ants are one of the most ferocious predators in the insect world [28]. Colonies of new world army ants may contain upwards of 600,000 members. In order to feed this veritable horde, army ants conduct raids almost daily, netting over 30,000 prey items. Raids may contain up to 200,000 attacking worker ants, all of which are almost completely blind. As with all swarm raiding phenomena, army ant raids are ordered complex organizational structures conducted without any centralized control.

While raiding in army ants is not completely understood, many models of these collective intelligence systems have been constructed [15, 18, 66]. A more recent model, which we follow in our simulations, suggest that the rates at which ants are able to turn act as control mechanisms for the formation of traffic lanes during raids [14], a phenomenon also observed in human crowds (see Sect. 4.9 for details). Individual ants within our simulation move on a discrete grid. They are oriented either along the vertical or horizontal axis, but can assume any position within a grid square. Grid sites are assumed to hold dozens of ants and are therefore much larger than a single ant. Each grid site is attributed with different kinds of pheromones, that are deposited by ants and decay over time. While moving from grid to grid, ants make their

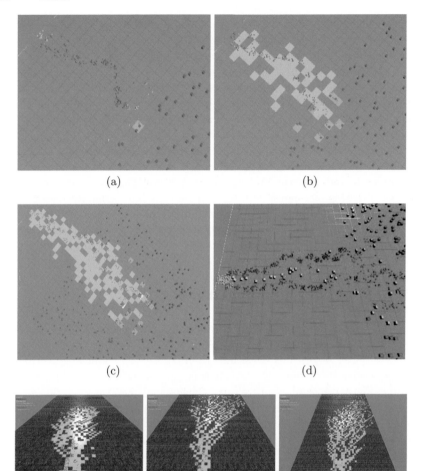

(a) (b) (c) (d)

(e) *E. burchelli*   (f) *E. hamatum*   (g) *E. rapax*

Fig. 4.7. Army ants raiding patterns: a swarm-based simulation models different species of *Eciton* army ants [69]. **(a)**–**(d)** Different stages during raiding visualize the ants' behaviors during raiding. Ants are color-coded dependent on their roles as follows: pioneers (red), followers (blue), primary recruiters (yellow), secondary recruiters (green), attackers (cyan), transporters (yellow). Cyan squares contain pheromone. Pink squares mark combat sites **(e)**–**(g)** Three different raiding patterns emerge dependent on food distribution and ant species

decisions depending on the amount of pheromone sensed in the three grid sites ahead of the ant's current orientation (Fig. 4.7).[5]

The simulation incorporates the following types of ants: *Pioneer ants* venture into unexplored territory at the front of the swarm. *Follower ants* detect and follow a trail of pheromones. Pioneers turn into *primary recruiters* upon their first encounter with a potential prey. Recruiters then initiate mass recruitment by signaling followers to change into *secondary recruiters*. Also drawn by pheromone signals, *attack ants* move towards a prey site and initiate prey capture. Finally, *transport ants* retrieve prey back to the nest.

Figure 4.7(a) shows a swarm of ants leaving the nest, which is located at the top left corner. Pioneer ants are in red, which leave pheromone trails (cyan squares in Fig. 4.7b) behind for other ants to follow. Pink squares indicate sites of combat with prey, which is distributed in a quasi-uniform fashion. Prey capture is introduced into the model by applying the principle of Lanchester's theory of combat [19]. Once the ants succeeded in overwhelming their prey, transport ants bring prey items back to the nest. As these transporting ants have reduced maneuverability due to their heavy loads, other ants tend to get out of their ways, which leads to the formation of distinct lanes to and from the nest site (Fig. 4.7d). The last row of images in Fig. 4.7 show typical exploration patterns of the ants that result from different ant parameters (e.g., speed and turning rate) and from varying the distribution of food items.

4.8 Herd Behaviors: Predators and Prey

This section takes us from insect societies to predator–prey systems in a spatial, two-dimensional world. We use this example to demonstrate how seemingly 'intelligent' strategies might not result from conscious strategic planning at all, but could as well be effects from collective interactions driven by simple rules. This predator–prey scenario also shows the role that the environmental setup plays when one tries to understand and interpret agent behaviors.

Figure 4.8(a) presents a simple predator–prey scenario. Seven predator agents (in red) are going to start off with random walks, until they encounter any of the prey agents (in cyan), which are arranged in a cluster near the top left corner. Both predators and prey are equipped with distance sensors. This way, predators become aware of prey, in which case they will go after the detected prey agent. On the other hand, prey agents are able to sense approaching predators and will turn away from any approaching predator. Prey agents also have an urge to stay close to their flockmates. Following these simple rules, once predators approach the group of prey agents, they retreat into a corner as if acting as a herd. In Fig. 4.8(b) the white prey agents are the ones 'seeing' the approaching predators. Notice that the prey group

---

[5] These sites are the one straight ahead of the ant's current site and the two sites diagonally ahead to the left and right.

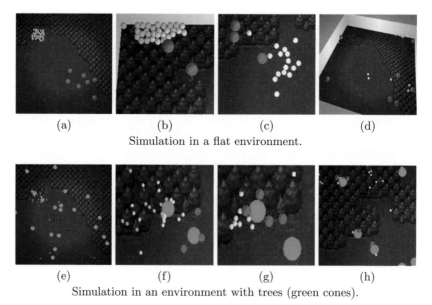

(a)   (b)   (c)   (d)

Simulation in a flat environment.

(e)   (f)   (g)   (h)

Simulation in an environment with trees (green cones).

Fig. 4.8. Predators and Prey: environmental cues influence system behaviors. Initially, prey are cyan and predators are red. Prey agents turn white when they become aware of a predator. Light green cones represent trees (bottom row)

sticks together (Fig. 4.8c). Eventually, as more predators happen to discover more prey agents and start following them, groups of predators become more scattered over the landscape (Fig. 4.8d). Eventually, most of the prey agents get caught and eliminated by the predators; this is built into the system, as our prey agents are slightly slower than the predators.

Now imagine the same scenario, but this time the landscape also contains a number of trees, which appear as green cones in Fig. 4.8(e). Both predator and prey agents follow exactly the same rules as previously. Predators will approach the prey agents, which tend to become more and more scattered into smaller groups (Fig. 4.8f). However, something interesting happens. Many of the prey agents seem to use the trees as hiding places. Both Figs 4.8(g) and (h) show examples of prey agents taking refuge from predators behind trees. In many cases, predators and prey engage in a chasing game around trees. Not knowing about the same underlying agent rules as in the scenario with trees, one might be misled to think that prey agents have discovered a successful survival strategy. Yet this is another example of emergent effects – this time, however, environmental factors play a key role, where interactions among agents and their surroundings combine into seemingly 'intelligent strategies.'

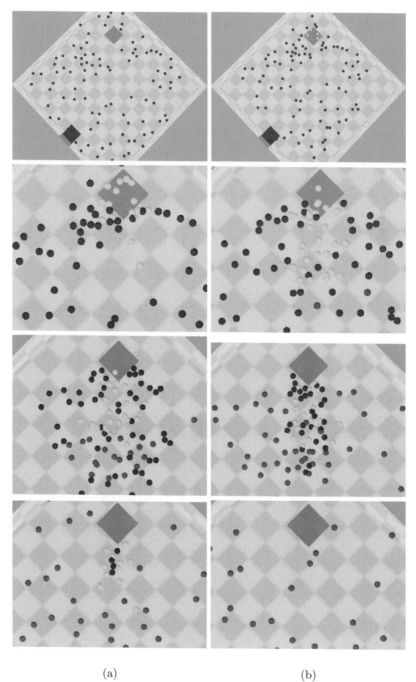

Fig. 4.9. Human crowds: sales table rush. **(a)** Low separation urge. **(b)** High separation urge. The corresponding snapshots are taken at the same simulation time steps

4.9 Human Crowds: Sales Table Rush

Looking at interactions in human societies, it is interesting to see how observations made among social insects, such as army ants, have parallels in the interaction patterns of human crowds. Here we explore this particular aspect in a simulation of a sales event, with customers rushing to and from a sales desk.

Figure 4.9 shows the scenario. One hundred agents are spread across an area confined by four walls. An exit door is located at the bottom left, marked by a blue square. The sales table area is the green square near the top corner. Agents in blue are still waiting to get into the sales area. Yellow agents have made their purchase recently. Once these agents are heading towards the exit door, they are displayed in red.

Two experiments are shown along the two columns in Fig. 4.9. On the left side, agents approach the sales area (yellow) and then move almost straight away from the sales table (yellow and red, Fig. 4.9a). On their way they encounter the blue agents that still have to get into the sales area. A downward stream of returning agents (yellow and red) seams to separate the blue agents. This is similar to the trail formation we have observed in the army ant raiding model (compare Fig. 4.7d).

In the second experiment (Fig. 4.9b) the separation urge of agents going to and leaving from the sales area is increased five-fold. This means that agents tend to separate from each other; they need more 'personal space'. This has the consequence that agents are pushing towards the sales table within a wider area. In fact, as agents approach the sales area from a wider range of directions – as they are pushed apart – they get their purchases faster and can return earlier. Eventually, only a lane of approaching (blue) and returning (yellow) agents in the centre remains.

4.10 SwarmArt.com: The Art of Swarm Intelligence

Swarm intelligence systems have also entered the artistic scene. Over the last few years, we have been working on interactive computer installations that promote emergent systems behaviors [9, 10, 53]. These exhibitions engage the user in a scientific exploration, but at the same time push the envelope for modern multi-media art. This section gives a brief overview of our *Swarm-Art.com* initiative, which resulted from a close collaboration between a Calgarian artist and computer scientists from the University of Calgary.[6]

---

[6] Gerald Hushlak, Department of Art; Jeffrey Boyd and Christian Jacob, Department of Computer Science.

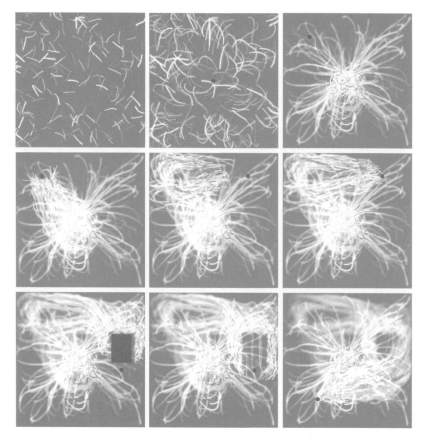

Fig. 4.10. Swarm art with BoidLand: Illustration of a swarm of paint brushes following a user-controlled red dot. An obstacle is illustrated as a grey square

The *SwarmArt.com* project[7] began in 2002 with a simple idea. Instead of having an artist direct his/her paint strokes with a single brush, we wanted to explore how a 'swarm of paint brushes' could be used to create art. We first started with a simulated two-dimensional canvas in a system called *BoidLand*, which later got extended to 3D canvas spaces [35, 46]. Figure 4.10 gives an illustration of the swarm brushes. A collection of swarming agents moves on a 2D canvas and leaves trails of paint behind. A user-controlled cursor, visualized by a red dot, acts as a focal point for the swarm agents. Hence, keeping the dot in the center of the canvas for a while, makes the swarm congregate in the middle. Continuing to move the cursor, the swarm follows. Obstacles

---

[7] The name is derived from the utilization of *swarm* intelligence techniques to *art*, implemented through computer and information processing technologies (*dotcom*).

are enveloped by the swarm agents. Once a former obstacle disappears, the canvas space is reclaimed by the swarms.

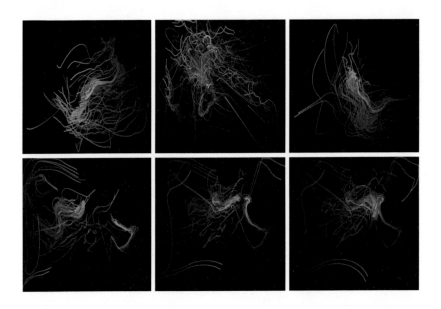

Fig. 4.11. BoidLand II: Examples of swarm paintings on a 2D canvas

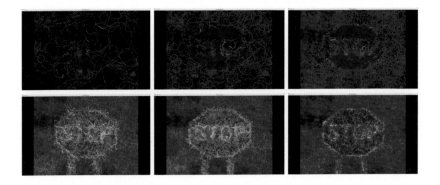

Fig. 4.12. BoidPainter: van Gogh-style swarm recreations of still images

Figure 4.11 shows a few examples of swarm art created with the second version of *BoidLand*. Here a user can either be present to direct the swarm brushes, or the system will create its own designs. In 2002, this system was extended to use live video input to direct the swarms, and was exhibited at the Nickle Arts Museum in Calgary (http://www.ucalgary.ca/~nickle/). An implementation of 3D canvas spaces and swarm choreography [46] led to another installation at the same gallery in 2003.

The following year, we not only included video-controlled, swarming agents but also aspects of bacterial chemotaxis (compare Sect. 4.5 and Fig. 4.3, in particular). We illustrate the basic idea of *BoidPainter* in Fig. 4.12. Again, a swarm of painter agents inhabits the canvas space. The canvas now consists of either a still image or a frame from a video stream. The agents pick up color information from each pixel they come across during their random walks and interactions on the canvas. Each agent remembers the color for a certain amount of time. Similar to pheromones, the agents then paint a single pixel square that they are visiting with that same color they have stored. At the beginning, color pheromones evaporate quickly, so have to be replenished often by the agents. Over time the color pheromones persist longer and thus the actual underlying image becomes more and more visible. After a predefined time, the cycle starts again with high pheromone evaporation rates. As these processes can be simulated quite fast, this technique can also be effectively applied to video streams. This allows one to directly interact with the swarm system, direct the agents, and also influence the composition and decomposition patterns created. An exhibition which applied the *BoidPainter* technique to live video streams was implemented both at the Nickle Arts Museum and in The Other Gallery at the Banff Centre (Banff, Alberta, Canada; http://www.banffcentre.ca/). More recent versions of our continued *SwarmArt.com* projects as well as a more detailed overview can be found in [33] and [10].

4.11 Conclusion and Outlook

In this chapter, we have given a glimpse into the world of massively parallel, decentralized information processing systems – with emergent properties from agents that determine the outcome of the overall system through their local interactions, and with essentially no central controller in charge. What have we learnt by looking at these example simulations which covered a wide range of scales? We went from gene regulation inside a bacterial cell, populations of cells engaged in competition for nutrients, and orchestrated blood-related biomolecular agents to army ant raiding patterns, seemingly strategic predator–prey interaction scenarios, and human crowd interaction around a sales event. A small excursion into the arts concluded our journey.

From a programming perspective, it turns out that many code libraries developed for one emergent computing application can be re-used in many others, regardless of the scale of the simulated system. Nature's systems exhibit analogous properties. Natural evolution is not possible without interactions of entities that scale from molecules to ecosystems [4]. Evolutionary computation [29] as well as more sophisticated *in silico* implementations of evolutionary processes (computational evolution) can be very beneficial for the design and fine tuning of swarm intelligence models [35, 46].

Our examples seem to be in strong support of a purely reductionist view, but it should be noted that we will only succeed with a more holistic perspective that integrates bottom-up design with downward causation [54]. Massively parallel, decentralized systems require a different approach to design, with feedback loops in both directions up and down the scales across all levels of organizational structures, following the design and control principles in living systems. Once we better understand swarm-based systems in combination with self-organization, regulatory and evolutionary mechanisms within this broader context, we are prepared to implement one of the next revolutions in engineering: self-assembling and self-adjusting structures that follow Nature's inspiring examples [6]. As mentioned in the introductory section, programming tools to explore such systems based on a large number of interacting agents are readily available, even for our youngest students [44]. Among those is the next generation of engineers and scientists that will look at our world and our technological systems from a different perspective: the world as a big swarm!

4.12 ESD Resources and Examples on the Web

Programming code and tools, as well as movies and further details about the models presented in this chapter are available online through our *Evolutionary & Swarm Design Laboratory* web site: http://www.swarm-design.org/.

Acknowledgments

Many students have come through my lab over my last few years at the University of Calgary. I am grateful to all of them for their enthusiasm and contributions. In particular I want to acknowledge the following students: Ian Burleigh (lac operon), Joanne Penner and Ricardo Hoar (bacterial chemotaxis), Garret Suen and Patrick King (army ants), Euan Forrester (BoidLand), Paul Nuytten (BoidLand II and BoidPainter). Some of the projects discussed here are based on class projects implemented by student groups. Especially I want to thank Sonny Chan, Matthew Templeton, Garth Rowe, Rennie Degraaf, Aarti Punj (blood clotting), Wilson Hui, Michael Lemczyk, David Robertson, Lia Rogers, Leslie Schneider, Vaughan van der Merwe (predator–prey), Brett Hardin, Linh Lam, Lillian Tran, Phil Tzeng, and Oscar Yeung (human crowds).

I am also grateful to my colleagues Jeffrey Boyd and Gerald Hushlak for their participation in SwarmArt.

References

1. Adamatzky, A., Komosinski, M. (eds.): Artificial Life Models in Software. Springer (2005)
2. Altaweel, M., Collier, N., Howe, T., Najlis, R., North, M., Parker, M., Tatara, E., Vos, J.R.: REPAST: Recursive porous agent simulation toolkit (2006). URL: http://repast.sourceforge.net/
3. Bak, P.: How Nature Works: The Science of Self-organized Criticality. Copernicus, Springer, New York (1996)
4. Banzhaf, W., Beslon, G., Christensen, S., Foster, J., Képès, F., Lefort, V., Miller, J., Radman, M., Ramsden, J.: From artificial evolution to computational evolution: a research agenda. Nature Reviews: Genetics 7 (2006)
5. Beckwith, J., Zipser, D. (eds.): The Lactose Operon. Cold Spring Harbor Laboratory Press, Cold Spring Harbor, NY (1970)
6. Benyus, J.: Biomimicry: Innovation Inspired by Nature. William Morrow, New York (1997)
7. Bonabeau, E., Dorigo, M., Theraulaz, G.: Swarm Intelligence: From Natural to Artificial Systems. Santa Fe Institute Studies in the Sciences of Complexity. Oxford University Press, New York (1999)
8. Bonner, J.: The Evolution of Complexity by Means of Natural Selection. Princeton University Press, Princeton, NJ (1988)
9. Boyd, J., Hushlak, G., Jacob, C.: SwarmArt: interactive art from swarm intelligence. In: ACM Multimedia. ACM (2004)
10. Boyd, J., Jacob, C., Hushlak, G., Nuytten, P., Sayles, M., Pilat, M.: SwarmArt: interactive art from swarm intelligence. Leonardo (2007)
11. Burleigh, I., Suen, G., Jacob, C.: DNA in action! A 3D swarm-based model of a gene regulatory system. In: First Australian Conference on Artificial Life. Canberra, Australia (2003)
12. Camazine, S., Deneubourg, J., Franks, N., Sneyd, J., Theraulaz, G., Bonabeau, E.: Self-Organization in Biological Systems. Princeton Studies in Complexity. Princeton University Press, Princeton (2003)
13. Corne, D., Dorigo, M., Glover, F. (eds.): New Ideas in Optimization. Advanced Topics in Computer Science. McGraw-Hill, London (1999)
14. Couzin, I., Franks, N.: Self-organized lane formation and optimized traffic flow in army ants. Proceedings of the Royal Society of London B(270), 139–146 (2002)
15. Deneubourg, J., Goss, S., Franks, N., Pasteels, J.: The blind leading the blind: modeling chemically mediated army ant raid patterns. Journal of Insect Behavior 2, 719–725 (1989)
16. Dittrich, P., Ziegler, J., Banzhaf, W.: Artificial chemistries: a review. Artificial Life 7(3), 225–275 (2001)
17. Dorigo, M., Gambardella, L.: Ant colony system: a cooperative learning approach to the traveling salesman problem. IEEE Transactions on Evolutionary Computation 1(1), 53–66 (1997)
18. Franks, N.: Army ants: collective intelligence. American Scientist 77, 139–145 (1989)
19. Franks, N., Partridge, L.: Lanchester battles and the evolution of combat in ants. Animal Behaviour 45, 197–199 (1993)
20. Gaylord, R., D'Andria, L.: Simulating Society: A Mathematica Toolkit for Modeling Socioeconomic Behavior. TELOS/Springer, New York (1999)

21. Gaylord, R., Wellin, P.: Computer Simulations with Mathematica: Explorations in Complex Physical and Biological Systems. Springer, New York (1995)
22. Gershenfeld, N.: The Nature of Mathematical Modeling. Cambridge University Press, Cambridge (1999)
23. Goodwin, B.: How the Leopard Changed Its Spots: The Evolution of Complexity. Touchstone Book, Simon & Schuster, New York (1994)
24. Gordon, D.: Ants at Work: How an Insect Society is Organized. The Free Press, New York (1999)
25. Hoar, R., Penner, J., Jacob, C.: Evolutionary swarm traffic: if ant roads had traffic lights. In: Proceedings of the IEEE World Congress on Computational Intelligence. IEEE Press, Honolulu, Hawaii (2002)
26. Hoar, R., Penner, J., Jacob, C.: Transcription and evolution of a virtual bacteria culture. In: Proceedings of the IEEE Congress on Evolutionary Computation. IEEE Press, Canberra, Australia (2003)
27. Holland, J.: Emergence: From Chaos to Order. Helix Books, Addison-Wesley, Reading, MA (1998)
28. Hölldobler, B., Wilson, E.: The Ants. Harvard University Press, Cambridge, MA (1990)
29. Jacob, C.: Illustrating Evolutionary Computation with Mathematica. Morgan Kaufmann, San Francisco, CA (2001)
30. Jacob, C., Barbasiewicz, A., Tsui, G.: Swarms and genes: exploring λ-switch gene regulation through swarm intelligence. In: Proceedings of the IEEE Congress on Evolutionary Computation (2006)
31. Jacob, C., Burleigh, I.: Biomolecular swarms: an agent-based model of the lactose operon. Natural Computing $3(4)$, 361–376 (2004)
32. Jacob, C., Burleigh, I.: Genetic programming inside a cell. In: T. Yu, R. Riolo, B. Worzel (eds.) Genetic Programming Theory and Practice III, pp. 191–206. Springer (2006)
33. Jacob, C., Hushlak, G.: Evolutionary and swarm design in science, art, and music. In: P. Machado, J. Romero (eds.) The Art of Artificial Evolution, Natural Computing Series. Springer-Verlag (2007)
34. Jacob, C., Litorco, J., Lee, L.: Immunity through swarms: agent-based simulations of the human immune system. In: Artificial Immune Systems, ICARIS 2004, Third International Conference. LNCS 3239, Springer, Catania, Italy (2004)
35. Jacob, C., von Mammen, S.: Swarm grammars: growing dynamic structures in 3D agent spaces. Digital Creativity $18(1)$ (2007)
36. Jacob, C., Steil, S., Bergmann, K.: The swarming body: simulating the decentralized defenses of immunity. In: Artificial Immune Systems, ICARIS 2006, 5th International Conference. Springer, Oeiras, Portugal (2006)
37. Jacob, F., Monod, J.: Genetic regulatory mechanisms in the synthesis of proteins. Molecular Biology 3, 318–356 (1961)
38. Johnson, S.: Emergence: The Connected Lives of Ants, Brains, Cities, and Software. Scribner, New York (2001)
39. Kasabov, N.: Foundations of Neural Networks, Fuzzy Systems, and Knowledge Engineering. MIT Press, Cambridge, MA (1998)
40. Kauffman, S.: At Home in the Universe: The Search for Laws of Self-Organization and Complexity. Oxford University Press, Oxford (1995)
41. KEGG: Bacterial chemotaxis – Escherichia coli K-12 MG1655 (path:eco02030) (2003). URL: http://www.genome.ad.jp/kegg/expression

42. Kelly, K.: Out of Control. Perseus Books, Cambridge, MA (1994)
43. Kennedy, J., Eberhart, R.: Swarm Intelligence. The Morgan Kaufmann Series in Evolutionary Computation. Morgan Kaufmann, San Francisco (2001)
44. Kloper, E., Begel, A.: StarLogo TNG (2006). URL: http://education.mit.edu/starlogo-tng/
45. Kühn, R., Menzel, R., Menzel, W., Ratsch, U., Richter, M., Stamatescu, I. (eds.): Adaptivity and Learning: An Interdisciplinary Debate. Springer (2003)
46. Kwong, H., Jacob, C.: Evolutionary exploration of dynamic swarm behaviour. In: Proceedings of the IEEE Congress on Evolutionary Computation. IEEE Press, Canberra, Australia (2003)
47. Levskaya, A., Chevalier, A., Tabor, J., Simpson, Z., Lavery, L., Levy, M., Davidson, E., Scouras, A., Ellington, A., Marcotte, E., Voigt, C.: Engineering Escherichia coli to see light. Nature **438**, 441–442 (2005)
48. Madigan, M., Martinko, J., Parker, J.: Brock Biology of Microorganisms, 10th edn. Prentice-Hall, Upper Saddle River, NJ (2003)
49. Minar, N., Burkhart, R., Langton, C., Askenazi, M.: The swarm simulation system: a toolkit for building multi-agent simulations. Working paper 96-06-042, Santa Fe Institute, Santa Fe, New Mexico, USA (1996). URL: http://www.swarm.org
50. Morowitz, H.: The Emergence of Everything: How the World Became Complex. Oxford University Press, Oxford (2002)
51. Morris, R.: Artificial Worlds: Computers, Complexity, and the Riddle of Life. Harper Collins Canada/Perseus Books (2003)
52. Müller-Hill, B.: The Lac Operon – A Short History of a Genetic Paradigm. Walter de Gruyter, Berlin (1996)
53. Nguyen, Q., Novakowski, S., Boyd, J., Jacob, C., Hushlak, G.: Motion swarms: video interaction for art in complex environments. In: ACM Multimedia. ACM, ACM (2006)
54. Noble, D.: The Music of Life. Oxford University Press (2006). URL: http://www.oup.com
55. Novartis-Foundation (ed.): Immunoinformatics: Bioinformatics Strategies for Better Understanding of Immune Function, vol. 254. Wiley (2003). URL: http://www.novartisfound.org.uk
56. Palsson, B.: The challenges of in silico biology. Nature Biotechnology **18**, 1147–1150 (2000)
57. Penner, J., Hoar, R., Jacob, C.: Swarm-based traffic simulation with evolutionary traffic light adaptation. In: L. Ubertini (ed.) Applied Simulation and Modelling, International Association of Science and Technology for Development, pp. 289–294. ACTA Press, Zurich, Crete, Greece (2002)
58. Penner, J., Hoar, R., Jacob, C.: Bacterial chemotaxis in silico. In: ACAL 2003, First Australian Conference on Artificial Life. Canberra, Australia (2003)
59. Ptashne, M.: A Genetic Switch: Phage λ and Higher Organisms, 2nd edn. Cell Press & Blackwell Scientific Publications, Palo Alto, CA (1992)
60. Ptashne, M., Gann, A.: Genes & Signals. Cold Spring Harbor Laboratory Press, Cold Spring Harbor, NY (2002)
61. Resnick, M.: Turtles, Termites, and Traffic Jams: Explorations in Massively Parallel Microworlds. Complex Adaptive Systems. MIT Press, Cambridge, MA (1997)
62. Resnick, M.: StarLogo (2006). URL: http://education.mit.edu/starlogo/

63. Schlosser, G., Wagner, G. (eds.): Modularity in Development and Evolution. University of Chicago Press, Chicago (2004)
64. Sherwood, L.: Human Physiology, 5th edn. Thomson (2004)
65. Simon, H.: Modularity: Understanding the Development and Evolution of Natural Complex Systems. MIT Press (2005)
66. Solé, R., Bonabeau, E., Delgado, J., Fernández, P., Marin, J.: Pattern formation and optimization in army ant raids. Artificial Life **6**, 216–227 (2000)
67. Spector, L., Klein, J.: Evolutionary dynamics discovered via visualization in the BREVE simulation environment. In: Artificial Life VIII. Addison-Wesley, Reading, MA (2002)
68. Strogatz, S.: Sync – The Emerging Science of Spontaneous Order. Theia Books, New York (2003)
69. Suen, G.: Modelling and simulating army ant raids. M.Sc. thesis, University of Calgary, Dept. of Computer Science, Calgary, Canada (2004)
70. Suen, G., Jacob, C.: A symbolic and graphical gene regulation model of the lac operon. In: Fifth International Mathematica Symposium, pp. 73–80. Imperial College Press, London, England (2003)
71. Suzuki, H.: An approach to biological computation: unicellular core-memory creatures evolved using genetic algorithms. Artificial Life **5**(4), 367–386 (1999)
72. Vertosick, F.: The Genius Within: Discovering the Intelligence of Every Living Thing. Harcourt, New York (2002)
73. Wolfram, S.: A New Kind of Science. Wolfram Media, Champaign, IL (2002)

Part II

Art

Evolutionary Design in Art

Jon McCormack

Centre for Electronic Media Art, Faculty of Information Technology, Monash University, Clayton 3800, Victoria, Australia
Jon.McCormack@infotech.monash.edu.au

Evolution is one of the most interesting and creative processes we currently understand, so it should come as no surprise that artists and designers are embracing the use of evolution in problems of artistic creativity. The material in this section illustrates the diversity of approaches being used by artists and designers in relation to evolution at the boundary of art and science. While conceptualising human creativity as an evolutionary process in itself may be controversial, what is clear is that evolutionary processes can be used to complement, even enhance human creativity, as the chapters in this section aptly demonstrate.

When it comes to using evolutionary methods for art and creative design, there are two related technical problems:

(i) the choice and specification of the underlying systems used to generate the artistic work;

(ii) how to assign fitness to individuals in an evolutionary population based on aesthetic or creative criteria.

In the case of (i) there are a multitude of possibilities, but a recurring theme seems to be the use of complex dynamical systems, probably due to the easy mapping of their complexity to visual or sonic output, combined with their ability to generate a large variety of aesthetically unusual output. These systems are, in computational terms, relatively simple. Yet they are able to produce highly detailed and complex images or sounds.

Two of the chapters that follow deal with fractal-based images as their underlying generative mechanism. Fractals have become a perennial obsession for many who work in evolutionary art. Mandelbrot's exploration of the 'fractal geometry of nature', as espoused in the book of the same name, triggered a revolution in thinking about the mathematical description of natural shape and form from the time it was first published in the 1970s. Fractal images seem to have a rich 'nature' of their own; one that is beguiling, cosmic, and endless. Fractals permit a vast phase-space of possibilities well-suited to evolutionary exploration. The chapter by Daniel Ashlock and Brooke Jamieson introduces the reader to this topic and illustrates ways in which evolution can be used to explore the complexity of the Mandelbrot set. The chapter by Jeffrey Ventrella uses an automated fitness function based on likeness to real images (including Mandelbrot himself!) to evolve fractal shapes reminiscent of real things, yet with their own unique aesthetic qualities – what might be termed 'a fractal reality'.

For problem (ii) listed above (evaluating fitness according to aesthetic criteria), the paradigm-setting breakthrough was the concept of the *interactive genetic algorithm* (IGA), first appearing in Richard Dawkins' *Biomorph* software, described in his 1986 book, *The Blind Watchmaker*. In this method the role of aesthetic fitness evaluation – a problem difficult to represent in algorithmic form – is performed by the user of the interactive evolutionary system, who selects individuals based on their own aesthetic preferences. Selected individuals become the parents for the next generation and the process is repeated until a satisfactory result is obtained or the user runs out of patience.

Following the pioneering work of artists such as Karl Sims and William Latham in the early 1990s, the IGA, or *aesthetic selection* as it is sometimes called, established the significant idea that creativity was no longer limited by the imagination of the creator. Artistic designs could be evolved through the simple act of selection, even if the artist doing the selecting had no underlying knowledge of how the system was producing the design! Indeed, a common observation for evolved complex systems is that the designer does not understand explicitly how a particular genotype produces the individual outcomes in the phenotype. Unlike engineering or biological applications, reverse engineering genotypes seems to offer little insight into the nature of what constitutes aesthetically pleasing images.

The use of the IGA in evolutionary art applications continues today as evidenced in my own chapter, where L-system grammars are interactively evolved using the IGA to create strange and bizarre hybrids of real plants. In the *E-volver* system described in the chapter by Erwin Driessens and Maria Verstappen, drawing agents are evolved based on their ability to modify their environment, which happens to be an image. Multiple interacting agents form a self-generating image that constantly undergoes change and renewal. The idea of conceptualising the artwork as an 'evolutionary ecosystem' seems to hold much promise for evolutionary art applications.

In artistic applications, we observe the importance of 'strange ontologies', whereby the artist forgoes conventional ontological mappings between simulation and reality in order to realise new creative possibilities. In scientific studies, the conceptualisation of models is based on our understanding of reality and we selectively try to preserve as much isomorphic mapping between reality and model as possible. Moreover, in implementing the model as a computer program, the programmer is forced to define basic ontological structures (or at least symbolic representations of them). We talk about 'individuals' and 'environments' interacting through well-defined channels of possibility. These concepts and interactions represent the ontology of the system, and must be explicitly mapped to data structures and algorithms.

In the most interesting creative works, however, conventional ontologies are changed and perverted in order to open new creative possibilities: an agent becomes a pixel in a image that is its environment; a plant grows in a way that defies biological convention, yet still maintains its own topological logic – two simple examples of the strange ontologies that appear in the chapters that follow. In this mode of artistic enquiry we are no longer trying to model reality, but to use human creativity to devise new relationships and interactions between components. These new relationships and interactions may be impossible, even illogical, yet through their impossibility they expose implicit thinking about the way we conceptualise our models in the first place. This kind of thinking may shed light on previously unconsidered assumptions in scientific models. Indeed, if we are lucky, it might even result in some enlightening artwork.

Jon McCormack
Art Area Leader

5

Natural Processes and Artificial Procedures

Erwin Driessens and Maria Verstappen

http://www.xs4all.nl/~notnot/
notnot@xs4all.nl

5.1 Introduction

The boundaries between nature, culture and technology are increasingly fading. Nature has become increasingly less natural in everyday life and technological applications have become more and more natural. Our relation to nature has rested on technical control and scientific and economic analysis for many years. Technology is at a stage where it is virtually invisible, having become a natural part of our life. New worlds open up as boundaries disappear. If technology itself sets the preconditions for survival of existing species and ecosystems, as well as the development of new and artificial ones, then this becomes an inseparable part of our experience of nature.

In addition to this shaping role, another aspect of technology is now coming to the fore, for it is deployed in a manner allowing for chance and stimulating self-organisation. Our work shows developments in the field of Artificial Life and Artificial Evolution in the form of machine generative systems, aiming for unpredictable behaviour and variety of form. This revives a century old desire: "to bring a work of art to life". These developments in art, encompassing technology, provide the contemporary phase of our culture with new possibilities of experiencing nature. As artists this challenges us to enlist vivid, artificial worlds in alliance with technology, and to design them in so refined a manner that they may awaken new experiences of beauty. Not merely in relation to the external forms themselves, but also in terms of the underlying processes which create it all.

5.2 A Brief Review

To sketch the background to our interest in Artificial Life and Artificial Evolution we will pay a brief visit to the end of the eighties and the beginning of the nineties. At the time, still studying at Rijksakademie Amsterdam, we found ourselves in a situation that forced us to question art practice. We were under the impression that a work of art seemed to be primarily a strategic instrument guaranteeing the continuity of the institutionalised art establishments. New artworks had to be shown every month, and production had to be kept up. The journals gave the most extensive and favourable reports to those galleries and art institutions buying big expensive pages of advertisements in their magazine. The so-called new and interesting seemed to be strongly intertwined with mutual commercial interest. We concluded that the art world was a generative system maintaining itself. In addition, post modernism stripped us of illusions, leaving us with the view that nothing is possible beyond the appropriation of images from other cultures and from the past: repetition, recombining, eclecticism, the end of history [2]. At the same time, we became interested in the theories of chaos and complexity. Scientists had discovered that through complex matter-energy flows, order could arise spontaneously from chaos. Under the right circumstances this dynamism allows new wholes to arise, which are more than just the sum of their various parts. Small modifications may sometimes lead to a dramatic turn around that is unpredictable. In spite of the fact that chaos research attempts to explain the origin of what is new, it remains uncertain what is going to occur in the future [10, 13]. As artists, we experienced that the associative tendency of our human mind may disturb the spontaneous development of new possibilities. By following our intuition, we often close off avenues that might have led to interesting results.

In this context the idea of automating art production arose. This would expose the underlying generative mechanism of the art world, and it would circumvent the cultural and the biological limitations of human art at the same time. Initially this idea was a rather nihilistic response to the powerless situation in which we seemed to find ourselves. However, this point of departure rapidly became an adventure when we realised how difficult and at the same time challenging it is to attempt an art in which spontaneous phenomena are created systematically, art that is not entirely determined by the subjective choices of a human being, but instead is generated by autonomously operating processes. These are self-organising processes able to create an endless flow of surprising results.

With the above in mind, we began to develop generative systems in which we attempted to formalise processes of creation. In addition to working with physical and chemical processes generating morphological transformations, we use the computer for the development of artificial worlds with self-organising properties. We wish to see what happens if you describe all the aspects of the development and growth processes. Not simulations of laws that are valid in our physical world, but instead the definition of an artificial nature,[1] with fic-

---

[1] The concept "Nature" is often used in the meaning of unspoiled land, an area that has not been touched or cultivated by Man. In a broader sense, however, nature means "disposition" or "character". From this point of view, Nature is the expression of the underlying laws that shape a world or entity. These fundamental laws of nature determine which development, growth and transformation processes are possible, and which are not.

tional laws constituting a complete world of its own: a parallel universe with intrinsic creative principles and spontaneous expressions. These generative systems are able to create very detailed images and forms, without precisely determining a priori how they will look and without attributing a specific meaning to them. We discovered that our approach fits the broader research field of Artificial Life. But whereas A-life science emphasises "knowing and understanding" of complex behaviour and self-organisation, we are more interested in "making and becoming", to give expression to the desire for an activity exploring the unseen, the unimagined and the unknown [16].

The specific properties of the computer enable it to be equipped with an autonomous, non-human way of "thinking". Unlike the associative tendency of the human mind, most software to date has no built-in preference for one solution or another. It is able to research the whole spectrum of possibilities consistently and systematically. Yet not all solutions within this infinite universe of possibility are equally interesting to human beings. Just as in the world around us, these artificial worlds have fascinating, strange but also tedious phenomena. We therefore need a powerful mechanism to enable us to explore the field of possibilities and which will enable further development of areas potentially interesting to us. In the world around us the process of evolution takes care of the multiformity of biological life and the creation of new species from old ones. The species become increasingly better adapted to the changing ecosystem by a process of mutation and selection. This is a mechanism that we can use. If we can link a mechanism of artificial evolution to a morphogenetic system (morphogenesis is the structural development of an organism) then we have an effective manner of giving direction to the image generating properties, without pre-determination of how the results deriving from this are going to look.

In the following, we will go into a few projects and describe the changing role of artificial evolution in relation to various morphogenetic processes. Two works that are more involved with the observation of existing processes introduce these projects. The theme of artificial evolution will emerge in another manner here.

5.3 Carrots and Tubers

Nowadays, science and industry are able to develop new and more attractive crops via direct intervention in genetic material, but the role of industry and technology have influenced the development of new biological species and variants for far longer than this. In particular, the species used for large-scale food production have been subject to an evolutionary process spurred on by industry for a long time, with far-reaching control of environmental factors. While the natural process of evolution usually engenders multiformity and diversity so that species remain strong and adaptive, industry manoeuvres the process as much as possible in a direction of efficiency, uniformity and

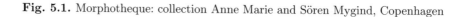

(a) Morphotheque#8   (b) Morphotheque#9

Fig. 5.1. Morphotheque: collection Anne Marie and Sören Mygind, Copenhagen

homogeneity. This results in many current species being extremely sensitive
to disease and degeneration; science then subsequently develops technological solutions for this. In short, species development and species preservation
(including humanity) are strongly correlated with technological management
and monitoring. This technology driven reciprocal dependency has become
our natural condition [12]. Morphotheque #8 (1994) and Morphotheque #9
(1997) (see Fig. 5.1) are two works reflecting this topic. Morphotheque #8 is
a collection of 28 elements, 1:1 copies of potatoes (cultivar Jaerla) executed
in painted aluminium. Morphotheque #9 is a collection of 32 elements, exact
copies of carrots executed in painted plaster.

Wild potatoes have relatively large diversity of form. Form characteristics
are in part genetically determined, but environmental factors (climate and
soil structure) influence the ultimate form at least as strongly. The industrial potato has become rather uniform due to Man's continual selection and
cultivation regarding form. That is also the intention, for peeling irregular
potatoes is impractical and even more important: it is very uneconomical.
Potatoes for consumption are reproduced via clones to ensure product characteristics remain constant. In the northern European industrial agriculture
there are strict checks on the growth quality of the seed-potatoes. The soil
structure is also maintained as homogeneously as possible and artificial fer-
tiliser ensures sufficient nutrients. Pesticides and chemicals ensure that no
other species can inhibit the culture of the potato plant and its tubers [19]. In
spite of that control, some tubers or roots do escape this imposed uniformity.
These amusing and suggestive forms stimulate our imagination. Local irregu-
larities in soil structure among other things (stones, roots or other tubers) or
sudden climate change (including a period of too much or too little moisture)
may cause these irregular forms of growth. Other causes are possible mutation
in the genes or accidents in the process of growth, which is not yet fully under control. Because these differently formed products do not meet standards,
they cannot be sold to consumers, and we are therefore never allowed to see

these variants in form. If by accident this does in fact happen, then they are associated with pathological deviation, disease, degeneration and ugliness. A farmer selling carrots and potatoes at an Amsterdam market experienced this as provocation when we asked him if he could put the irregular, strangely formed variants aside for us: "How can you even think that I would grow inferior and malformed products!" Normally speaking, these deviant growths disappear into potato starch or they serve as cattle fodder. Sorting takes place in large distribution centres and on location, we made a selection out of a great number of rejected products, representing the variety and diversity in form within the species. We thought it was important to record the forms in their three-dimensionality, and not only by means of photography. For this reason a copy was made of each form in a durable material and it was then painted with acrylic paint so that the colour impressions would also match the original.

We also researched whether the diversity in shape of the Morphotheque #8 potatoes was in fact determined by external factors. The following year we planted the potatoes as seeds in a piece of land, where they grew under circumstances comparable to those in the industry. Without exception, this yielded "normal", round progeny. We suspect that a varied environment and variable circumstances may stimulate diversity in form, but we did not carry out the experiment in reverse (planting uniform seeds in an uneven soil, no removal of weeds, no spraying and fertilising, etc.). In Morphotheque #9, one of the carrots stands out immediately. Here it is the straight form that is different, and in fact it is the only carrot from the supermarket.

5.4 Breed

Is it possible to breed industrial products via a technological route whereby multiformity is achieved instead of uniformity? In 1995, we began the project titled Breed with this question as a starting point, which ultimately led to a system that automates both the design and the production of three-dimensional sculptures. Breed makes a "mass production of unique artefacts" possible, with a minimum of human intervention.

To attain multiformity without designing each individual form in detail, we focussed on a system – in fact a recursive cellular automaton – able to generate a large number of morphogenetic rules independently. The rules prescribe how a detailed spatial form arises from a simple elementary form, or rather "grows". However, not all rules lead to a spatial object that is executable as sculpture. The system therefore comprises an evolutionary component that seeks solutions for this problem completely independently.

Fig. 5.2. Breed, stages of morphogenic development

5.4.1 Morphogenesis

The underlying principle of the Breed morphogenesis is division of cells.[2] One initial cell, a cube, engenders throughout successive stages of cell division a complex, multi-cellular "body". Morphogenetic rules determine how the division of a cell occurs, dependent on its situation between the cells surrounding it. Every potential situation has a separate rule, so a cell surrounded on all sides by other cells may divide differently from a cell that only has a neighbour on the left and underneath, or a cell with nothing at all in the vicinity, et cetera. Each rule is coded in a gene, and the complete set of rules forms the genotype of the growth.

A parent cell divides by halving the cubic space that it takes up according to length, width and depth. This creates eight smaller spaces, each of which either contain a child cell or remain empty, according to the rule applicable to the parent cell. The new child cells again function as parent cells during the following division stage, the empty spaces remain empty. Each division refines the cells and differentiates the form further, until growth ceases after a number of stages (see Fig. 5.2). The ultimate form, the phenotype, is not specified at a high level in its genotype, but is the result of the recursive application of simple rules on the lowest organisation level, that of the cell. A spatial characteristic is thus not directly specified, but is the result of a local chain reaction of events. The ultimate appearance of such a virtual Breed object may take many forms (see Fig. 5.3). The majority of these are comprised of many parts floating separately from each other in space. This is no problem while it remains a computer model on the screen where gravity does not count, but turned into real material, subject to gravity, such an incoherent structure would collapse. If we wish to produce such a computer designed object, then it must form a physical whole. If the model is to be feasible as sculpture there can be no loose cells, or groups of cells that do not connect to the rest. When the morphogenetic system was eventually operational, we made protracted experiments with randomly composed genotypes, in order to see which phenotypes this would produce. By doing a large number of samples we realised how huge the room for possible forms of manifestation was, and

---

[2] We often borrow terms from the domain of biology such as organism, cell, genotype, phenotype, ecosystem, morphogenesis and evolution, but we use these concepts in an abstract manner. This concerns three-dimensional forms in the case of Breed, thus volume-element or voxel is intended by the term cell, and voxel-multiplication process by cell division.

that only a smaller section of that space met the conditions of constructibility we considered necessary. Yet this smaller section would still be able to run to an enormous number, far too great to generate within a human life, let alone examine.

(a) Breed 1.2 #e234   (b) Breed 1.2 #e365   (c) Breed 1.2 #f315

Fig. 5.3. ProMetal models, approx. 80 × 80 × 80 mm, stainless steel/bronze. Courtesy gallery VOUS ETES ICI, Amsterdam

5.4.2 Selection and Mutation

To automate the search for constructable results, it is necessary to establish objective and measurable preconditions for constructibility. One crucial condition is that the phenotype should be completely coherent, i.e. consisting of a single part. The computer can compare two models with each other by counting the number of separate parts the model contains. The model with fewer parts satisfies the criterion of "coherence" more, or rather is "fitter" than the model with the higher number of parts. We are now able to implement a computerised process of trial and error that incrementally evolves in the direction of potential solutions. The simplest method is already effective: take a randomly composed genotype as base, generate the phenotype and test it for fitness; mutate the base genotype, generate the phenotype and test it for fitness; compare both results with each other, and take the result with the highest fitness as the new base. Repeat the mutations until the result satisfies the stated requirements. It is not of course certain that this process of evolution will always lead to a solution of the constructibility problem. This minimal mechanism (the two-membered evolution strategy) has the characteristic of firmly clinging to a turning it has already taken. Favourable mutations in general become rarer, and sometimes even impossible, to the extent that fitness increases. As Mitchell Whitelaw put it: "... the structure of the evolutionary process mirrors the morphogenetic process; rather than searching for a single absolute goal, this simple stepwise evolution uses only a local comparison. As it forms a sequence of incrementally fitter forms, the process paints itself into a corner; the final optimal form is in fact only the most optimal form that the specific sequence of random alterations has produced." [17]

Breed carries out one evolution after the other, always from a different initial genotype. The number of mutations per evolution is limited; if no suitable solution presents itself within the maximum number of mutations allowed, Breed begins a new evolution from a subsequent starting point. By approaching the problem from different sides, all sorts of solutions reveal themselves. The first evolution experiments had a surprising result. It was apparent that virtually all constructable solutions consisted merely of one single cell (the path of least resistance also seems to occur here). This unexpectedly simple result, however valid, did not meet our attempt at multiformity! We therefore added a "volume" and "surface area" criterion alongside the "coherence" criterion. All these criteria determine the objective characteristics of the object, but at the same time leave open how the form is going to turn out. We only fix essential limits, yet there are billions of possibilities within this.

5.4.3 Selection and Execution

Breed evolves forms based on a fixed set of selection criteria, that remain the same during the run of the process or in the subsequent breeding sessions. Innumerable forms of manifestation may show up. Some forms appeal more to the imagination than others, but this plays no role at all within the system. Evolution has the role of constructor here, taking no account of the aesthetic quality of the result. Ultimately, we specified which models should or should not be produced based on our personal preferences. We paid attention to significant differences in form, style and structure in the selection – just as for the Morphotheques – in order to reflect the form diversity of the system.

An initial series of six sculptures was executed manually in plywood. The tracing, sawing and gluing was a time-consuming technique and in any case, we wanted an industrial procedure. It was not possible to produce such complex forms via computer controlled Rapid Prototyping techniques until the end of the nineties. Only with the arrival of the SLS (Selected Laser Sintering) technology did it become possible to computerise the whole line from design to execution. A second series of nine samples was realised with this technique in DuraForm nylon. Recently, a series of six sculptures were produced with the ProMetal technique. These objects are made out of stainless steel, infiltrated with bronze. The results became smaller and more detailed so that the separate cells of the objects began to be absorbed into the total form.

5.5 E-volver

We wondered if we could design an artificial evolutionary process that would not be evaluated based on a set of objective criteria hard-coded within the system, but on the basis of different subjective selection criteria introduced by user interaction. In this way one obtains a very open system, where one can explore a great many different paths within the gigantic realm of possibilities. The project E-volver has expanded this idea into a working system. E-volver encompasses an alliance between a breeding machine on the one hand and a human breeder on the other. The combination of human and machine properties leads to results that could not have been created by either alone. The cultivated images show that it is possible to generate lively worlds with a high degree of detail and often of great beauty.

In this system, a population of eight virtual creatures is active within the screen to construct an ever-changing picture. Each pixel-sized creature is an individual with its own characteristic behaviour that literally leaves its visible traces on the screen. Because each creature is incessantly interacting with its environment, the picture remains continually changing. Figure 5.4 shows snapshots of such pictures. Artificial evolution is the means used to develop the characteristics of the individual creatures and the composition of the population as a whole. The user directs the process based on visual preferences, but they can never specify the image in detail, due to the patterns and unpredictabilities within the system. We will now give a detailed description of the system, first of the morphogenetic and then of the evolutionary component.

5.5.1 Generative Process

The E-volver software generates a colourful dynamic flow of images in real-time. The successive images are cohesive because each new image only differs a little from the previous one. Due to this rapid succession of separate images, like frames in a film, you experience continuity of movement. In contrast to a film (where the images once filmed and recorded can be repeated over and over again), there is no question of filming and recording E-volver – for the images are created at the moment of presentation. E-volver will never repeat earlier results; it always exists in the present and gives a immediate, life-like experience. In order to generate images that are coherent in space and time, we use a development process that is also active in space and time. In this case, the space consists of a right-angled grid of "picture elements" corresponding to the pixels of the screen. The essential characteristics of the picture elements are their individual colours and positions in the grid. Time, just like space, is divided up into equal pieces. Time passes in discrete steps, each as short as the other. Thousands of such time steps pass per second.

The pixel grid forms the environment – the habitat – of eight artificial "monocellular" creatures, who each have the size of a pixel. They are simple,

Fig. 5.4. E-volved culture, two stills, inkjet on canvas, 600 × 300 cm. Courtesy LUMC, Leiden and SKOR, Amsterdam

autonomous operating machines, mobile cellular automata, which operate according to a fixed, built-in program. The creatures are able to change the colour of the pixel they stand on, and move to a neighbouring pixel. Their program describes how they do this, dependent on local circumstances: colours from the immediate vicinity. Each creature is provided with its own control program, endowing it with a specific and individual behaviour, reminiscent of Langton's work [8]. During each successive time step all the creatures execute their program, one after the other; they change their environment by leaving visible traces in their habitat. Innumerable scenarios for interaction and emergence are possible because one creature's traces can influence the other creature. The habitat functions here as a common buffer, which holds history and shapes the present with it. The collective activity of all the creatures in the habitat forms an artificial ecosystem, developing gradually in time, and constantly brought into view. In principle, all the actions of the creatures are visible; E-volver does not employ any complicated visualisation techniques or hidden layers; the image plane is the world itself and not a representation of something else. The image plane, the habitat, is unlimited in the eyes of the creatures. When a creature bursts through the edge of the plane, it reappears on the opposite side. All the picture elements are equal to the creature in the sense that they all have a complete environment: each pixel is surrounded by eight neighbouring pixels (Moore neighbourhood). There are no edge or corner pixels impeding its perception and movement – its notion of space is continuous.

At a superficial glance, pixels on a screen each have a colour. On closer inspection, however, we see that each pixel is actually constructed from three smaller sub-pixels, a red and a green, and a blue one. By controlling the intensity of the three sub-pixels, a pixel can arouse the impression of any colour at all. RGB (Red, Green, Blue) is therefore the standard for machines producing coloured light, such as monitors and projectors, and a standard for colour specifications in computers. It is not so much the RGB components that are important to human perception and the artificial beings in E-volver, as the interpretation in the components Hue, Saturation, Lightness (HSL). These elementary properties of colour are perceptually more relevant, because they are more revealing than the RGB values. E-volver deploys both colour models: HSL for the actual habitat, RGB for its reproduction in coloured light. Both colour models can easily be converted into each other arithmetically. In this habitat of coloured pixels the creature exists. The creature is able to perceive its immediate environment – in the form of HSL colour components. The creature is able to alter the colour components at its present standpoint, and take a step towards a neighbouring pixel. It is a simple creature, its state is only given by a colour, a position, and a direction of looking and moving. The memory of the creature does not extend further than the colour and direction it had during the previous time step. Its control program propels the creature; this determines what the creature's following state will be, given the present state and the local context. The control program converts input into output within a single time step. The input is the creature's present colour, position and direction, and the colour of the underlying and bordering pixels. The output is the creature's new colour and direction. A control program is thus constructed of two parts, one part taking care of the colour determination, and one part deciding on the direction to be taken. One random example of a control program in pseudo-code follows:

- Colour determination:
  - view the light values of the surrounding eight pixels;
  - choose the value that most resembles your current light value, and adopt it;

- if the value is greater than the light value of the pixel you are on, then increase your own light value;
- mix your hue and saturation with the hue and saturation of the pixel you are on.

- Direction determination:
  - view the saturation values of the surrounding pixels;
  - choose the least saturated pixel, and turn in its direction;
  - if this least saturated pixel is as saturated as your own colour, then make a 90 degree turn to the left.

The creature subsequently colours the pixel it is on in its new colour, and jumps to the following position in its new direction. This winds down all activities in this time step, and the creature is ready for the following time step, that only begins when all other creatures in the habitat have had their turn.

We do not write the control programs for E-volver creatures ourselves, the computer automatically designs and assembles them from building blocks. The building blocks are code fragments forming a working whole by linking them up together, somewhat like Koza's genetic programming [6]. The building blocks are simple for they encode basic mathematical operations and flow control. There are different types of building blocks, and each type in its turn has different sub-types. For example, there is a type of building block to extract a colour component value from the environment. Each sub-type of this "extractor" performs this operation in a specific way: one taking the minimum, one taking the maximum, one taking the average, one taking the median, one taking a value randomly, one taking the value most similar to the current creatures component value, one taking the least similar, et cetera. Each creature has a set of thirteen genes that prescribe which building blocks are to be used, and how these will be combined. This whole set of thirteen genes, the genotype, is a creature's compact symbolic building scheme.

Two pipelines are constructed, one for the specification of the new colour, and one for the specification of the new direction. In the present version of E-volver each pipeline is constructed out of four building blocks at most, each responsible for a stage of processing. Further, a pipeline is accompanied by a building block that specifies which of the colour components (hue, saturation or lightness) serve as the input for the processing stages. Finally, there are three more building blocks in effect, which potentially modify the operation of every building block forming the pipeline. In the present version, E-volver uses more than a hundred different sorts of building blocks when compiling the artificial creatures' control programs. In total, there are 2859,378,278,400 different possible programs for controlling a creature. E-volver therefore has a wide choice when creating a single creature; when putting together an ecosystem where several creatures are present next to each other the choice is even considerably greater. For a community of eight creatures the number of possibilities increases to 4.46861e+83, a number composed of 84 digits.

5.5.2 Evolutionary Component

Evolution is in essence a stochastic process of trial and error – nothing has been decided beforehand about where it leads to. Evolution is not necessarily progressive and does not lead per se to ever greater complexity. Evolution is, however, accumulative. A favourable step taken is a basis for further development. As long as variants are still to hand of which the "worse ones" (less adapted) are suppressed, or rather the "better ones" (more adapted) are stimulated and as long as variants shape more variants of themselves, a directed development will occur – an evolution.

The E-volver software is divided into a generative component, that in principle offers as large a space of potential cultures as possible, and an evolutionary component that within that space concentrates attention around fields of special interest. The human being standing outside the software system forms the link between the two components. He/she is the one who evaluates and selects the output of the generative component, and thus furnishes the evolutionary component with input. The human being functions as a breeder who continually makes a selection from a diverse supply of results. The system itself does not prefer specific cultures – it remains impartial here. In contrast, a spectator has the tendency to prefer specific cultures above other ones and is thus able to make personal choices from the supply of various cultures. The reasons for these choices are of no significance to the system. We only posit that cultures similar to cultures previously chosen have more chance of re-selection than random other cultures. This point of departure offers the possibility of exploring the enormous potential of cultures in an individual manner, without tracing demarcated paths beforehand. Repeatedly selecting from a varied supply puts an evolution into motion that can lead to the development of cultures increasingly fitting the preferences, interests, sensitivities, affinities and peculiarities of the spectator/user. In this way, the system is reminiscent of Dawkins' Blind Watchmaker program [3].

Four separate cultures are presented on the screen at the same time, as in Fig. 5.5, over and over again. Each culture consists of a habitat containing a community of eight creatures. By picturing the cultures next to each other an immediate visual comparison is possible. You can end a culture at any moment, after which a new culture is initiated automatically. The ending of one culture, also signifies the "survival" of the three other growths. Each time that a culture survives a selection it earns a point, and its fitness score increases. Growths that are terminated quickly have a low score, cultures that keep going for a long time collect many points. Each culture tried out is added to a database that records the culture's name, the final fitness score, and the genotype of each creature in that culture. The evolution component uses this data when it puts together new cultures. Characteristics of higher scoring cultures have a greater chance of being deployed than characteristics from lower scoring cultures.

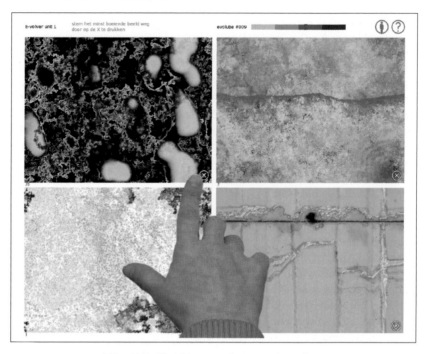

Fig. 5.5. E-volver, touch screen interface

The properties of a culture are spread over two levels. The higher level, of the culture as a whole, is characterised by the specific collection of creatures that form a community. The lower level, of the individual creature, is characterised by the specific composition of its genotype. Properties of both levels may be involved when composing a new culture, thus variation can arise both in the composition of the community, and in the composition of the genotypes of the creatures that form a community. In order to come to a new culture, E-volver uses three sorts of evolutionary operators: mixing, crossing and mutation. "Mixing" composes a new culture from creatures that have already appeared in previous cultures. It chooses a number of cultures, proportional to their fitness score, and makes a random selection of creatures from these cultures. These are then assembled in the new culture. "Crossing" composes a new culture from an earlier culture, chosen in proportion to its fitness score. It crosses a small number of creatures from the culture by cutting their genome in two pieces, exchanging the parts with each other and assembling them again into a complete genome. "Mutating" composes a new culture from the creatures of an earlier culture, whereby the genome from some creatures is subject to variation. Each gene from the genome from a mutating creature has a small chance of randomly changing into another possible variant of this gene. By applying these three evolutionary operations to already evaluated cultures, E-volver is able to constantly supply new variants for evaluation and selection, variants that continue to build on selected properties but at the same time also introduce new properties.

At the start of a new E-volver breeding session no evaluations are known yet, and so no evolutionary operations can be performed. A breeding session therefore begins with a "primordial soup" phase. Cultures are formed out of creatures that each have a completely randomly composed genome. Only after a large number of these "primordial soup cultures" have been developed and evaluated, and the E-volver database is filled with a range of cultures with varying fitness scores, does the system begin to apply the evolutionary operators in order to evolve further on the basis of the above.

5.5.3 Cultures and Interpretation

The E-volver creatures are embedded in their habitat, which influences the creature's behaviour; the behaviour of the creatures in turn influences the habitat. The creatures have no notion of each other's existence, and do not enter into direct relationships with each other. Nevertheless, because all eight creatures share their habitat with each other and give it form, a stigmergic network of indirect relationships arises. Here the habitat functions as the reflection of a shared history and as an implicit communications medium for all the creatures present. Each event that leaves its traces may later lead to other events [4]. The habitat, which is homogeneously grey at the start of each new culture, develops in the course of time into a heterogeneous whole. It is a dynamic and differentiated landscape that, in spite of its complexity, still forms an organic unity. The evolution is not restricted to the evolution of separate artificial life forms, but particularly includes the evolution of the culture as a whole. It is not an evolution of one individual in isolation, but a co-evolution of individuals within their mutual connection. So when we evaluate a culture, we do not so much evaluate the individual creatures, the habitat at a specific moment of development, but rather the continuing interplay between them. The selection criteria applied have a bearing on how the dynamic of the culture allows itself to be read and interpreted as a visual performance. The colourful abstract animations arouse associations with microscopic observations, rasters, geological processes, noise, cloud formations, modern art, fungus cultures, organ tissues, geometry, satellite photos etc., but ultimately they still avoid any definitive identification. The pixel creatures are in fact elementary image processors. Klaas Kuitenbrouwer puts it thus: "The system can also be interpreted as a machine which does nothing more than present the process of representation itself. Imaging without an object: we see a mutating process of visualisation, in which local image features operate and feedback on themselves. ... Another way to appreciate the work is as a system, an entirety of relationships between forces and processes. The beauty is then in the amazing connection between the simplicity of the pixel creatures and the complexity and endless variation that they create. The images therefore look like living processes because similar processes are actually taking place in the work! . . . The work as a system automatically includes the viewers, who have a very important role to play in the whole thing. It is they who give the work its direction, who appreciate the differences that occur, and thus give it meaning. . . . However you view the work, whatever way you choose to interpret it, there will be different reasons for deleting the images. The choices made by the viewer will also select the type of meaning and these meanings are made manifest in the work by the viewer's choice." [7]

5.6 Directed Versus Open

While we applied artificial evolution in Breed as an optimising technique, in E-volver it has become an inherent content aspect of the work. The aim of the evolutionary process is not described, but any imaginable fitness criterion can make surprising and significant properties of the system visible. Through the evolution process the users of the software become conscious of the influence they have on the development of the generative system, because personal preferences are progressively reflected in the changing visual patterns. This in itself is not new (for example Richard Dawkins' Biomorphs [3], Karl Sims' Genetic Images [14] and William Latham's Evolutionary Art [15]). It is particularly the linking of artificial evolution to a powerful and elegant morphogenetic process that throws new light on our quest for a system of self-organisation that is as open as possible. A user of E-volver confronted with the continually changing patterns several times a day over a long period, builds up a profound relationship with the work. In the first instance, the spectator is inclined to impose his personal taste on the system. Gradually we see that fathoming of the generative system more and more interferes with taste and judgement. We discover which sort of images occur more generally and which are rarer. It can indeed become a challenge to develop and sublimate a more unusual phenomenon, by adapting the evolutionary strategy that was used. However, what was initially exceptional becomes ordinary again after a little while, and thus attention keeps moving, always in the direction of the unseen and the unknown. The selection criteria themselves are thus subject to change, they co-evolve in interaction with the E-volver system.

The initial idea of automating art production, "art that is not made or developed by the subjective choices of the human being but is instead generated by independently operating processes", does not have to exclude human subjectivity completely. We have established a collaboration of typically human characteristics and the specific qualities of the machine.

5.7 Generative Art and Present Day Experience of Nature

Ultimately, it is important that the results of our research be presented in order to communicate with the public. When presenting Breed we prefer to show a number of materialised objects next to each other. This enables the objects to be compared with each other, seeing the similarities and differences. They clearly form a close family with their own characteristic style features. In the first instance, the constructions remind one of enlarged models of complicated molecules or crystals. Or the reverse – scaled down models of a grotesque futuristic architecture. But usually, it becomes progressively clearer that a process that is not immediately obvious, must have been at work here. For a human being would not be easily inclined to think up something so complex, and neither does the Nature we know produce this sort of technical construction as a matter of course. So there is conjecture, or at least it sets people thinking. Even if the creation process cannot be perceived in detail in the static final results, it does give rise to speculations about the process underlying it all. A small series of objects is often sufficient to summon the potential of possibilities indirectly.

Fig. 5.6. E-volver image breeding unit installed at LUMC Research Labs, Leiden. Photography Gert Jan van Rooij

In 2006, E-volver was installed in the form of an interactive artwork in the new building of the LUMC Research Labs in Leiden (see Fig. 5.6). Research within the laboratories is concentrated on Human Genetics, Clinical Genetics, Anatomy, Embryology, Molecular Cell Biology, Stem Cells and many other areas of medical DNA research. Four interactive image-breeding-machines are to be found in the common areas of the laboratories, which the scientists use to direct a collective evolution process via a touch-screen. If desired, one can log on to a personal culture environment and perform an individual evolution process. Each breeding unit has a large wide-screen monitor, where you can see the cultures that score highly. From time to time, these top cultures are replaced, as the process produces even higher scores. The evolution is terminated after a fixed number of selections, and the whole process starts again from the beginning – a primordial soup once again, where new possibilities lie enshrined which may subsequently reach expression. Five large prints hang in other places in the building, random snapshots from the most unusual growth processes that have been developed at high resolution. The permanent installation E-volver in the LUMC allows the scientists to experience by means of interaction how intriguing images can evolve out of a primordial soup of disconnected lines and spots. The underlying generative processes have full scope, with the scientists in the role of force of nature practising pressure of selection. Wilfried Lentz, director SKOR has said "While the scientists are concerned with the biochemistry, genetics and evolution of biological life, the artists address the potential offered by the underlying mechanisms of these processes for art, by implementing them as a purely visual and image-generating system. In this way, E-volver mirrors the complexity of the biochemical universe and humankind's desire for knowledge and understanding. Self-organising processes such as growth and evolution – which from a theoretical point of view may be well understood, but which in daily life are never directly observable – can with E-volver be experienced at a sensory level. It is at the same time a source of great aesthetic pleasure." [9]

5.7.1 Concrete Approach

The visual structures of Breed and E-volver do not represent anything but they arise from a logical and direct use of the medium and its formal visual means. This testifies to an approach that is related to concrete art, which has its origin in early Modernism. The artwork itself is the reality. Jean Arp said "We do not wish to copy Nature; we do not wish to reproduce, but to produce. We want to produce as a plant produces its fruit. We wish to produce directly, and no longer via interpretation. ... Artists should not sign their works of concrete art. Those paintings, statues and objects ought to remain anonymous; they form a part of Nature's great workshop as do trees and clouds, animals and people ..." [1] Concrete art does not reject the increasing prevalence of technology and industrialisation, but it wishes to provide new times with a fitting visual language. We share this approach but in contrast to modernistic artwork – that attempts to reveal the underlying harmony of reality by rational ordering and reduction of visual means – we are actually striving for complexity and multiformity in the final result. In our case the reductionist approach is not manifest in the visible results but on the other hand, we do use a form of minimalism in the design of the underlying generative process. The harmony model has been replaced in our case by the conviction that chance, self-organisation and evolution order and transform reality. Thus, new scientific insights and technologies contribute to a continuation and adjustment of older modernistic and expressionistic ideals. The concrete and formal approach can now enter into a union with new possibilities of vivification – vivification in the sense of bringing a work of art literally to life. These new possibilities are enshrined in the procedural character of the computer. By developing artificial life/evolution programs we unlock worlds each having their own generative principles and spontaneous expressions. These expressions make visible the internal structures of an artificial nature in all its variation, with a high degree of detail and often of great beauty.

5.7.2 Computational Sublime

Our A-life/evolution projects contribute to a new possibility of experiencing Nature. Jon McCormack introduced the term "the computational sublime", during the Second Iteration conference in Melbourne in 2001 [11]. It is an important concept in relationship to our work. He states that an important aspect of the sublime is the tension between pleasure and fear. Here, the pleasure is that we can be conscious of that which we cannot experience; fear is that things exist too great and too powerful for us to experience. In the dynamic sublime it is about the incomprehensible power of Nature [5], in the computational sublime such feelings are aroused by an uncontrollable process taking place in the computer. The underlying generative process is not directly comprehensible but we are in fact able to experience it through the machine. By real time generation and visualising of the internal processes, the spectator can become overwhelmed by a sense of losing control, and at the same time enjoy the spectacle of this artificial Nature. What Romantic Painting in the nineteenth century could only arouse via figurative representation, can now in fact be generated and experienced live with Artificial Life and Artificial Evolution techniques. To quote Mitchell Whitelaw again: "Evolution, an idea that has become the most powerful organising narrative of contemporary culture, appears to unfold on a screen. A-life proposes not a slavish imitation of this or that living thing but, at it strongest, an abstract distillation of aliveness, life itself, re-embodied in voltage and silicon" [18, p. 5].

Acknowledgments

Thanks to The Netherlands Foundation for Visual Arts, Design and Architecture.

References

1. Arp, J.: Abstract art, concrete art. In: Art of This Century. New York (1942)
2. Baudrillard, J.: Les Stratégies Fatales. Grasset & Fasquelle, Paris (1983)
3. Dawkins, R.: The Blind Watchmaker. W. W. Norton & Company (1986)

4. Grassé, P.: La réconstruction du nid et les coördinations interindividuelles. La théorie de la stigmergie. Insectes Sociaux **6**, 41–84 (1959)
5. Kant, I.: Kritiek der Urteilskraft (Critique of Judgement).   Lagarde und Friederich, Berlin (1790)
6. Koza, J.: Genetic Programming – On the Programming of Computers by Means of Natural Selection. MIT Press (1992)
7. Kuitenbrouwer, K.: E-volver. In: E-volver, SKOR, Foundation Art and Public Space, pp. 7–22. Amsterdam (2006)
8. Langton, C.: Studying artificial life with cellular automata. Physica **22D**, 120–149 (1986)
9. Lentz, W.: Art and scientific research. In: E-volver, SKOR, Foundation Art and Public Space, pp. 4–6. Amsterdam (2006)
10. Lewin, R.: Complexity: Life at the Edge of Chaos. Macmillan, New York (1992)
11. McCormack, J., Dorin, A.: Art, emergence, and the computational sublime. In: Second Iteration – Proceedings of the Second International Conference on Generative Systems in the Electronic Arts, pp. 67–81. CEMA, Monash University, Melbourne (2001)
12. Pollan, M.: The Botany of Desire. Random House (2001)
13. Prigogine, I., Stengers, I.: Order out of Chaos – Man's New Discourse with Nature. Flamingo (1984)
14. Sims, K.: Artificial evolution for computer graphics. In: Proceedings of SIGGRAPH, pp. 319–328. ACM Press, Las Vegas, Nevada (1991)
15. Todd, S., Latham, W.: Evolutionary Art and Computers. Academic Press (1992)
16. Whitelaw, M.: Tom Ray's hammer: emergence and excess in A-life art. Leonardo **31**(5), 377–381 (1998)
17. Whitelaw, M.: Morphogenetics: generative processes in the work of Driessens & Verstappen. Digital Creativity **14**(1), 45–53 (2003). Swets & Zeitlinger
18. Whitelaw, M.: Metacreation, Art and Artificial Life. MIT Press (2004)
19. van der Zaag, D.: Die Gewone Aardappel (That Ordinary Potato). Wageningen University (1996)

6

Evolutionary Exploration of Complex Fractals

Daniel Ashlock[1] and Brooke Jamieson[2]

[1] University of Guelph, Department of Mathematics and Statistics, 50 Stone Road East, Guelph, Ontario N1G 2W1, Canada `dashlock@uoguelph.ca`

[2] University of Guelph, Department of Mathematics and Statistics, 50 Stone Road East, Guelph, Ontario N1G 2W1, Canada `bjamieson@uoguelph.ca`

6.1 Introduction

A *fractal* is an object with a dimension that is not a whole number. Imagine that you must cover a line segment by placing centers of circles, all the same size, on the line segment so that the circles just cover the segment. If you make the circles smaller, then the number of additional circles you require will vary as the first power of the degree the circles were shrunk by. If the circle's diameter is reduced by half, then twice as many circles are needed; if the circles are one-third as large, then three times as many are needed. Do the same thing with filling a square and the number of circles will vary as the second power of the amount the individual circles were shrunk by. If the circles are half as large, then four times as many will be required; dividing the circle's diameter by three will cause nine times as many circles to be needed. In both cases the way the number of circles required varies is as the degree to which the circles shrink to the power of the *dimension* of the figure being covered. We can thus use shrinking circles to measure dimension.

For a figure whose dimension is to be computed, count how many circles of several different sizes are required to just cover the figure. Fit a line to the log of the number of circles required as a function of their diameter. The slope of this line is the dimension of the figure. For a line segment the value will be close to 1, while for a square the value will be close to two. This procedure approximates the *Hausdorff* dimension [9] of the objects. For Mandelbrot and Julia sets, the fractals that are the focus of this chapter, the boundaries of the sets will have dimensions strictly between 1 and 2. Examples of these fractals appear in Figs 6.1 and 6.2. Objects as diverse as clouds, broccoli, and coastlines exhibit the property of having non-integer Hausdorff dimension.

Fractals appeared in art before their defining property of fractal dimension was formalized. A nineteenth century painting, taken from a series of 36 views of Mt. Fuji, depicts a great wave threatening an open boat. This picture appears in [9] as an example of a depiction of a natural fractal. The wave in the picture is an approximation to a self-similar fractal. This picture was

Fig. 6.1. An example of a view of the Mandelbrot set

painted long before the 1868 birth of Felix Hausdorff, who is credited with defining the notion of fractal dimension.

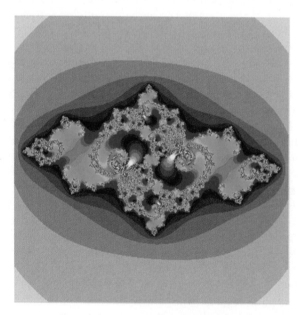

Fig. 6.2. An example of a Julia set

Fractals are sometimes taken as a type of art in their own right [1,3,4,6,7]. They may also be incorporated as elements of a piece of visual art [2,12,13]. Fractals are also used as models that permit nature to be incorporated more easily into artistic efforts [5,9,11]. Examples of evolution of fractals including systems that use genetic programming [14] and which optimize parameters of (generalized) Mandelbrot sets to generate biomorphs [16]. Fractals that are located by evolutionary real parameter optimization to match pictures of faces appear in [17]. Finally, fractals can be used as a way of making the beauty of math apparent to an audience whose last contact with math may have been in the context of a less-than-enjoyed class [8,10]. The American Mathematical Society's web page on mathematics and art (http://www.ams.org/mathimagery/) displays several efforts in this direction.

This chapter explores fractals as a form of art using an evolutionary algorithm to search for interesting subsets of the Mandelbrot set and for interesting generalized Julia sets. The key problem is to numerically encode the notion of "interesting" fractals. Using an evolutionary algorithm automates the search for interesting fractal subsets of the complex plane. The search proceeds from a simple form of arithmetic directions given by the artist. After this direction is given, the process runs entirely automatically.

The *quadratic Mandelbrot set* is a subset of the complex plane consisting of those complex numbers z for which the *Mandelbrot sequence*

$$z, \; z^2 + z, \; (z^2 + z)^2 + z, \; ((z^2 + z)^2 + z)^2 + z, \; \ldots \tag{6.1}$$

does not diverge in absolute value. A *quadratic Julia set* with parameter $\mu \in \mathbb{C}$ is a subset of the complex plane consisting of those complex numbers z for which the *Julia sequence*:

$$z, \; z^2 + \mu, \; (z^2 + \mu)^2 + \mu, \; ((z^2 + \mu)^2 + \mu)^2 + \mu, \; \ldots \tag{6.2}$$

fails to diverge in absolute value. These two fractals are related. The defining sequences for both fractals involve iterated squaring. The difference is that, when checking a point z in the complex plane for membership in the fractal, the Mandelbrot sequence adds the value z under consideration after each squaring, while the Julia sequence adds the fixed parameter μ. The overall appearance of a Julia set matches the local appearance of the Mandelbrot set around μ. Good Julia set parameters μ are thus drawn from *near* interesting regions of the Mandelbrot set.

The indexing of Julia sets by the Mandelbrot set is demonstrated in Fig. 6.3. Notice in the lower picture how the Julia sets changes as their parameter moves across a small region of the Mandelbrot set. This indexing property of the Mandelbrot set means that the search for interesting Julia sets can be undertaken as a search of the Mandelbrot set. Because of this we use a generalization of the Julia set we discovered.

We tested two versions of the evolutionary algorithm. The first searches the Mandelbrot set for interesting views. A *view* of the Mandelbrot set is a

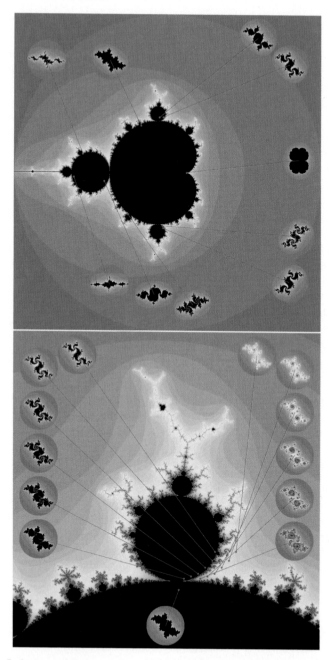

Fig. 6.3. Indexing of Julia sets by the Mandelbrot set. Julia sets are displayed as small insets in the Mandelbrot set (above) and a subset of the Mandelbrot set (below). Julia set parameters are drawn from the location of the red dots connected to the inset thumbnails

square subset of the complex plane specified by three real parameters: x, y, and r. The square has the complex number $x + iy$ as its upper left corner and has side length r. The second version of the evolutionary algorithm searches for the parameters of a generalized Julia set. The generalization of Julia sets is indexed by a four-dimensional form of the Mandelbrot set and so is in greater need of automatic search methods than the standard two-dimensional (with one complex parameter) space of Julia sets. The generalization of the Julia set used in this study requires two complex parameters, ω_1 and ω_2. As with the Mandelbrot set and standard Julia sets, the members of the generalized Julia set are those complex numbers z for which a sequence fails to diverge. The sequence is:

$$z \tag{6.3}$$
$$z^2 + \omega_1 \tag{6.4}$$
$$(z^2 + \omega_1)^2 + \omega_2 \tag{6.5}$$
$$((z^2 + \omega_1)^2 + \omega_2)^2 + \omega_1 \tag{6.6}$$
$$(((z^2 + \omega_1)^2 + \omega_2)^2 + \omega_1)^2 + \omega_2 \tag{6.7}$$
$$\cdots \tag{6.8}$$

Rather than adding a single complex parameter after each squaring, the two complex parameters are added alternately. This generalization originated as a programming error a number of years ago. It encompasses a large number of fractal appearances that do not arise in standard Julia sets. One clear difference is this. A standard Julia set is a connected set if its parameter is a member of the Mandelbrot set and is a fractal dust of separate points if its parameter is outside of the Mandelbrot set. The generalized Julia sets include fractals that have multiply connected regions.

   The remainder of the chapter is structured as follows. Section 6.2 reminds those readers who have not used complex arithmetic recently of the details as well as explaining how the fractals are rendered for display. Section 6.3 defines the fitness function used to drive the evolutionary search for Mandelbrot and generalized Julia sets. Section 6.4 gives the experimental design, specifying the evolutionary algorithm and its parameters. Section 6.5 gives the results, including visualizations. Section 6.6 gives possible next steps.

6.2 Complex Arithmetic and Fractals

The *complex numbers* are an extension of the familiar real numbers (those that represent distances or the negative of distances) achieved by including one "missing" number, $i = \sqrt{-1}$, and then closing under addition, subtraction, multiplication, and division by non-zero values. The number i is called the *imaginary unit*. A *complex number* z is of the form $z = x + iy$ where x and y

are real values. The number x is called the *real part* of z, and y is called the *imaginary part* of z. The arithmetic operations for complex numbers work as follows:

$$(a + bi) + (c + di) = (a + c) + (b + d)i \tag{6.9}$$

$$(a + bi) - (c + di) = (a - c) + (b - d)i \tag{6.10}$$

$$(a + bi) \times (c + di) = (ac - bd) + (ad + bc)i \tag{6.11}$$

$$\frac{a + bi}{c + di} = \frac{ac + bd}{\sqrt{c^2 + d^2}} + \frac{bc - ad}{\sqrt{c^2 + d^2}}i. \tag{6.12}$$

One of the properties of the complex numbers is that they place an arithmetic structure on points (x, y) in the Cartesian plane so that arithmetic functions over the complex numbers can be thought of as taking points (x, y) (represented by $x + yi$) in the plane to other points in the plane. Because the complex numbers have this one-to-one correspondence with the points of the Cartesian plane, they are also sometimes referred to as the *complex plane*. Complex fractals are easier to define when thought of as consisting of points in the complex plane. The *absolute value* of a complex number $z = x + yi$ is denoted in the usual fashion $|z|$ and has as its value the distance $\sqrt{x^2 + y^2}$ from the origin.

Suppose we are examining a point z in the plane for membership in a Mandelbrot or a generalized Julia set. The *iteration number* for a point is the number of terms in the relevant sequence (Equation 6.1 for the Mandelbrot set, Equation 6.2 for standard Julia sets, or Equations 6.3–6.7 for generalized Julia sets) before the point grows to an absolute value of 2 or more. The reason for using an absolute value of 2 is that the point in the Mandelbrot set with the largest absolute value is $z = -2 + 0i$. For any value z with an absolute value greater than 2, its square has an absolute value exceeding 4 and so the absolute value of the sequence must increase. Points not in the fractal thus have finite iteration numbers, while points in the fractal have infinite iteration numbers. Iteration numbers are used for rendering visualizations of the fractal.

This chapter uses a coloring scheme with a cyclic continuously varying palette for rendering divergence behaviors. The software uses an RGB representation for color, with each color represented by a red, green, and blue value in the range $0 \le r, g, b \le 255$. The palettes used are of the form:

$$R(k) = 127 + 128 \cdot \cos(A_R \cdot k + B_R) \tag{6.13}$$

$$G(k) = 127 + 128 \cdot \cos(A_G \cdot k + B_G) \tag{6.14}$$

$$B(k) = 127 + 128 \cdot \cos(A_B \cdot k + B_B) \tag{6.15}$$

where the A_* values control how fast each of the three color values cycle, the B_* values control the phase of their cycle, and k is the iteration number of the point being colored. The palette used in this chapter uses the color-cycle values: $A_r = 0.019$, $A_G = 0.015$, $A_B = 0.011$ (a slowly changing palette). The phase-shift values were set to $B_R = B_G = B_B = 0.8$ so that the first color, for $k = 0$, is a uniform gray. Thus, when the cosine waves that control each color are in phase, the color is some shade of gray. Since the parameters A_* controlling speed differ, the hue changes with k. The apparent differences in palettes between pictures are caused by the fact that the scale of the views is different (some have far smaller side lengths than others), and so the colors displayed in the view are sampled from different parts of the periodic palette.

When checking for membership in a complex fractal, it is necessary to set an upper limit M_{max} on the number of terms of the sequence checked. In this chapter, $M_{max} = 200$. Points with M_{max} terms of the sequence with absolute value less than two are taken to be within the set. The color value for these points is calculated using $k = M_{max}$ in the periodic palette.

6.3 Fitness Function Design

The fitness function we used to locate interesting complex fractals transforms the three real parameters defining a view of the Mandelbrot set or the four parameters that specify a particular generalized Julia set into a scalar fitness value. The fitness function gives the artist control over the appearance of the fractals located by the algorithm. A fitness function that can do this is presented in [1]. A grid of points, either 11×11 or 15×15 is placed on a square subset of the complex plane. When searching for views into the Mandelbrot set, this square is the square defined by the view, while for generalized Julia sets it is the square with corners $(0.8, 0.8)$ and $(-0.8, -0.8)$.

For a given set of parameters defining a complex fractal, the iteration values for the points in the grid are computed. A *mask* is used to specify desired iteration values on the grid points. Fitness is the average, over the grid points, of the squared error of the true iteration value and the desired value specified by the mask. The average is used, rather than the total, to permit comparison between masks with different numbers of points.

The mask specifies where points with various approximate iteration values are to be in the evolved images. This ability to specify the behavior of the fractal on the grid gives the artist control over the appearance of the resulting complex fractals. The masks used for the generalized Julia sets are shown in Fig. 6.4, while those used to search for Mandelbrot views are shown in Figs 6.5 and 6.4. The evaluation square of $(-0.8, -0.8)$ to $(0.8, 0.8)$ used for the generalized Julia sets was chosen by experimentation. When too large a square is used, the behavior far from the origin of the complex plane dominates the fitness function, and the fractals found look like simple lagoons. When too small a square is used, the overall appearance of the Julia set becomes visually unrelated to the mask. The rendered views of the Julia sets have a side length of 3.6 units with the square used for fitness evaluation centered in the rendered

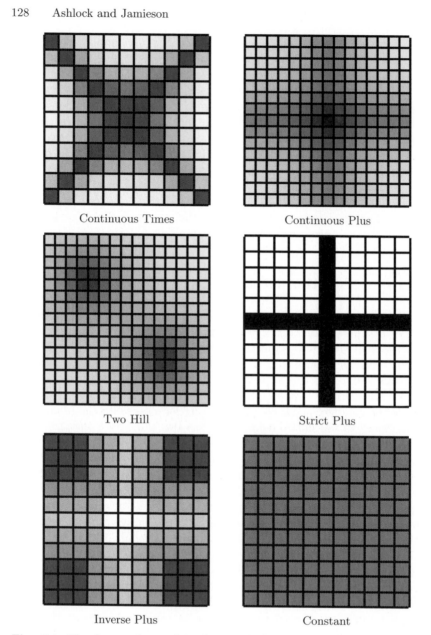

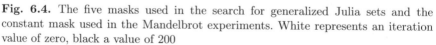

Fig. 6.4. The five masks used in the search for generalized Julia sets and the constant mask used in the Mandelbrot experiments. White represents an iteration value of zero, black a value of 200

view. The squares used for rendering Mandelbrot views are exactly the square associated with the view.

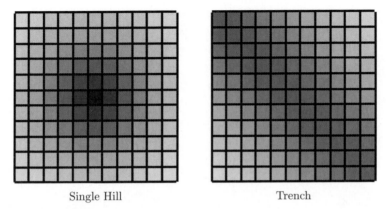

Single Hill   Trench

Fig. 6.5. Two of the masks used in the search for views into the Mandelbrot set. The constant mask appearing in Fig. 6.4 was also used in the Mandelbrot experiments. White represents an iteration value of zero, black a value of 200

6.4 Specification of Experiments

The evolutionary algorithm for both sorts of complex fractal operates on a population of 800 structures. The initial population for the Mandelbrot experiments is generated by placing (x, y) uniformly at random in the square region with corners $(-2, -1.5)$ and $(1, 1.5)$. This is a relatively small square including all of the Mandelbrot set. The value of r is initialized in the range $0.0001 \leq r \leq 0.01$ using a distribution whose natural log is uniformly distributed. Variation operators for the Mandelbrot views are as follows. The y-values are exchanged, and the r-values are exchanged 50% of the time. After the exchange of y-values and r-values, the new structures are mutated. Mutation has two parts. The first consists of adding a displacement in the plane to (x, y) in a direction selected uniformly at random and with magnitude given by a normal random variable with mean zero and standard deviation $\sigma = 0.002$. For the small range of zooms used in this chapter, this constant value for σ was acceptable but a greater range of potential zoom would require that this parameter be made to adapt to the current level of zoom. After the corner of a view has been displaced in this fashion, the side length is multiplied by $1.1e^{N(0,1)}$ where $N(0, 1)$ is a standard normal random variable. The effect of this is to multiply r by a value near 1. These mutation operators were chosen to permit incremental exploration of view-space with as little mutational bias as possible. The multiplicative mutation operator used for r makes small adjustments whose scale does not depend on the current value of the side length. Twenty independent runs were performed for each mask.

Parameters for the initial population of generalized Julia sets are stored as an array (a, b, c, d) where $\omega_1 = a + bi$ and $\omega_2 = c + di$. These parameters are initially chosen uniformly at random in the range $-2 \leq a, b, c, d \leq 2$. The

Fig. 6.6. Selected thumbnails of best-of-run Mandelbrot views for the fitness function using the constant mask

variation operators for the Julia sets consist of two-point crossover operating on the array of four reals and a single point mutation that adds a Gaussian with a variance of 0.1 to one of the four real parameters selected uniformly at

Fig. 6.7. Selected thumbnails of best-of-run Mandelbrot views for the fitness function using the single hill mask

random. Thirty-six independent runs were performed for each mask for the Julia sets.

The model of evolution for all experiments uses size seven single tournament selection. A group of seven distinct population members is selected at

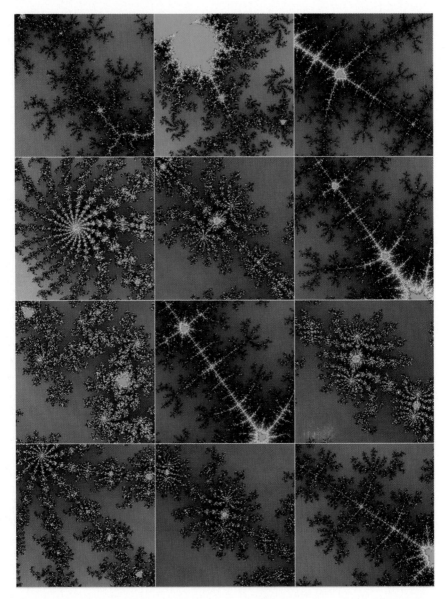

Fig. 6.8. Selected thumbnails of best-of-run Mandelbrot views for the fitness function using the trench mask

random without replacement. The two best are copied over the two worst, and then the copies are subjected to the variation operators defined for the respective sorts of fractals. The evolutionary algorithm is a steady state [15] algorithm, proceeding by mating events in which a single tournament is pro-

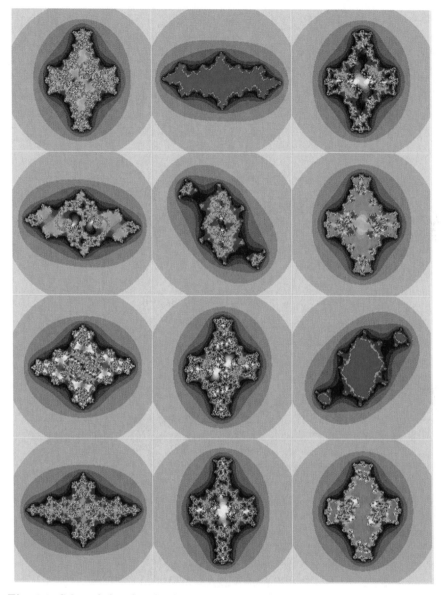

Fig. 6.9. Selected thumbnails of the best-of-run generalized Julia sets for the fitness function using the continuous plus mask

cessed. Evolution continues for 100,000 mating events for the Mandelbrot views and for 50,000 mating events for the generalized Julia sets. The most fit (lowest fitness) individual is saved in each run. The fitness functions used are described in Section 6.3.

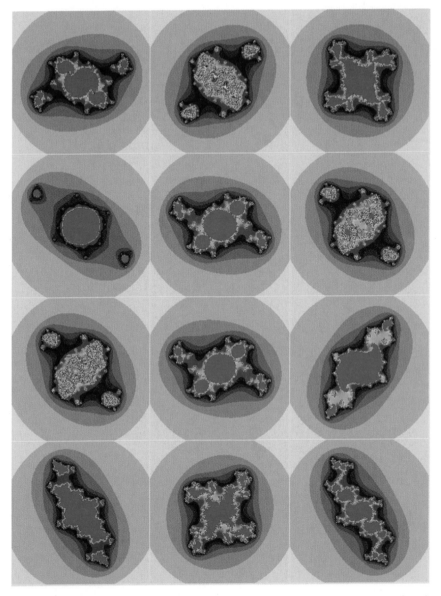

Fig. 6.10. Selected thumbnails of the best-of-run generalized Julia sets for the fitness function using the continuous times mask

6.5 Results and Discussion

For all the complex fractals studied in this chapter, final fitnesses varied considerably within the runs done for each of the masks. This suggests that all eight fitness landscapes used are pretty rugged. The thumbnails of the fractals located also show a substantial diversity of appearance within each fitness function. The fitness landscapes are themselves fractal. In the search for Mandelbrot views the fitness function is a smoothed version of the iteration number function for the Mandelbrot set. The mask defines an average of the iteration numbers over its sample points, acting as a smoothing filter. Since this iteration function makes arbitrarily large jumps in arbitrarily small spaces, a finite amount of smoothing will not erase its rugged character.

The fitness landscape for the generalized Julia sets is fractal for a similar reason. The parameters represent a point in a four-dimensional Mandelbrot set that specifies the complexity of the resulting generalized Julia set. This four-dimensional Mandelbrot set, alluded to earlier in the manuscript, can be understood as follows. Each set of four parameters (a, b, c, d) yields a generalized Julia set. If we fix $\omega_1 = a + bi$ and then check the iteration value of $\omega_2 = c + di$ using ω_2 as both the point being tested and the second Julia parameter, then the points that fail to diverge are a (nonstandard) Mandelbrot set. The points tested depend on all the parameters a, b, c and d, and so the set is four dimensional. The role of ω_1 and ω_2 can be interchanged, but this yields a different parameterization of the same four-dimensional Mandelbrot set. Much as the fitness function for Mandelbrot views filters the Mandelbrot set, the Julia parameters are a filtered image of the four-dimensional Mandelbrot set.

Figs 6.6–6.13 give selected thumbnails for the runs performed for the eight different masks. One of the goals in the design of the mask-based fitness function was to give the artist control over the appearance of the fractals located. Comparing the within-fitness-function variation in appearance with the between-fitness-function variance, it is clear that the different masks grant substantial control over the character of the fractals located. Each fitness function found similar but not identical fractals (with the exception of the constant mask). The thumbnails shown in the figures show the diversity of fractals found after eliminating fractals with substantially similar appearance.

The results presented represent a successful controlled search of both the Mandelbrot set and the 4-space of generalized Julia sets. The success lies in providing an automated search tool that, via the use of the mask-based fitness function, is still controlled by the user. The mask fitness functions give the artist input while still leaving a rich space of fractals that are near optima (probably local optima) of the fitness function. The difference between the continuous plus mask, Fig. 6.9 and the strict plus, Fig. 6.12, is substantial. The softer plus-shaped mask locates "softer" fractals. All eight of the masks used in the study yield markedly different appearances.

The most diverse collection of fractals located were the Mandelbrot views associated with the constant mask. The location of these views on the Mandelbrot set is scattered in space and their appearances have less in common than those found using any of the other masks. The constant mask was intended as a control, and for the generalized Julia sets it serves as one, locating two Julia sets that are both quite simple and mirror images of one another (data not

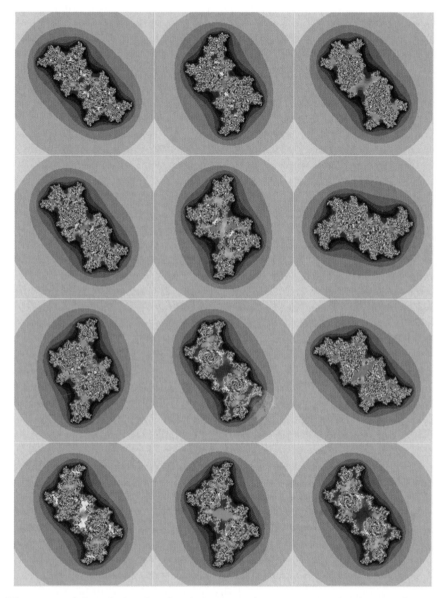

Fig. 6.11. Selected thumbnails of the best-of-run generalized Julia sets for the fitness function using the twin hill mask

shown). In retrospect it is not surprising that the Mandelbrot views located with the constant mask are interesting. The constant mask tries for iteration numbers that are 75% of the maximum number of iterations (150 out of 200 in this case). This requires that a fit view have high iteration numbers on its

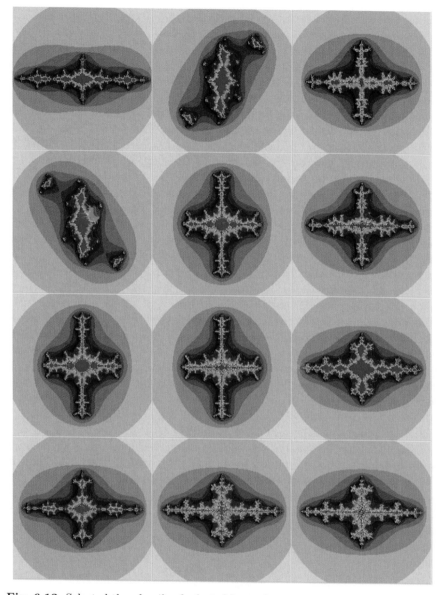

Fig. 6.12. Selected thumbnails of selected best-of-run generalized Julia sets for the fitness function using the strict plus mask

sample points, and that it not contain points in the interior of the Mandelbrot set. Thus, the constant mask looks for action near the boundary of the set where most of the interesting fractals reside. The choice of 75% of maximum iteration value for the constant mask was made arbitrarily. Other constant masks might also yield interesting appearances.

The single hill mask for Mandelbrot views turned out to be good at locating a well-known feature of the Mandelbrot set – the *minibrot*. A minibrot (see Fig. 6.7) is a smaller copy of the Mandelbrot set within the Mandelbrot set. There are an infinite number of minibrots within the set. Since a minibrot gives a central collection of high iteration number, it is intuitive that the single hill mask would tend to locate minibrots. The second most common type of view located by the single hill mask is a radiation collection of strips of high iteration value. Several examples of these appear in Fig. 6.7.

The continuous plus and continuous times masks used for the Julia set evolution are quite similar; the continuous plus mask is slightly lower resolution, but both are twin intersecting ridges when viewed as functions. The images in Figs 6.9 and 6.10 form two clearly distinct groups. The twin hill mask (Fig. 6.11) provides a collection of superficially similar fractals – showing the influence of the mask – but close examination reveals that, in terms of spiral structure, connectivity, branching factor, and other properties of Julia sets, these fractals are actually quite diverse. The masks provide some control but yield a substantial diversity of appearances.

An interesting aspect of the type of evolutionary search presented here is the new role of local optima. The mask-based fitness functions do *not* clearly specify the exact, desired outcome. They are more like "guidelines." This means that the local optima of these fitness functions may have greater appeal than the global optimum. This is very different from a standard evolutionary optimization in which local optima can be an absolute bane of effective search. This is good news given the complexity of the fitness landscape; locating a global optimum would be very difficult.

The search for interesting complex fractals presented here is an example of an automatic fitness function for locating artistically interesting images. Enlargements of some of the thumbnail views appear in Figs 6.14–6.16. This contrasts with the process of using a human being as the fitness function. It gives the artist a tool to guide the search, in the form of the masks, but does not require the artist's attention throughout evolution.

6.6 Next Steps

The fractal evolution techniques used here are intended as a technique for use by artists. Various evolved art contests held at the Congress on Evolutionary Computation and the EvoMusArt conferences have stimulated a good deal of interest in evolved art as a technique. With this in mind several possible directions for the extension of this research follow. Potential collaborators are welcome to contact the first author to discuss possible variations and extensions of this work.

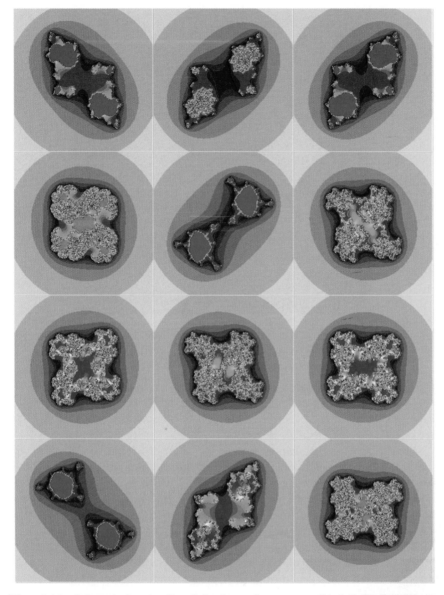

Fig. 6.13. Selected thumbnails of the best-of-run generalized Julia sets for the fitness function using the inverse plus mask

The time to run the evolutionary search for complex fractals grows in direct proportion to the number of sample points. This is why the number of sample points used here have been restricted to an 11×11 or 15×15 grid. Larger grids could cause tighter groupings of the appearances than occurred here. In

Fig. 6.14. An enlargement of the third thumbnail from the Mandelbrot views located with the constant mask

addition, there is the potential for sparse masks that have an indifference value at some locations. The minimization of squared error would simply ignore those points, permitting more rapid search on what are, in effect, non-square masks.

Another obvious extension of this line of research lies in the fact that quadratic Julia sets are simply one of an infinite number of types of Julia sets. Each possible power (cubic, quartic, etc.) yields its own Mandelbrot set, indexing a corresponding collection of Julia sets. Similarly, Mandelbrot and Julia sets can be derived from transcendental functions such as e^z, $\sin(z)$, et cetera. Each of these is its own domain for evolutionary search.

An obvious extension comes from continuing the generalization process for Julia sets. The generalized Julia sets presented here use, alternately, two complex parameters when generating the sequence used to test for set membership; a standard Julia set uses one. An n-fold, rather than two-fold, alternation of constants could be used. Likewise a set of constants could be added into terms of the series in a pattern rather than in strict alternation or rotation. Each of these variations has a simple realization as a real-parameter optimization problem using the mask fitness functions.

The masks tested in this study are a small set, designed using the intuition of the authors. A vast number of other masks are possible. One possibility is to used an existing complex fractal as a source for a mask. The artist would

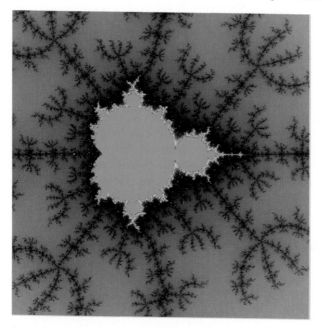

Fig. 6.15. An enlargement of the tenth thumbnail from the Mandelbrot views located with the single hill mask

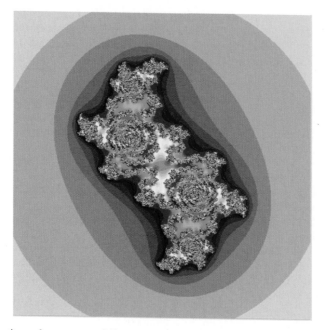

Fig. 6.16. An enlargement of the tenth thumbnail from the generalized Julia sets located with the twin hill mask

hand select a set of parameters that yield an interesting complex fractal. The iteration values for this fractal would then be used as a mask, permitting evolutionary generalization of the hand-selected fractal. This proposed technique would require a parameter study on the number of sample points used in the mask. Too few points and the resulting fractals will likely bear no resemblance to the original one. Too many sample points and the search will locate collections of nearly identical Julia sets.

Finally we feel that automating the selection of a coloring algorithm is a possibly difficult but rewarding avenue for future research. This might be a natural place for evolution with a human serving as the evaluator, or it might be possible to select a feature set and make chromatic analogies with existing pictures thought to be chromatically balanced.

Acknowledgments

The authors would like to thank the University of Guelph Department of Mathematics and Statistics for its support of this research.

References

1. Ashlock, D.: Evolutionary exploration of the Mandelbrot set. In: Proceedings of the 2006 Congress on Evolutionary Computation, pp. 7432–7439 (2006)
2. Ashlock, D., Bryden, K., Gent, S.: Creating spatially constrained virtual plants using L-systems. In: Smart Engineering System Design: Neural Networks, Evolutionary Programming, and Artificial Life, pp. 185–192. ASME Press (2005)
3. Ashlock, D., Jamieson, B.: Evolutionary exploration of generalized Julia sets. In: Proceedings of the 2007 IEEE Symposium on Computational Intelligence in Signal Processing, pp. 163–170. IEEE Press, Piscataway NJ (2007)
4. Barrallo, J., Sanchez, S.: Fractals and multi-layer coloring algorithm. In: Bridges Proceedings 2001, pp. 89–94 (2001)
5. Castro, M., Pérez-Luque, M.: Fractal geometry describes the beauty of infinity in nature. In: Bridges Proceedings 2003, pp. 407–414 (2003)
6. Fathauer, R.: Fractal patterns and pseudo-tilings based on spirals. In: Bridges Proceedings 2004, pp. 203–210 (2004)
7. Fathauer, R.: Fractal tilings based on dissections of polyhexes. In: Bridges Proceedings 2005, pp. 427–434 (2005)
8. Ibrahim, M., Krawczyk, R.: Exploring the effect of direction on vector-based fractals. In: Bridges Proceedings 2002, pp. 213–219 (2002)
9. Mandelbrot, B.: The Fractal Geometry of Nature. W. H. Freeman and Company, New York (1983)
10. Mitchell, L.: Fractal tessellations from proofs of the Pythagorean theorem. In: Bridges Proceedings 2004, pp. 335–336 (2004)
11. Musgrave, F., Mandelbrot, B.: The art of fractal landscapes. IBM J. Res. Dev. **35**(4), 535–540 (1991)

12. Parke, J.: Layering fractal elements to create works of art. In: Bridges Proceedings 2002, pp. 99–108 (2002)
13. Parke, J.: Fractal art – a comparison of styles. In: Bridges Proceedings 2004, pp. 19–26 (2004)
14. Rooke, S.: Eons of genetically evolved algorithmic images. In: P. Bentley, D. Corne (eds.) Creative Evolutionary Systems, pp. 339–365. Academic Press, London, UK (2002)
15. Syswerda, G.: A study of reproduction in generational and steady state genetic algorithms. In: Foundations of Genetic Algorithms, pp. 94–101. Morgan Kaufmann (1991)
16. Ventrella, J.: Creatures in the complex plane. IRIS Universe (1988)
17. Ventrella, J.: Self portraits in fractal space. In: La 17 Exposicion de Audiovisuales. Bilbao, Spain (2004)

7

Evolving the Mandelbrot Set to Imitate Figurative Art

JJ Ventrella

http://www.ventrella.com/
Jeffrey@Ventrella.com

7.1 Introduction

This chapter describes a technique for generating semi-abstract figurative imagery using variations on the Mandelbrot Set equation, evolved with a genetic algorithm. The Mandelbrot Set offers an infinite supply of complex fractal imagery, but its expressive ability is limited, as far as being material for visual manipulation by artists. The technique described here achieves diverse imagery by manipulating the mathematical function that generates the Set.

The art of portraiture includes many art media and many styles. In the case of self-portraiture, an artist's choice of medium is sometimes the most important aspect of the work. A love for math and art, and an irreverence concerning the massacring of math equations for visual effect, inspired the medium described in this chapter. It evolves manipulations of the Mandelbrot equation to mimic the gross forms in figurative imagery, including digital photographs of human and animal figures. To exploit the Mandelbrot Set's potential for this, the technique emphasizes shading the "flesh" (the interior of the Set), rather than the outside, as is normally done. No image has been generated that looks exactly like a specific figure in detail, but nor is this the goal. The images have an essence that they are "trying" to imitate something, enough to evoke a response in the viewer. As an example, Fig. 7.1 shows evolved images that are all based on a single digital image of the author's head. The technique for generating these images will be explained near the end of the chapter.

The technique is partly inspired by two movements in Modernist painting: abstract expressionism and surrealism. But in this case, the effect is not achieved by the physics of paint on canvas and an artist's dialog with an emerging image. Instead, the effect is achieved by dynamics in the complex plane (the canvas upon which the Mandelbrot Set is painted) and an artist's algorithmic searching methods.

After developing many artworks by painstakingly adjusting numerical parameters, an interactive interface was added to help automate the selective

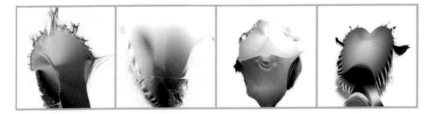

Fig. 7.1. Four examples based on an image of the author's head

process and to eliminate the need to control the numbers directly. Then a question came up: is it possible to evolve these kinds of images automatically using a genetic algorithm and an image as the objective fitness function? Also, given a specific image as a fitness function, what are the limits of the Mandelbrot Set – and the array of parameters added to increase its dimensionality and plasticity – to imitate images? How evolvable are these mathematically-defined forms? And for the sake of image-making, are not other parametric schemes more evolvable – such as Koch fractal construction, IFS, chaotic plots, L-systems, cellular automata, et cetera? General questions like this will be threaded throughout this chapter, touching upon the nature of the Mandelbrot Set, and the notion of evolutionary art as imitation of natural form.

The Mandelbrot Set has a special place in the universe of visual materials available to artists for manipulation – it has a peculiar complexity all its own. And the fascination with its arbitrary symmetry and beauty can reach near-religious levels. Roger Penrose believes, as do many mathematicians, that there is a Platonic truth and universality to mathematics, and that the Mandelbrot Set demonstrates this. It was not, and could never have been, invented by a single human mind [13]. It could only have been discovered. A different approach to mathematics is presented by Lakoff and Nunez [9] who claim that mathematics springs out of the *embodied* mind. It expresses our physical relationship with the world, and the subsequent metaphors that have evolved. Math is not universal truth, but a language of precision, contingent upon the nature of the human brain and the ecology from which it evolved.

We will not try to address this debate in this chapter. Only to say that this technique takes a rather un-Platonic approach to manipulation of the math, which may be unsavory from a mathematical standpoint, but from an artistic standpoint it breaks open the canvas of the complex plane to a larger visual vocabulary.

Even when pulling the function out of the realm of complex analysis, a manipulated Mandelbrot Set still possesses remarkable properties, and it appears not to be as plastic and malleable as a lump of sculptor's clay. It is more like an organism whose entire morphology, at every scale, is determined by a specific genetic code and the constraints of its iterative expression. For instance, rotational behavior is characteristic of the complex plane (from the multiplication of two complex numbers). Curvilinear, rotational, and spiral-like features are common, and they are still present when the Mandelbrot function has been manipulated, as seen in the detail at right of Fig. 7.2.

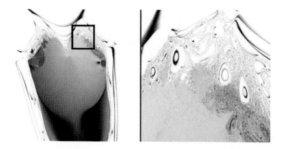

Fig. 7.2. Curvilinear forms remain upon manipulation of the Mandelbrot equation

And so, while this approach may be irreverent mathematically, what can be learned about the nature of this particular artistic canvas elevates admiration for the magic of the Mandelbrot Set, and its extended family of related fractals.

Like the playful and evocative variations on complex plane fractals created by Pickover [14] and others, the technique described here is heavy on the "tweaking". It is not focused on finding interesting regions in the pure unaltered Set (as in "Mandelzoom" [4]). It is more like sculpting than nature photography. Or perhaps it is more like genetic engineering than painting. For this reason, the technique is called "Mandeltweak". Images produced by this technique are also called "Mandeltweaks".

7.1.1 Genetic Space

Inspired by a metaphor which Dawkins uses to describe the genes of species as points existing within a vast multi-dimensional "genetic space" [3], Mandeltweaks are likewise considered as existing within a genetic space. Figure 7.3 illustrates this. In each image, two genes vary in even increments, one in the horizontal dimension, and the other in the vertical dimension. The values are default in the middle of each space, and this is where the Mandelbrot Set lies. The image on the left is of a large nine-panel photo series shown in various gallery settings.

7.1.2 Genetic Parameters

The number of ways one can alter a digital image is practically infinite. Consider the number of plug-in filters available for software tools like Photoshop. But while an arbitrary number of filters, distortions, layerings, etc. could have been applied with parameters for manipulating Mandelbrot images in the pixel domain, the choice was to keep all variability to within the confines of the mathematical function. The game is to try to optimize the parameters

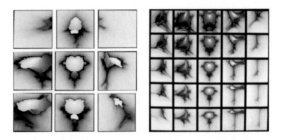

Fig. 7.3. Mandelbrot Set in the middle of two 2D genetic spaces

of the manipulated Mandelbrot equation so that the resulting images resemble an ideal form (or in some cases an explicit target image). This is a challenge because the mapping of genetic parameters to image attributes (genotype to phenotype) is non-trivial, and unpredictable.

7.2 Background

The Mandelbrot Set has been called "the most complex object in mathematics" [4]. It is like the mascot of the chaos and complexity renaissance – replicated in popular science books like a celebrity. When looking at it in its whole, it looks like a squashed bug – not pretty. But its deep remote recesses reveal amazing patterns that provoke an aesthetic response. Is the Mandelbrot Set a form of abstract art? No. Not if you consider abstractionism as human-made art that is "abstracted" from nature, with human interpretation. The Platonic stance claims that the Mandelbrot Set "just is". It had been hiding in the folds of complex mathematics until humans and computers revealed it. But consider the canvas upon which the Mandelbrot Set is painted. We can alter our mathematical paint brush from $z = z^2 + c$ to something else which is not so clearly defined, and make a departure from its Platonic purity. We can render images with the kind of interpretation, imprecision, and poetry that distinguishes art from pure mathematics.

It is possible that the Mandelbrot Set was discovered a handful of times by separate explorers, apparently first rendered in crude form by Brooks and Matelski [13]. Benoit Mandelbrot discovered it in the process of developing his theory of fractals, based on earlier work on complex dynamics by Fatou and Julia [10]. His work, and subsequent findings by Douady [6], helped popularize the Set, which now bears Mandelbrot's name.

Briefly described, the Set is a portrait of the complex function, $z = z^2 + c$, when iterated in the two-dimensional space known as the complex plane, the space of all complex numbers. When the function is applied repeatedly, using its own output as input for each iteration, the value of z changes in interesting ways, characteristically different depending on c (i.e., where it is being applied in the plane). The dynamics of the function as it is mapped determines the colors that are plotted as pixels, to make images of the Set. Specifically, if the magnitude of z grows large enough (> 2) and escapes to infinity, it is considered outside of the Set. The inside is shown as the black shape on the left of Fig. 7.4.

The Set has become a favorite subject for computer art. On its boundary is an infinite amount of provocative imagery. The most common visual explorations involve zooming into remote regions and applying color gradations on the outside of the Set near the boundary. This is somewhat like simple point-and-shoot photography, with a bit of darkroom craft added. Some techniques have been developed to search-out interesting remote regions, including an evolutionary algorithm developed by Ashlock [1]. A particle swarm for converging on the boundary, under different magnifications, was developed by Ventrella [21].

Deeper exploration into the nature of the Set is achieved by manipulating the mathematics, to reveal hidden structures. Pickover [14] has fished out a great wealth of imagery by using varieties of similar math functions. Peitgen et al. [12] describe a variety of complex plane fractals, with ample mathematical explanations. Dickerson [5], and others, have explored higher-order variations of the function to generate other "Mandelbrot Sets". According to Dickerson, the equation can be generalized to: $z = a \times f(z) + c$, where a is a scale constant and $f(z)$ is one of a broad range of functions, such as higher powers of z, polynomials, exponentials and trigonometric functions. As an example of a higher-order function: $z = z^3 + c$ creates the shape shown at the right in Fig. 7.4. This "Mandelbrot cubed" function is used in some experiments described below.

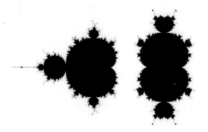

Fig. 7.4. Mandelbrot Set (left), and Mandelbrot "cubed" (right)

The algorithms used in Mandeltweak can also be considered as a generalization from $z = z^2 + c$ to $z = a \times f(z) + c$, only in this case, the complex nature of the number z is violated: the real and imaginary components of the number, as exposed in the software implementation, are manipulated. A collection of fractals created by Shigehiro Ushiki [18], are generated in a similar manner. More examples of separate treatment of the real and imaginary parts of the equation are emerging. These are described in various web sites on the Internet, including a technique developed by eNZed Blue [2].

7.2.1 Evolutionary Art

One way an artist can approach mathematically-based image-making is to identify a number of variables that determine visual variations and to tweak them to suit his/her own aesthetic style. The more variables available for tweaking, the more the artist can potentially tweak to reach some level of personal expression. The problem is that in most cases the variables are interdependent, and it is hard to predict the effects of variables in combination. Besides, most artists would rather not use numbers to manipulate visual language. Evolutionary computation addresses this problem.

In evolutionary art, a computer software program becomes a creative collaborator to the artist. The most common process is *interactive evolution* (also called "aesthetic evolution" or "aesthetic selection"). In contrast to the standard genetic algorithm (GA) [7], the selection agent is not determined by an objective fitness function, but rather by a human observer/participant (the artist), whose aesthetic choices guide the direction of evolution in a population of variations of an artwork. McCormack [11] outlines a number of problems that remain open as we articulate and refine the tools for evolutionary art. Among these are the problem of finding interesting and meaningful phenotypes, which are capable of enough variation to allow for artistic freedom.

Among the earliest examples of evolutionary art are the work of Latham [17]. Sims [16] has applied genetic programming [8] to various visual realms. Rooke [15], Ventrella [19], and others have developed evolutionary techniques for generating visual art and animation.

7.3 Technique

The standard black-and-white figure of the Mandelbrot Set is created as follows: on a rectangular pixel array (i, j), determine whether each pixel lies inside or outside of the Set. If inside, color the pixel black, otherwise, color it white. This 2D array of pixels maps to a mathematical space (x, y) lying within the range of -2 to 1 in x, and -1 to 1 in y. The function $z = z^2 + c$ is iterated. The value c is a complex number $(x + yi)$ – (the 2D location in the complex plane corresponding to the pixel), and z is a complex number which starts at $(0 + 0i)$ and is repeatedly fed back into the function. This is repeated until either the number of iterations reaches a maximum limit, or the magnitude of z exceeds 2. If it exceeds 2, it is destined to increase towards infinity, and this signifies that it is outside of the Set, as expressed in Algorithm 1.

This is not implemented as optimally as it could be, but it exposes more variables for manipulation – which is part of the Mandeltweak technique.

maxIterations could be any positive number, but higher numbers are needed to resolve a detailed definition of the boundary, which is especially important for high magnifications. Mandeltweak does not require high magnifications, and so this value ranges from 30 to 50. The mapping of (i, j) to

---

Algorithm 1 Plot Mandelbrot

For each pixel (i, j)
    map screen pixel values (i, j) to real number values (x, y)
    $zm = 0$
    $zx = 0$
    $zy = 0$
    $timer = 0$
    while ($zm < outsideTest$ AND $timer < maxIterations$)
        $z1 = y + zy \times zy + zx \times -zx$
        $z2 = x + 2.0 \times zx \times zy$
        $zm = z1 \times z1 + z2 \times z2$
        $zx = z2$
        $zy = z1$
    end while
    if ($zm < outsideTest$)
        set pixel color black
    else
        set pixel color white
    end if
    plot pixel (i, j)
end for loop

(x, y) determines translation, rotation, and scaling in the complex plane. Complex number z is represented by zx and zy. The variable zm is the squared magnitude of z. The variable $outsideTest$ is set to 4 for normal Mandelbrot plotting, but it could be set to other values, as explained below.

Note that swapping x and y rotates the Set 90 degrees. The Mandeltweak approach is to make the real number axis vertical, orienting the Set as if it were a fellow entity with left-right bilateral symmetry, as shown in Fig. 7.5.

7.3.1 Coloration

The typical coloration scheme for Mandelbrot Set images applies a color gradient on the outside of the Set which maps to the value of $timer$ after iteration. A smoother version of the outside gradient, described by Peitgen [12] can be achieved by replacing the value $timer$ with: $0.5 \times \log(z)/2^{timer}$. Setting $outsideTest$ to higher values, such as 1000, makes it smoother. Mandeltweak uses this technique, and in most cases, the background is rendered with a light color, which shifts to a dark color very close to the boundary. The effect is a mostly-light colored background, which is darker near the complex boundary, as seen in Fig. 7.5(c) and (d).

The inside of the Set is colorized with gradient $(zm/outsideTest + za)/2$, where za is a special value determined by analyzing the orbit of z during iteration. As the value of z jumps around the complex plane, the change in angle between vectors traced by each consecutive pair of jumps is saved, and when iteration is complete, the average angle is calculated. This is normalized to create za. In addition, both zm and za are modulated by sine waves whose frequencies and phases are evolvable. The result is that a variety of coloration scenarios are possible, sometimes accentuating hidden features in the *flesh*.

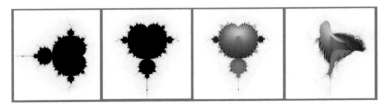

Fig. 7.5. (a) The default set, (b) rotated, (c) colorized, (d) tweaked

One artistic post-process is added to give the images a sepia tone quality with a subtle blue shadow effect. While keeping pure black and white at the extremes, a slight blue shift in the darker range, and a slight orange shift in the lighter range, are applied.

7.3.2 Morphological Tweakers

The kernel of the Mandelbrot equation is in the two lines in Algorithm 1 that express the complex number equation $z = z^2 + c$ in real number terms:

$$z1 = y + zy \times zy + zx \times -zx \tag{7.1}$$
$$z2 = x + 2.0 \times zx \times zy. \tag{7.2}$$

Arbitrary morphological tweakers are inserted. Following is an example of a typical set of tweakers:

$$z1 = y + (zy + p1) \times (zy \times p2) + (zx \times p3) \times ((zx \times p4) + p5) \times p6 \tag{7.3}$$
$$z2 = x + p7 \times (zx + p8) \times (zy + p9) \tag{7.4}$$

where $p1$ through $p9$ are real number variables. Their default values are as follows: $p2$, $p3$, and $p4$ are set to 1; $p6$ is set to -1; and $p7$ is set to 2. The rest are set to 0. Each tweaker can deviate from its default value within a range (extending in both the negative and positive directions). These ranges are unique for each tweaker – most of them are around 1 or 2. Each tweaker controls a unique visual behavior. For instance, $p2$ is responsible for the distortion shown in Fig. 7.5(d).

In addition to these morphological tweakers, the following lines:

$$zm = z1 \times z1 + z2 \times z2$$
$$zx = z2$$
$$zy = z1$$

are expanded as follows:

$$zm = z1 \times z1 + z2 \times z2$$
$$zm = zm \times (1 - p10) + z2 \times p10$$
$$zx = z2 \times (1 - p11) + z1 \times p11$$
$$zy = z1 \times (1 - p12) + z2 \times p12$$

where $p10$, $p11$, and $p12$ are 0 by default. Also, before the iterative loop, the lines:

$$zx = 0$$
$$zy = 0$$

are expanded to:

$$zx = p13$$
$$zy = p14$$

where $p13$ and $p14$ are set to 0 as default.

Since the kernel of the Mandelbrot equation could also be expressed as:

$$z1 = y + zy \times zx + zx \times zy \tag{7.5}$$
$$z2 = x + zx \times zx - zy \times zy \tag{7.6}$$

a different set of tweakers could be applied as follows:

$$z1 = y + (zy \times t1 + t9) \times zx \times t2 + t10)$$
$$+ (zx \times t3 + t11) \times zy \times t4 + t12) \tag{7.7}$$
$$z2 = x + (zx \times t5 + t13) \times zx \times t6 + t14)$$
$$- (zy \times t7 + t15) \times zy \times t8 + t16). \tag{7.8}$$

This is a more orderly expansion, and it provides a larger genetic space than the $p1$–$p9$ shown in the original expression (Equations 7.1 and 7.2). New and intriguing forms exist in this space as well.

The example of tweakers just shown is not the whole story. To describe all the variations that have been tried could take potentially many more pages. These examples should give a general sense of how the technique is applied. The reader, if interested in trying out variations, is encouraged to choose a unique set of tweakers appropriate to the specific goal. Mandeltweaking is more an Art than a Science.

These tweakers are stored in an array as real number values within varying phenotype-specific ranges (the phenotype array). They are generated from an array of genes (the genotype array). The genotype holds normalized representations of all the tweakers, in the range 0–1. Before an image is generated, the genotype is mapped to the phenotype array, taking into consideration the tweakers' default values and ranges. This normalized genotype representation will come in handy as explained later when a genetic algorithm is applied to the technique.

Besides morphology tweaking, the entire image can be transformed. There are four tweakers for this which determine:

1. angle of rotation;
2. magnification (uniform scaling in both x and y);
3. translation in x; and
4. translation in y.

These can be thought of as transformations of the digital microscope that views the complex plane.

When these tweakers are set to their default values, the function produces the Mandelbrot Set. Any offset from a default value will push it out to a superset of the Mandelbrot Set, called *Mandeltweak*, a dimension that does not obey the normal rules of complex numbers. All experiments can be considered as deviations from the true Set – deviations from its home in the complex plane. This is the control point – the point of registration from which to build a visual vocabulary describing this multidimensional genetic space.

7.3.3 Interactive Evolution for Artistic Breeding

Originally, the Mandeltweak software was given an interface to generate images which could be repeatedly reviewed and altered, until interesting and provocative forms were resolved. These sessions sometimes involved hundreds of adjustments. Considering the accumulated memory in the artist's mind of the variations being explored, you could say that a sort of wetware genetic algorithm was being run, resulting in convergence towards a desired image. These sessions were sometimes very long. These final images were stored mostly as photographs, and exhibited in art shows and on the web [20]. Figure 7.6 shows six examples of images created with this process.

This experience set the stage for developing a modified genetic algorithm with an interactive evolution interface, whereby the artistic choices could be stored in a population of tweak settings, and re-circulated within the population to offer up combinations of favorite images. This was a great improvement – a natural application of evolutionary programming to the problem domain. This interactive evolution scheme is described below.

1. Initialization

A population of genotypes is generated, with their genes initialized randomly, distributed evenly in the range (0–1). Population size is usually set to around 100. The genotypes are associated with initially random fitness values ranging from 0 to 1 (which is meaningless at first, but as explained below, fitness

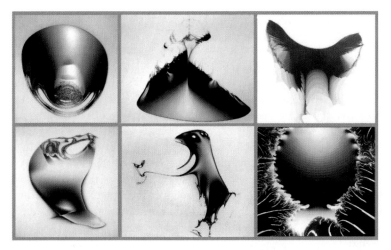

Fig. 7.6. Some early examples of Mandeltweaks

values will gradually change to take on meaning).

2. Iteration

The iterative loop has three basic components: (a) mating, (b) evaluation, and (c) death. This is explained below.

(a) Mating via Tournament Selection

Two random, relatively fit genotypes are chosen as parents, each by way of a competition for relative fitness, as follows:

1. Two genotypes are randomly chosen, and their fitness values are compared. The one with the highest fitness is labeled "parent 1".
2. Another competition is run with two other randomly-chosen genotypes, and the winner is chosen as "parent 2".

The two resulting parent genotypes mate to produce one offspring genotype using crossover, with some chance of mutation. During mating, standard crossover and mutation techniques are used, with crossover rate $C = 0.2$, and mutation rate m ranging from 0.01 to 0.1 in most experiments. While parent genotypes are being read to generate the offspring genotype, gene-by-gene, there is C chance of the parent genotype which is being copied to the offspring to swap to the other parent. And there is m chance that the gene being read will mutate. If mutated, a random number is added to the gene value, which ranges between -1 and 1, and is weighted towards zero. If after mutation, the gene value falls out of the normal interval 0 to 1, it wraps around to keep the value normalized.

(b) Evaluation

The resulting offspring genotype is then mapped to a phenotype array to determine the tweakers for generating a new Mandeltweak image. The user evaluates this image by giving it a value in the range of (0–1). Different versions have been explored as far as inputting this value, including binary (0 = bad vs. 1 = good); a three-choice scheme (bad-medium-good); and continuous (clicking the mouse on the screen, with the location from left to right or bottom to top determining a value from 0 to 1). There are pros and cons to each of these input techniques, which will not be covered here – what's important is that a value ranging from 0–1 is provided by the user.

(c) Death

Once a fitness value has been provided for this image, the associated genotype replaces the least-fit genotype in the population.

This process is iterated indefinitely. In the beginning, the user experiences no progress, especially in the case of large populations (like more than 100), but in time, the choices that the user has been making start to effect the population, and images begin to come up that are preferable, having visual qualities that the user has been responding positively to.

In some experiments, the selection of parent genotypes is set to not take relative fitness into consideration, and so any two genotypes can become parents. In this case, the only driving force for evolution is the fact that the least-fit genotype is killed off to make room for the offspring. This causes slower convergence, but more thorough exploration of the genetic space. It appears not to have a major impact on the outcome.

7.3.4 One Image at a Time, One Mating at a Time

A common design in interactive evolution schemes is to present the user with a collection of images with variation, and for the user to compare these and make some selection based on that comparison. A major difference in the technique described here is that the user is presented with only one image at a time, and uses visual memory to compare with other images seen. This interaction design is intended to enhance the experience of perceiving an individual image as a work of art. It allows the aesthetic sense to operate more like viewing art than shopping for a product. This interface is meant to allow the process of interactive evolution to be pure and direct – an evolving dialog between the artist's visual memory and a progression of fresh images, with an arc of aesthetic convergence that threads through the experience.

Many genetic algorithm schemes use *generational selection*: all the genotypes in the population are sized-up for fitness in one step, and then the entire population is updated to create a new generation of genotypes, which selectively inherit the genetic building blocks from the previous generation. For software implementation, a backup population is required in order create each new generation. In contrast, this scheme uses *steady-state selection*: it keeps only one population in computer memory, and genetic evolution is performed on that population one genotype at a time. This is slower than generational selection, but it has the effect of preserving the most fit genotypes at all times. And the grim reaper only visits the least fit.

7.3.5 Fitness Decay

At the start of the process, fitness values are randomly distributed from 0 to 1. Since relatively-fit genotypes are chosen to mate and offer up Mandeltweak images for evaluation, the first images reviewed are essentially random in their relative quality. But this soon begins to change as the user provides meaningful fitness values. In addition to this, a global decay scalar d is applied to all fitness values at each iteration ($d =$ just under 1, typically 0.99). The effect of d at the beginning of the process is that genotypes that were randomly initialized with high fitness values decrease as new genotypes rise to the top of the fitness range as a result of positive user selection. The distribution of fitness values begins to reflect user choice.

As the initial random distribution of fitness values gives way to meaningful values from user interaction, the decay operator then begins to serve a different purpose: that of allowing for meandering aesthetic goals. Genotypes that were once considered fit are allowed to decay. The decay effect roughly corresponds to memory. This avoids having the highest fit genotypes continually dominate the population and prohibit new discoveries to take the lead. If there were no fitness decay, the population would lose any flexibility to respond to changes in the aesthetic trajectory. The fitness decay operator is like the evaporation of ant pheromones – chemicals released by ants for communicating which build up in the environment and permit ants to establish trails for foraging. If pheromone scent never decayed, ant colonies would not be able to adapt to changing distributions of food, and their collective behavior would become rigid. Same with this fitness decay: it gives the user a chance to push the population in new directions when an aesthetic dead-end is reached.

The value d is sensitive. If it is set too low (like, 0.9 – causing fitness to decay too quickly), then the results of the user's choices will not stay around long enough to have an effect on the general direction of evolution. If it is too weak (like 0.999 – decreasing too slowly) then inertia sets in: user selections that were either "mistakes" (or choices that are no longer relevant) will stick around too long and make evolution inflexible. This value is sensitive to population size, user psychology, the nature of the evolvable imagery, mutation rate, and other factors.

The interactive evolution technique just described has a few notable properties:

1. it always preserves the most fit genotypes (imagine looking at a picture, liking it, and choosing to keep it in a box for future use – you can rely on it being there for a long time);
2. it always overwrites the least-fit genotype, which has the effect of increasing average fitness over time (note that the least-fit genotype fell to its place either because the user put it there directly by selection, or else it slowly "faded into the remote past" as a result of fitness decay – in both cases, it is appropriate to replace it);
3. it allows user aesthetics to change direction over time, and to redirect the population.

7.3.6 Using a Digital Image as a Fitness Function

The latest stage in this progression towards automating the process is to use an image as an objective fitness function. Instead of a user providing the fitness of a Mandeltweak based on aesthetics, the Mandeltweak is compared to an ideal image to determine similarity. The genetic algorithm for this scheme is the same as the interactive evolutionary scheme described above, except for three important differences:

1. The human user is replaced by an image-comparison algorithm, which uses a single ideal image.
2. There is no fitness decay operator. Fitness decay is a psychological mechanism, and is not needed in this case. Since the ideal image is static (as opposed aesthetic whim), there is no need for the flexibility that decay affords.
3. Instead of setting all fitness values randomly at initialization, the initial genotypes are used to generate an initial population of Mandeltweaks to establish meaningful fitness values. This becomes the starting point for the iterative loop.

The ideal image is either painted in Photoshop, pulled off of a website, or snapped with a digital camera and then post-processed with Photoshop. Only gray-scale images are used, for three reasons:

1. it reduces the complexity of the experiment to fewer variables;
2. the addition of color was not found to contribute significantly to the perception of figurative form in the images; and
3. it was an artistic choice: to encourage the resulting images to resemble black-and-white portrait photography.

All ideal images have a white background, to simplify the technique and to disambiguate figure vs. ground.

7.3.7 Image Resolution

It was found that comparing a Mandeltweak image with the ideal image could be done adequately using a resolution r of only 50 pixels. So, the ideal image consists of r^2 (50 × 50 = 2500) pixels, where each pixel color is a shade of gray ranging from black to white in $g = 256$ possible values ($0 \le g \le 1255$). When a Mandeltweak is generated so as to compare with the ideal image to calculate a fitness value, it is rendered at the same resolution. The images are compared, pixel-by-pixel, and so there are r^2 pixel value differences used to determine the difference between the images, and thus the fitness.

Note that even though the Mandeltweak is rendered at a specific resolution for comparison, this does not mean that it could not be re-rendered at a higher resolution. In fact, since the genotype is what constitutes the representation of the image (not the pixel values), it could be re-rendered at any arbitrary resolution. What r represents, then, is the amount of image detail that could potentially be compared. And since the mimicking ability of Mandeltweak is limited only to general forms and approximate gray-scale values, it is not necessary to present it with high-resolution ideal images – the extra detail would be wasted.

7.3.8 Image Comparison

To illustrate how the comparison scheme works, consider the ideal image in Fig. 7.7 of a black disk against a white background.

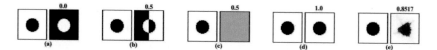

Fig. 7.7. Five examples showing images compared to an ideal image (black disk)

Compare each image to the right of the black disks in examples (a) through (e). In example (a), every pixel value is as different as possible (covering the complete range of pixel difference: 255), and so the resulting fitness is 0. In example (b), half of the pixel values are identical, and the other half are as different as possible, and so fitness is 0.5. In example (c), the image is filled entirely with gray value 128. In this case, fitness is 0.5, since the difference between a gray pixel and either black or white is 128. In example (d), all pixels are identical, and so fitness is 1.0. Example (e) shows a Mandeltweak that resulted from evolution in a population using the black disk as the fitness function. It was able to approach the ideal, reaching a fitness value of 0.8517.

Let's define p, ($-255 \le p \le 255$) as the difference in value of a pixel in the Mandeltweak image and its corresponding pixel in the ideal image. The normalized pixel difference is $|p/g|$. P is the sum of all normalized pixel differences. Thus, the fitness f of a Mandeltweak image is $f = 1 - P/r^2$, $(0 \le f \le 1)$.

This technique uses a simple pixel-wise comparison. A few variations have been explored in order to encourage sensitivity to certain features. But nothing conclusive has come of this. There is certainly a lot of research, and many techniques, for feature-based image comparison, and it would make for an interesting enhancement to this technique. But for the preliminary purposes of these experiments, this simple scheme is sufficient.

7.4 Experiments

To help visualize the evolution of a population of Mandeltweaks using an ideal image as the fitness function, each genotype is plotted as a row of rectangles. The gray value of a rectangle corresponds to the value of its associated gene, with the range 0–1 mapped to a gray scale from black to white. An example is illustrated in Fig. 7.8.

Fig. 7.8. Visualization of a genotype with gene values mapped to grayscale

This genotype visualization in used in Fig. 7.9, which shows a population of 1000 genotypes evolving to imitate an ideal image of the author's face (upper left). Four stages of the evolution are plotted, at times 0, 1000, 10,000, and 50,000. Fitness is visualized as the vertical height of the genotype, and convergence is revealed as similarity in genotype coloration. The Mandeltweak with the highest fitness in the population is shown at the top of each plot, and its fitness value is shown at the left of the plot. The final Mandeltweak is enlarged at right.

At initialization genotypes are randomized and their associated images are compared to the ideal image to determine fitness. In this particular experiment, the fitness values range from just under 0.5 to 0.827 at initialization. This distribution of fitness values is believed to be due to the three following factors:

1. the characteristics of the genotype-to-phenotype mapping, and thus the resulting Mandeltweak images;
2. the ideal image; and
3. the nature of the fitness comparison scheme.

The graph shows that after 1000 iterations the lower end of the fitness range has raised. This is because the least-fit genotype is always replaced

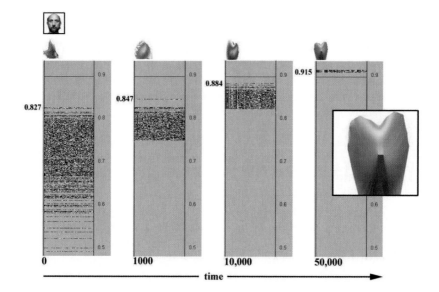

Fig. 7.9. Plotting fitness and genetic convergence in a population of 1000 genotypes

with the newly-created genotype for each step, and since each new genotype is the product of two relatively-fit genotypes, it usually has higher fitness. This is especially the case in the beginning. By time 50,000 we see that the highest fitness is 0.915. Notice also that the genotypes cover a much smaller range of fitness and that they have converged considerably (revealing visible bands of similar colors across the population).

Figure 7.10 shows the results of six experiments, each using a different ideal image as the fitness function.

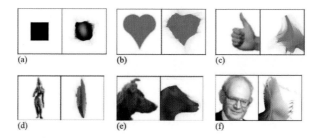

Fig. 7.10. Six examples of ideal images and evolved Mandeltweaks

In these examples, the ideal images are shown to the left of their associated Mandeltweaks. The ideal images may appear jagged or pixelated because of their low resolution. The Mandeltweaks, in contrast, are shown at a higher resolution – recall that pixel resolution is arbitrary in Mandeltweaks, as the encoding of the image is parametric.

In all of these experiments, population size was set to 1000, and mutation rate was set to 0.05 except for examples (a) and (c), in which population size was set to 100 and mutation rate was set to 0.1. In all cases, the highest fitness achieved was in the approximate range of 0.95. The number of iterations in each case ranged, averaging around 2000.

7.4.1 Range of Genetic Variation

Each tweaker has a unique range within which it can deviate from its default value, as explained earlier. To manipulate this range, a global range scale s was created so that the whole array of range values could be scaled at once. In most experiments, s is set to 1 (resulting in the normal tweak ranges as originally designed). But s can be varied to explore Mandeltweak's imitative performance over different genetic ranges. Figure 7.11 shows the results of 11 experiments with the ideal image set to a portrait of painter Francis Bacon.

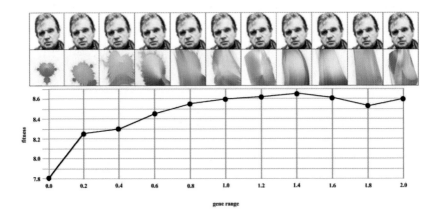

Fig. 7.11. Varying tweak ranges in a series of experiments

In each experiment, population size was set to 1000, and mutation rate was set to 0.05. When s is set to 0.0, the result is that when genotypes are mapped to phenotypes, the values of the tweakers are clamped to their defaults. And so the image at the left-most side of Fig. 7.11 has the Mandelbrot morphology. As s is increased by increments of 0.2, we see that fitness increases on average, because there is an increasingly larger genetic space available for searching. The reason the default Mandelbrot image has a fitness of 0.78 has to do with the nature of the ideal image, the nature of the default Mandeltweak settings, and the comparison technique. What is of interest, though, is not this value, but the rate at which fitness increases, and then reaches what appears to be an upper limit. All experiments indicate a similar limitation, including when they are run for many more iterations and with much larger populations.

Artistically-speaking, one might find the visual "sweet-spot" to occur before fitness has reached its maximum. In fact, the images in Fig. 7.1 were intentionally evolved using a slightly smaller range. They also used smaller populations and were run for fewer iterations – this allowed the peculiar vestiges of Mandelbrot features to remain, making the head-like shapes more intriguing and ambiguous.

7.4.2 Imitating ... The Mandelbrot Set?

Since the vast genetic space of Mandeltweak contains the Mandelbrot Set at the point in the space where all values are at their default settings, it is possible that an initial population of random Mandeltweaks can converge on the Mandelbrot Set. But in a number of experiments with different population sizes and mutation rates, this was not achieved. Instead, the population converged on other regions of the space which are similar to the shape of the Set. Figure 7.12 shows the most fit Mandeltweak in a population in multiple stages of evolution, starting at time 0, then 2000, and then doubling the time intervals, up to 128,000. It reached a maximum fitness of 0.895. The image at right is the Mandelbrot Set (a high-res version of the ideal image used) to show what it was trying to imitate.

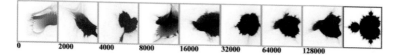

Fig. 7.12. Mandeltweak imitates the Mandelbrot Set, only backwards

In this and other similar experiments, the Mandeltweak got stuck on a local hill in the fitness landscape, which corresponds roughly to the shape, but it is rotated almost 180 degrees! The fitness landscape is very large, and the shapes in the initial random population are too varied from the original shape – and so a common protrusion resulting from tweaking (such as the one shown in Fig. 7.5d), ends up being a proxy for the main bulb. As a test, a critical gene, the "angle" gene (responsible for varying the rotation of the shape in the complex plane), was not allowed to vary. The population was then able to more easily converge on the Mandelbrot Set. This supports intuition that the angle gene enlarges the fitness landscape considerably.

7.4.3 Using Higher-Order Mandelbrot Functions

The shape created by the higher-order function, $z = z^3 + c$, is shown in Fig. 7.4. This uses more variables in the software implementation, to which tweakers can be attached, and so it was considered as another base function to explore. It can be expressed as follows: replace the kernel of the expanded Mandelbrot function shown in (7.7) and (7.8) with:

$$a = zx$$
$$b = zy$$
$$z2 = a \times zx - b \times zy$$
$$z1 = b \times zx + a \times zy$$
$$zx = z2$$
$$zy = z1$$
$$z2 = a \times zx - b \times zy$$
$$z1 = b \times zx + a \times zy$$
$$z1 = z1 + x$$
$$z2 = z2 + y$$
$$zx = z2$$
$$zy = z1$$

and tweak like this:

$$a = zx$$
$$b = zy$$
$$z2 = (a \times p1 + p2) \times zx \times p3 + p4) - (b \times p5 + p6) \times zy \times p7 + p8)$$
$$z1 = (b \times p9 + p10) \times zx \times p11 + p12) + (a \times p13 + p14) \times zy \times p15 + p16)$$
$$zx = z2$$
$$zy = z1$$
$$z2 = (a \times p17 + p18) \times zx \times p19 + p20) - (b \times p21 + p22) \times zy \times p23 + p24)$$
$$z1 = (b \times p25 + p26) \times zx \times p27 + p28) + (a \times p29 + p30) \times zy \times p31 + p32)$$
$$z1 = z1 + x$$
$$z2 = z2 + y$$
$$zx = z2$$
$$zy = z1.$$

Figure 7.13 shows the results of five experiments in which the normal Mandeltweak algorithm is compared to the one which uses the cubed algorithm just described. Population was set to 1000, and mutation rate was set to 0.05. Each experiment was run until the population converged significantly, and the

Fig. 7.13. Comparing normal versus cubed algorithm

number of iterations, while varying among pairs of tests, was kept constant for each algorithm in the pair.

The cubed algorithm doesn't appear to be much better at imitating the ideal image, and it might even be inferior, if the stars placed next to the fitness values in the figure are any indication. In the case of both algorithms, there is an inability to imitate local features – notice that the fingers of the hand and the legs of the horse are not picked up very well using either algorithm. The reasons for this are not clear. A few possibilities are:

1. the image comparison scheme is not feature-based;
2. the population is too small;
3. the genetic algorithm is not designed appropriately; or
4. these mathematical equations are simply not able to conjure up these particular shapes, even though the genetic space is very large.

There is more exploration to be done with higher-order Mandelbrot functions, as well as the other varieties of fractal equations.

7.5 Conclusion

Math-based computer art created using evolutionary computation is often non-objective or abstract – not meant to represent anything in particular. The Platonic Mandelbrot Set and its kin are neither art, nor are they abstract art. But the curious animal-like nature of Mandeltweak, as far as how it behaves upon manipulation, invites one to read it as an organic entity, and thus it enters into an interpretive space, making it more pregnant as an art medium. This was part of the initial motivation behind the technique described. Its ability to imitate explicit images is limited. But this tension – the tension between being the Mandelbrot Set and being coerced into a representational form – is part of the game. It is a kind of conceptual art. The question of how evolvable an image-making scheme can be is a common problem in evolutionary art: is the phenotype space large enough? – and can a subset of it map to the artist's aesthetic space? The Mandeltweak technique was created to take on this question. In the process of asking these questions, and to better understand its limits, the artist's aesthetic (and mathematical) vocabulary has grown larger.

References

1. Ashlock, D.: Evolutionary exploration of the Mandelbrot set. In: Proceedings of the 2006 Congress On Evolutionary Computation, pp. 7432–7439 (2006)
2. eNZed Blue: The Koruandelbrot (2005). http://www.enzedblue.com/Fractals/Fractals.html
3. Dawkins, R.: The Blind Watchmaker – Why the Evidence of Evolution Reveals a Universe Without Design. W. W. Norton and Company (1986)
4. Dewdney, A.: Computer recreations: a computer microscope zooms in for a look at the most complex object in mathematics. Scientific American pp. 16–25 (1985)
5. Dickerson, R.: Higher-order Mandelbrot fractals: experiments in nanogeometry (2006). http://mathforum.org/library/view/65021.html
6. Douady, A., Hubbard, J.: Etude dynamique des polynomes complexes, I and II. Publ. Math. Orsay (1984, 1985)
7. Goldberg, D.: Genetic Algorithms in Search, Optimization, and Machine Learning. Addison-Wesley (1989)
8. Koza, J.: Genetic Programming: On the Programming of Computers by Means of Natural Selection. MIT Press (1992)
9. Lakoff, G., Núnez, R.: Where Mathematics Comes From – How the Embodied Mind Brings Mathematics Into Being. Basic Books (2000)
10. Mandelbrot, B.: The Fractal Geometry of Nature. W. H. Freeman and Company (1977)
11. McCormack, J.: Open problems in evolutionary music and art. Lecture Notes in Computer Science **3449**, 428–436 (2005)
12. Peitgen, H., Saupe, D. (eds.): The Science of Fractal Images. Springer-Verlag (1988)
13. Penrose, R.: The Road to Reality, A Complete Guide to the Laws of the Universe. Knopf (2004)
14. Pickover, C.: Computers, Pattern, Chaos, and Beauty – Graphics From an Unseen World. St. Martins Press (1990)
15. Rooke, S.: Eons of genetically evolved algorithmic images. In: P. Bentley, D. Corne (eds.) Creative Evolutionary Systems. Morgan Kaufmann, San Francisco, CA (2001)
16. Sims, K.: Artificial evolution for computer graphics. In: Proceedings of the 18th annual conference on Computer graphics and interactive techniques, pp. 319–328 (1991)

17. Todd, S., Latham, W.: Evolutionary Art and Computers. Academic Press (1992)
18. Ushiki, S.: Phoenix. IEEE Transactions on Circuits and Systems **35**(7), 788 (1988)
19. Ventrella, J.: Explorations in the emergence of morphology and locomotion behaviour in animated characters. In: R. Brooks, P. Maes (eds.) Artificial Life IV – Proceedings of the 4th International Workshop on the Synthesis and Simulation of Living Systems, pp. 436–441. MIT Press (1994)
20. Ventrella, J.: Mandeltweaks (2004). `http://www.ventrella.com/Tweaks/MandelTweaks/tweaks.html`
21. Ventrella, J.: Mandelswarm – particle swarm seeks the boundary of the Mandelbrot Set (2005). `http://www.ventrella.com/Tweaks/MandelTweaks/MandelSwarm/MandelSwarm.html`

Evolutionary L-systems

Jon McCormack

Centre for Electronic Media Art, Faculty of Information Technology, Monash University, Clayton 3800, Victoria, Australia
Jon.McCormack@infotech.monash.edu.au

> Any "novelty," ...will be tested before all else for its compatibility with the whole of the system already bound by the innumerable controls commanding the execution of the organism's projective purpose. Hence the only acceptable mutations are those which, at the very least, do not lessen the coherence of the teleonomic apparatus, but rather further strengthen it in its already assumed orientation or (probably more rarely) open the way to new possibilities.
>
> Jacques Monod, *Chance and Necessity* [17, p. 119]

8.1 Introduction

The problem confronting any contemporary artist wishing to use technology is in the relationship between algorithmic and creative processes. This relationship is traditionally a conflicting one, with the artist trying to bend and adapt to the rigour and exactness of the computational process, while aspiring for an unbounded freedom of expression. Software for creative applications has typically looked to artforms and processes from non-computational media as its primary source of inspiration and metaphor (e.g. the photographic darkroom, cinema and theatre, multi-track tape recording, etc.).

The process implicitly to be advanced in this chapter is that we should turn to natural processes, including *computation itself* as a creative source of inspiration. Adopting this approach removes the traditional conflict between algorithmic and creative processes that typifies most software used in creative applications, liberating the artist to explore computational processes as a source of inspiration and artistic creativity.

This chapter details the interactive evolution[1] of string rewriting grammars, known as *L-systems*. The method is a powerful way to create complex,

---

[1] Also known as *aesthetic evolution, interactive aesthetic selection,* or the *interactive genetic algorithm* (IGA).

organic computer graphics and animations. This chapter describes an inter-active modelling system for computer graphics in which the user is able to evolve grammatical rules. This system has been used by the author for more than 15 years to produce a number of creative works (see [14] for details). Starting from any initial timed, parametric D0L-system grammar (defined in Section 8.2.4), evolution proceeds via repeated random mutation and user selection. Sub-classes of the mutation process depend on the context of the current symbol or rule being mutated and include mutation of: parametric equations and expressions, development functions, rules, and productions. As the grammar allows importation of parametric surfaces, these surfaces can be mutated and selected as well. The mutated rules are then interpreted to cre-ate a three-dimensional, time-dependent model composed of parametric and polygonal geometry. L-system evolution allows a novice user, with minimal knowledge of L-systems, to create complex, 'life-like' images and animations that would be difficult and far more time-consuming to achieve by writing rules and equations explicitly.

8.1.1 L-systems and Graphics

Modern computer graphics and computer aided design (CAD) systems allow for the creation of three-dimensional geometric models with a high degree of user interaction. Such systems provide a reasonable paradigm for modelling the geometric objects made by humans. Many organic and natural objects, however, have a great deal of complexity that proves difficult or even impos-sible to model with surface or CSG based modelling systems. Moreover, many natural objects are statistically similar, that is they appear approximately the same but no two of the same species are identical in their proportions.

L-systems have a demonstrated ability to model natural objects, particu-larly branching structures, botanical and cellular models. However, despite the flexibility and potential of L-systems for organic modelling (and procedural models in general), they are difficult for the non-expert to use and control. To create specific models requires much experimentation and analysis of the ob-ject to be modelled. Even an experienced user can only create models that are understood and designed based on their knowledge of how the system works. This may be a limitation if the designer seeks to explore an open-ended design space. The method presented here gives the artist or designer a way of ex-ploring the "phase space" of possibilities offered by L-system models without an explicit understanding of how to construct the grammar that generates a given model.

8.1.2 Related Work

Interactive evolution is a well-explored method to search through combina-torial computational spaces using subjective criteria. As with all Evolution-ary Computing (EC) methods, it is loosely based on a Darwinian evolution metaphor, with the fitness of each generation explicitly selected by the user. The method was first described by Richard Dawkins in his 1986 book *The Blind Watchmaker* [4]. Dawkins demonstrated simulated evolution by evolving *biomorphs* – simple two-dimensional structures with a lateral symmetry resembling organic creatures in form. The survival of each generation of biomorphs is selected by the user who evolves features according to their personal selection.

In the 20 years since *The Blind Watchmaker* was published, a number of applications of the technique have been published. An extensive overview can be found in [24]. Applying evolutionary methods to generate L-systems has also been extensively researched, [12] provides a summary.

The next section provides a brief overview of the interactive evolution process. Section 8.2 looks at the syntax and structure of L-systems used for interactive evolution. Section 8.3 shows how the strings produced by L-systems may be interpreted into complex geometric structures, suited to the automated modelling of organic form. Section 8.4 explains the mutation process necessary for the evolutionary process to find new genotypes. Finally, Sect. 8.6 summarises the results of the system, including a case study of a recent commissioned artwork. The artwork was produced using software that implements the methods described in this chapter.

8.1.3 Overview of the Interactive Evolution Process

In nature, genetic variation is achieved through two key processes: mutation of the parent genotype; and crossing over of two parental genotypes in the case of sexual species. In the system described here, 'child' genotypes are created by mutating a single parent genotype. Such mutations cause different phenotypes to result. Changes may affect structure, size, topology and growth. Through a repeated process of selection by the user and mutation by the computer, aesthetic characteristics of the resultant forms can be optimised (the genetic algorithm can be thought of as an optimisation process).

Typically, the user of the system begins with an existing L-system that they wish to evolve. This *germinal genotype* becomes the first parent from which offspring are mutated. A library of existing L-systems is available for the user to select from to begin the evolutionary process. It is also possible to evolve 'from scratch', i.e. with only implicitly defined identity productions in the germinal genotype.

At each generation, 11–15 mutated offspring from the parent genotype are created. Each phenotype is a time-variant, three-dimensional structure that can be manipulated in real-time on screen. The user examines each offspring, selecting the child with the most appealing traits to become the new parent. The process is repeated for as long as necessary.[2]

---

[2] Normally until a satisfactory form is found, or the user runs out of patience.

Following the evolutionary process, the user has the option of saving the mutated genotype in the library. Alternatively, the phenotype (three-dimensional geometric data, shading and lighting information) can be exported to software or hardware based rendering systems for further visualisation and processing. The dataflow of the system is outlined in Fig. 8.1.

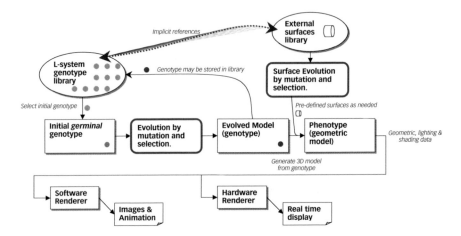

Fig. 8.1. The initial genome is selected from an existing library to undergo the evolutionary processes described in this chapter. Following evolution the genotype can be (optionally) saved into the library or output to rendering systems for visualisation

The interactive evolution process is a useful technique for the aesthetic exploration of parameterised computational systems, however is not without problems, both technical and creative [5, 15].

8.2 L-systems

Since the original formulation by Lindenmayer in 1968 [10], L-systems have been adapted to modelling a diverse range of phenomena, including herbaceous plants [19], neural networks [9] and the procedural design of cities [18]. This has necessarily involved many different formulations and extensions beyond Lindenmayer's original formalism. Fundamentally however, all L-systems are iterative symbol rewriting systems, where strings of symbols are replaced *in parallel* according to a set of rules, known as *productions*. For detailed descriptions, the reader is referred to [19, 21, 22].

In the following sections, I will briefly review a number of L-system variants: *deterministic, stochastic, parametric* and *timed*. Only *context-free* L-systems (0L-systems) will be considered here. The variants described can be subject to evolutionary algorithms, detailed in Sect. 8.4.

8.2.1 0L-systems

0L-systems are the simplest class of L-system, being context-free, interaction-less, or zero-sided. Following the terminology and presentations found in [7, 8, 19, 21], a formal description is presented below.

A context-free, 0L-System is defined as the ordered triple $G = \langle V, \omega, P \rangle$, where:

- $V = \{s_1, s_2, \ldots, s_n\}$ is an *alphabet* composed of a set of distinct *symbols*, s_i;
- $\omega \in V^+$, a non-empty word (sequence of symbols) over V, known as the *axiom*;
- $P \subset V \times V^*$ an endomorphism defined on V^*, known as the finite set of *productions*.

A production $(s, \chi) \in P$ is written in the form $s \to \chi$, where the symbol s is known as the *predecessor* and the word $\chi \in V^*$ the *successor* of the production. Where there is no specific production for a symbol s, the *identity production* $s \to s$ is assumed.

Deterministic L-systems (D0L-systems) have at most one production for each symbol. A 0L-system is *deterministic* if and only if for each $s \in V$ there is exactly one $\chi \in V^*$ such that $s \in \chi$.

The L-system, G, applies a production to a symbol, s, in a word when the predecessor for that production matches s. The notation $s \mapsto \chi$ means module s produces a word χ as a result of applying a production in G.

Let $\mu = s_1 s_2 \ldots s_k$ be an arbitrary word over V. The word $\nu = \chi_1 \chi_2 \cdots \chi_m$ is generated by G from μ, denoted $\mu \Rightarrow \nu$, iff $s_i \mapsto \chi_i \; \forall i : i = 1, 2, \ldots, k$. G generates a *developmental sequence* of words $\mu_0 \mu_1 \ldots \mu_n$ by applying the matching productions from P at each iteration, beginning with the axiom, ω. That is $\mu_0 = \omega$ and $\mu_0 \Rightarrow \mu_1 \Rightarrow \ldots \Rightarrow \mu_n$. A word ν is a *derivation of length* n if there exists a developmental sequence such that $\nu = \mu_n$.

Here is a simple D0L-system example:

$$V = \{F, R, L, [,]\}$$
$$\omega : F$$
$$p_1 : F \to FFR[RFLFLF]L[LFRFRF] \tag{8.1}$$

which generates the sequence of words:

$$\omega = \mu_0 = F$$
$$\mu_1 = FFR[RFLFLF]L[LFRFRF]$$
$$\mu_2 = FFR[RFLFLF]L[LFRFRF]FFR[RFLFLF]L[LFRFRF]R$$
$$[RFFR[RFLFLF]L[LFRFRF]LFFR[RFLFLF]L[LFRFRF]$$
$$LFFR[RFLFLF]L[LFRFRF]]L[LFFR[RFLFLF]L[LFRFRF]$$
$$RFFR[RFLFLF]L[LFRFRF]RFFR[RFLFLF]L[LFRFRF]].$$

As can be seen, the size of the string increases rapidly under such a production, illustrating the generative nature of productions. Moreover, it is easy to see recursive patterns developing in the produced strings, leading to self-similarity in the visualisations of the string (Fig. 8.5 below).

8.2.2 Stochastic L-systems

In the D0L-system defined in the previous section, application of productions was deterministic, meaning the L-system will always generate an identical developmental sequence. One way of introducing variation into produced strings is to incorporate a probabilistic application of productions.

A *stochastic 0L-system* is an ordered quadruplet $G_\pi = \langle V, \omega, P, \pi \rangle$ where:

- the alphabet, V, axiom ω and set of productions, P are as defined for D0L-systems in Sect. 8.2.1;
- the function $\pi : P \rightarrow (0, 1]$ is called the *probability distribution* and maps the set of productions to a set of production probabilities.

Let $\hat{P}(s) \subset P$ be the subset of productions with s as the predecessor symbol. If no production for s is specified the identity production, $s \rightarrow s$ is assumed. Each production $p_i \in \hat{P}(s)$ has with it an associated probability $\pi(p_i)$, where

$$\sum_{p_i \in \hat{P}(s)} \pi(p_i) = 1. \tag{8.2}$$

The derivation $\mu \Rightarrow \nu$ is known as a *stochastic derivation* in G_π if for each occurrence of s in the word μ, the probability of applying production p_i is equal to $\pi(p_i)$. In a single word, different productions with the same predecessor may be applied in a single derivation step. Selection is weighted according to the probabilities associated with each production with the appropriate predecessor.

A production in a stochastic L-system is notated $p_i : s \xrightarrow{\pi(p_i)} \chi$.

8.2.3 Parametric L-systems

Parametric L-systems were proposed to address a number of shortcomings in discrete L-system models, particularly for the realistic modelling of plants. As string rewriting systems, L-systems are fundamentally discrete in both production application and in relation to the symbolic representation of the strings themselves. This discrete nature makes it difficult to model many continuous phenomena or accurately represent irrational ratios. The application of parametric L-systems to plant modelling was extensively developed by Hanan [7], the definition below is based on this work.

Parametric L-systems associate a vector of real-valued *parameters* with each symbol, collectively forming a *parametric module*. A module with symbol $s \in V$ and parameters $a_1, a_2, \ldots, a_n \in \mathbb{R}$ is written $s(a_1, a_2, \ldots, a_n)$. Strings of parametric modules form *parametric words*. It is important to differentiate the real-valued *actual* parameters of modules, from the *formal* parameters specified in productions. In practice, formal parameters are given unique[3] identifier names when specifying productions.

A parametric 0L-system (p0L-system) is defined as an ordered quadruplet $G = \langle V, \Sigma, \omega, P \rangle$ where

- $V = \{s_1, s_2, \ldots, s_n\}$ is an alphabet composed of a set of distinct symbols, s_i;
- Σ the set of formal parameters;
- $\omega \in (V \times \mathbb{R}^*)^+$, a non-empty parametric word known as the axiom; and
- $P \subset (V \times \Sigma^*) \times \mathcal{C}(\Sigma) \times (V \times \mathcal{E}(\Sigma)^*)^*$ the finite set of productions.

$\mathcal{C}(\Sigma)$ and $\mathcal{E}(\Sigma)$ are the sets of correctly constructed logical and arithmetic expressions with parameters from Σ. Logical expressions evaluate to Boolean values of TRUE and FALSE, arithmetic expressions evaluate to real numbers. Expressions are composed of relational operators (e.g. $<, >, \leq, \geq, =, \neq$), logical operators !(not), &(and), |(or), arithmetic operators (e.g. $+, -, \div, \times$, unary $-$, etc.), trigonometric, stochastic and other functions, and parenthesis '(', ')'.

Productions $(\underline{s}, C, \chi)$ are denoted $\underline{s} : C \rightarrow \chi$ where the formal module $\underline{s} \in V \times \Sigma^*$ is the predecessor, the logical expression $C \in \mathcal{C}(\Sigma)$ is the *condition* and $\chi \in (V \times \mathcal{E}(\Sigma)^*)^*$ is the formal parametric word known as the *successor*. A formal parameter appears once in the predecessor. The number of formal parameters must be consistent for any given symbol and match the number of actual parameters for the same symbol. A production without a condition has an implicit condition constant value of TRUE.

A production is applied to a module in a parametric word if the production *matches* that module. The necessary conditions for matching are: if the module and production predecessor symbols *and* parameter counts match; the condition statement, C, evaluates to TRUE when the module's actual parameters are bound to the formal parameters as specified in the predecessor module. When a module is matched it is replaced by the successor word, χ, whose formal parameters are evaluated and bound to the corresponding actual parameters.

Parametric L-system Examples

This parametric L-system

$$\omega : A(1, 1)$$
$$p_1 : A(x, y) \rightarrow A(y, y + x) \tag{8.3}$$

generates the derivation sequence

$$A(1, 1) \Rightarrow A(1, 2) \Rightarrow A(2, 3) \Rightarrow A(3, 5) \Rightarrow \cdots$$

---

[3] Within the scope of the associated production.

calculating the Fibonacci sequence. This parametric L-system

$$\omega : A(1,1)$$

$$p_1 : A(x,c) : x \times x \neq c \rightarrow A\left(\frac{(x + \frac{c}{x})}{2}, c\right) \tag{8.4}$$

uses the conditional application of p_1 to compute the square root of the second parameter of A (labelled c) using Newton's method:

$$A(1,2) \Rightarrow A(1.5,2) \Rightarrow A(1.416667,2) \Rightarrow A(1.414216,2) \Rightarrow \cdots$$

For parametric L-systems to be *deterministic* no two productions can match the same module in a derivation word by the rules above. However, ensuring determinism by these criteria can be difficult to prove for all possible combinations of parameter values, hence a practical solution is to *order* the set of productions and apply the first production that matches in the list. If there are no matching productions, the identity production is assumed and the parameters for that module remain unchanged.

8.2.4 Timed L-systems

The development of parametric L-systems addressed problems with D0L-systems in representing irrational ratios and adding continuous components to individual modules. However, development still proceeds in discrete, incremental steps. While it is possible to simulate continuous development using very fine increments, this complicates the grammar, requiring the developer to fight against the elegance and simplicity of the basic L-system formalism.

The symbolic and discrete nature of an L-system alphabet is inherently suited to a stepwise topological and structural description, whereas the developmental aspects may be more suited to a continuous model. It is conceptually elegant to separate model development (which is continuous) from model observation (which is discrete). With these points in mind, the concept of *timed L-systems* was introduced [19, Chap. 6], the authors seeing an analogy with the theory of morphogenesis advanced by Thom [25] where development is a piecewise continuous process, punctuated by *catastrophes*.

The description that follows is based on [11]. We assume the definition of parametric 0L-systems in Sect. 8.2.3. Timed L-systems include continuously variable module *ages* that permit continuous temporal and spatial development that is difficult or impossible to achieve with conventional 0L- or nL-systems.

In this case, modules consist of *timed symbols* with associated parameters. For each module, the symbol also carries with it an *age* – a continuous, real variable, representing the amount of time the module has been active in the derivation string. Strings of modules form *timed, parametric words*, which can be interpreted to represent modelled structures. As with parametric L-systems, it is important to differentiate between *formal modules* used in production specification, and *actual modules* that contain real-valued parameters and a real-valued age.

Let V be an alphabet, $\mathbb{R}$ the set of real numbers and $\mathbb{R}_+$ the set of positive real numbers, including 0. The triple $(s, \lambda, \tau) \in V \times \mathbb{R}^* \times \mathbb{R}_+$ is referred to as a *timed parametric module* (hereafter shortened to *module*). It consists of the symbol, $s \in V$, its associated parameters, $\lambda = a_1, a_2, \ldots, a_n \in \mathbb{R}$ and the *age* of s, $\tau \in \mathbb{R}_+$. A sequence of modules, $x = (s_1, \lambda_1, \tau_1) \cdots (s_n, \lambda_n, \tau_n) \in (V \times \mathbb{R}^* \times \mathbb{R}_+)^*$ is called a *timed, parametric word*. A module with symbol $s \in V$, parameters $a_1, a_2, \ldots, a_n \in \mathbb{R}$ and age τ is denoted by $(s(a_1, a_2, \ldots, a_n), \tau)$.

A timed, parametric 0L-system (tp0L-system) is an ordered quadruplet $G = \langle V, \Sigma, \omega, P \rangle$ where:

- V is the non-empty set of symbols called the alphabet;
- Σ is the set of formal parameters;
- $\omega \in (V \times \mathbb{R}^* \times \mathbb{R}_+)^+$ is a non-empty, timed, parametric word over V called the axiom; and
- $P \subset (V \times \Sigma^* \times \mathbb{R}_+) \times \mathcal{C}(\Sigma) \times (V \times \mathcal{E}(\Sigma)^* \times \mathcal{E}(\Sigma))^*$ is a finite set of productions.

A production $(\underline{a}, C, \underline{\chi})$ is denoted $\underline{a} : C \to \underline{\chi}$, where the formal module $\underline{a} \in V \times \Sigma^* \times \mathbb{R}_+$ is the predecessor, the logical expression $C \in \mathcal{C}(\Sigma)$ is the condition, and the formal timed parametric word $\underline{\chi} \in (V \times \mathcal{E}(\Sigma)^* \times \mathcal{E}(\Sigma))^*$ is called the successor.

Let $(s, \underline{\lambda}, \beta)$ be a predecessor module in a production $p_i \in P$ and $(s_1, \underline{\lambda}_1, \underline{\alpha}_1) \cdots (s_n, \underline{\lambda}_n, \underline{\alpha}_n)$ the successor word of the same production. The parameter $\beta \in \mathbb{R}_+$ of the predecessor module represents the *terminal age* of s. The expressions, $\underline{\alpha}_i \in \mathcal{E}(\Sigma)$, $i = 1, \ldots, n$ sets the initial or *birth age*. Birth age expressions are evaluated when the module is created in the derivation string. This nomenclature is illustrated in Fig. 8.2.

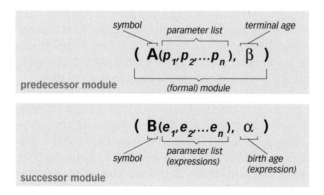

Fig. 8.2. Nomenclature for predecessor and successor modules in a timed L-system

As with parametric 0L-systems, if the condition is empty the production can be written $\underline{s} \rightarrow \chi$. Formal and actual parameter counts must be the same for any given symbol.

Here are some example productions:

$$\big(A(j,k), 3.0\big) : j < k \rightarrow \big(B(j \times k), 0.0\big)\big(C(j+1, k-1), 0.5\big) \tag{8.5}$$

$$\big(A(t), 5.0\big) \rightarrow \big(A(t+1), 5.0/t\big). \tag{8.6}$$

It is assumed:

- For each symbol $s \in V$ there exists as most one value of $\beta \in \mathbb{R}_+$ for any production $p_i \in P$ where $(s, \underline{\lambda}, \beta)$ is the predecessor in p_i. If s does not appear in any production predecessor then the terminal age of s, $\beta_s = \infty$ is used (effectively the module never dies).
- If $(s, \underline{\lambda}, \beta)$ is a production predecessor in p_i and $(s, \underline{\lambda}_i, \underline{\alpha}_i)$ any module that appears in a successor word of P for s, then $\beta > \alpha_i$ when $\underline{\alpha}_i$ is evaluated and its value bound to α_i (i.e. the lifetime of the module, $\beta - \alpha_i > 0$).

Development proceeds according to some global time, t, common to the entire word under consideration. Local times are maintained by each module's age variable, τ (providing the relative age of the module). A *derivation function* is used to obtain the derivation string at some global time, t (see [11] for details of this function). The derivation function guarantees a derivation string for any value of t, thus elegantly providing a continuous development that can be discretely sampled at arbitrary time intervals. This sampling has no effect on module or derivation string development.

Growth Functions

Any module in the development string may have associated with it a *growth function*, $g_s : (\mathbb{R}^* \times \mathbb{R}_+) \rightarrow \mathbb{R}$. The purpose of the growth function is to relate a module's age to changes in its parameters. For example a module may represent a developing sub-branch of a tree, whose radius and length change as the segment ages. The growth function for this module relates the module's age with parameters representing these attributes.

The growth function may involve any of the module's parameters, the current age, τ, and the terminal age β of s (determined by the predecessor of the production acting on s, the unique symbol of the module under consideration). Thus g_s is a real valued function that can be composed of any arithmetic expression $E(\lambda_s, \tau_s, \beta_s)$. In addition to the formal parameters supplied, expressions can include the operators, numeric constants, and a variety of functions, some of which are illustrated in Fig. 8.3. The development function returns a real value, which is then used as a scaling factor for the actual parameter vector λ. That is:

$$\lambda' = g_s \cdot [\lambda]. \tag{8.7}$$

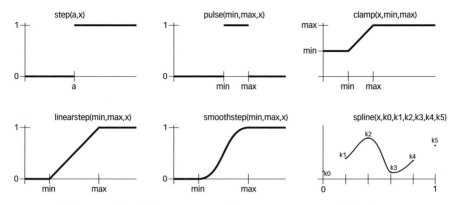

Fig. 8.3. Sample functions used for composing growth functions

The growth function is evaluated whenever a module requires turtle interpretation (explained in Sect. 8.3), with parameter vector λ' sent to the turtle, rather than λ as is the case with parametric L-systems. No constraints are placed on the range or continuity of g_s, however if continuity is required when a production is applied (such as the accurate modelling of cellular development), g_s must be monotonic and continuous. These constraints extend to the development functions for those symbols that are related as being part of the successor definition of s.

8.3 Turtle Interpretation

Having defined the mechanics of various types of L-systems, the reader might be curious as to how such formalisms may be turned into pictures. In the case of the system described in this chapter, we are interested in building three-dimensional geometric models, which develop over time and may be rendered as still images or animated sequences. This is achieved by a process known as *turtle interpretation*.

The derivation strings of developing L-systems can be interpreted as a linear sequence of instructions (with real-valued parameters in the case of parametric L-systems) to a 'turtle', which interprets the instructions as movement and geometry building actions. The historical term *turtle interpretation* comes from the early days of computer graphics, where a mechanical robot turtle (either real or simulated), capable of simple movement and carrying a pen, would respond to instructions such as 'move forward', 'turn left', 'pen up' and 'pen down'. Each command modifies the turtle's current position, orientation and pen position on the drawing surface [1]. The cumulative product of commands creates the drawing.

In the system described in this chapter, the simulated turtle operates in three-dimensions and maintains an extensive *state*, which includes:

- the current *position*, **t**, a position vector in three-space;
- the current *orientation*, an orthogonal basis matrix, composed of three vectors [**H**, **L**, **U**] representing the turtle's *heading*, *left* and *up* directions respectively;
- a homogeneous transformation matrix, **T**, used for local transformations on geometry created by the turtle;
- the current *drawing material*, a context dependent value determined by the type of geometric output;
- current drawing parameters such as line width, cylinder radius, generalised cylinder profile shape (see Sect. 8.3.1);
- a normalised *level-of-detail* value, used to control the complexity of output geometry.

In addition, the turtle maintains a *first-in-last-out* (FILO) *stack*. The current turtle state may be *pushed* onto and *popped* from the stack, permitting the easy construction of complex structures, such a branches. The turtle position and orientation co-ordinate system is shown in Fig. 8.4.

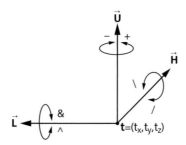

Fig. 8.4. Turtle position and orientation

The turtle reads the derivation string from left to right. Specific symbols from the alphabet, V, are interpreted by the turtle. If the symbol has associated parameters, the value(s) of the parameter(s) may be used to provide continuous control of the command. For example the symbol f is interpreted as 'move forward by the default move distance' by the turtle. The parametric module f(1.5) is interpreted as 'move forward 1.5 units'. In the case of tp0L-systems, the module's growth function would first be applied (Sect. 8.2.4). A subset of the turtle commands is summarised in Table 8.1. Other commands instance geometric primitives (e.g. spheres, discs, boxes), or change the turtle state (e.g. modifying the transformation matrix, **T**).

Recalling the simple D0L-system (8.1) and its derivation (Sect. 8.2.1): the turtle interpretation of this L-system for μ_4 is shown in Fig. 8.5. Note that, even though this image looks 'tree like' it is only two-dimensional. Three-dimensional examples will be given in the sections that follow.

Table 8.1. Summary of turtle commands. l_d and θ_d are global default values

Command	Default	Description
$\mathtt{f}(l)$	$l = l_d$	Move l units in the current heading direction, **H**.
$\mathtt{F}(l)$		Draw a line (2D) or cylinder (3D) of length l, in the direction of **H**, updating the current turtle position to be l units in the direction of **H**
$\mathtt{!}(r)$	$r = l_d/10$	Set line width (2D) or cylinder radius (3D) to r.
$\mathtt{+}(\theta)$	$\theta = \theta_d$	Turn left θ degrees.
$\mathtt{-}(\theta)$		Turn right θ degrees.
$\mathtt{\&}(\theta)$		Pitch down θ degrees.
$\mathtt{\wedge}(\theta)$		Pitch up θ degrees.
$\mathtt{\backslash}(\theta)$		Roll left θ degrees.
$\mathtt{/}(\theta)$		Roll right θ degrees.
$\mathtt{|}(n)$	$n = 1$	Turn around n times – equivalent to $+(180n)$.
$\mathtt{[}$		Save the current turtle state onto a first-in, last-out stack.
$\mathtt{]}$		Restore the turtle state from the top of the stack.
$\mathtt{\%}$		If this is the first time this particular symbol has been encountered: interpret current derivation string, μ, up to the position of this symbol without generating any output. Store turtle reference frame with this symbol. Subsequent readings of the symbol set the turtle reference frame to the value stored with the symbol.

Fig. 8.5. Turtle interpretation of L-system (8.1) in Sect. 8.2.1

8.3.1 Advanced Geometry with Generalised Cylinders

The turtle commands discussed in the previous section enable the generation of simple two- and three-dimensional shapes, constructed with lines (2D) or cylinders (3D). This is suitable for generating crude 'tree like' forms, but unsuited to more complex organic shapes such as horns, limbs, leaves, stems or tentacles. Previous solutions to this problem involved designing particular shapes in an external modelling system and reading them in as pre-defined surfaces [19]. Special symbols representing the surfaces were included in the grammar and the geometry instanced by the turtle when the symbol was encountered in the derivation string. This limits the complexity and variety of possible forms and is less elegant and compact than a grammar specification that generates all the geometry, rather than relying on external pre-computed surfaces.

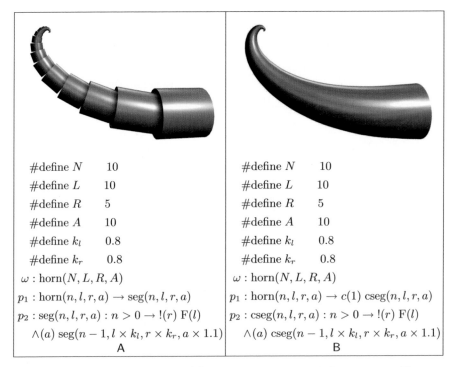

#define N	10	#define N	10
#define L	10	#define L	10
#define R	5	#define R	5
#define A	10	#define A	10
#define k_l	0.8	#define k_l	0.8
#define k_r	0.8	#define k_r	0.8

$\omega : \mathrm{horn}(N, L, R, A)$

$p_1 : \mathrm{horn}(n, l, r, a) \to \mathrm{seg}(n, l, r, a)$

$p_2 : \mathrm{seg}(n, l, r, a) : n > 0 \to \,!(r)\,\mathrm{F}(l)$

   $\wedge(a)\,\mathrm{seg}(n - 1, l \times k_l, r \times k_r, a \times 1.1)$

A

$\omega : \mathrm{horn}(N, L, R, A)$

$p_1 : \mathrm{horn}(n, l, r, a) \to c(1)\,\mathrm{cseg}(n, l, r, a)$

$p_2 : \mathrm{cseg}(n, l, r, a) : n > 0 \to \,!(r)\,\mathrm{F}(l)$

   $\wedge(a)\,\mathrm{cseg}(n - 1, l \times k_l, r \times k_r, a \times 1.1)$

B

Fig. 8.6. A simple horn defined (A) using cylinders, which leaves noticeable gaps where the radius and angle of the cylinder changes. In B, this problem is fixed with the use of generalised cylinders. The parametric L-system generating each model is shown below the image

The Biologist Stephen Wainwright suggests that the cylinder has found general application as a structural element in plants and animals [26]. He sees the cylinder as a logical consequence of the *functional morphology* of organisms, with the dynamic physiological processes (function) considered dependent on form and structure over time. Wainwright distinguishes the cylinder as a natural consequence of evolutionary design based on the physical and mechanical properties of cylindrical structures.

The problem of describing more complex cylindrical shapes can be solved using *generalised cylinders*, originally developed by Agin for applications in computer vision [2]. Generalised cylinders have found wide application in botanical and biological visual modelling [3, 13, 16, 20]. An example is shown in Fig. 8.6.

The basic principle for creating a generalised cylinder is to define a series of *cross-sections*, possibly of varying shape and size, distributed over some continuous curve, known as the *carrier curve*. The cross-sections are connected to form a continuous surface. This is illustrated in Fig. 8.7.

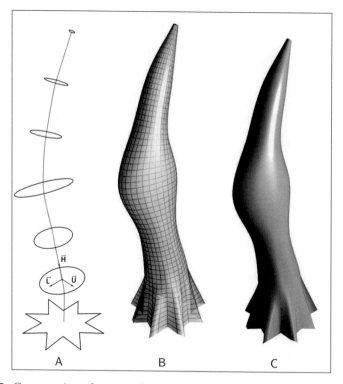

Fig. 8.7. Construction of a generalised cylinder defined by a turtle. A shows the cross-sectional elements and how they are oriented according to the turtle reference frame at cross-section instantiations. The path traced out through the cross sections represents the path taken by the turtle as it moves between cross-sections, thus creating a carrier curve. B shows the polygonalisation of the cylinder, and C a rendered version

As the turtle moves through space, it receives commands to 'drop' particular cross-section curves (also defined by the turtle and stored in an array of curves which forms part of the turtle state). Each cross-section is transformed to the turtle reference frame at the time it was dropped. The path of the turtle as it moves through space forms an implicit carrier curve for the cross-sections. Cross-sections are automatically swept along the carrier curve, undergoing possible interpolation, and forming a solid generalised cylinder as shown in Fig. 8.7. The method of automated turtle instantiation of generalised cylinders provides a powerful modelling capability for modelling organic form (Fig. 8.8).

#define N 10	#define N 10
#define R 10	#define R 10
#define L 50	#define L 50
$\theta_d = \pi/4$	$\theta_d = \pi/4$
ω : tree	ω : tree
p_1 : tree $\rightarrow$ seg(1)	p_1 : tree $\rightarrow$!(R) cseg(1)
p_2 : seg(n) : $n < N$ $\rightarrow$!(R/n)	p_2 : cseg(n) : $n < N$ $\rightarrow$!(R/n)
/ F(L/n) [+ seg($n + 1$)]	/ C(L/n) [+ cseg($n + 1$)]
[− seg($n + 1$)]	[− cseg($n + 1$)]
p_3 : seg(n) : $n \geq N \rightarrow$ F(L/n)	p_3 : cseg(n) : $n \geq N \rightarrow$ C(L/n)
A	B

Fig. 8.8. A simple tree-like structure with texture (A) using isolated, simple cylinders, and (B) using generalised cylinders. The L-system generating each form is shown. The generalised cylinder method requires only one extra symbol and the same number of productions as the isolated cylinder version

Having now covered a variety of L-systems, turtle interpretation of produced strings, and modelling of complex shapes using generalised cylinders, let us now turn to the mutation of L-system rules. These mutations form the basis of an evolutionary system to explore the possibilities for form generation using L-systems.

8.4 Mutation of L-system Rules

In order for the structure and form of an L-system generated model to change, its productions and module parameters must be changed. Small changes in genotype are usually preferred so as not to completely alter the form on which it is based. However, for radical change larger amounts of mutation are required.

There are three principal components of an L-system grammar to which mutation can be applied:

- mutation of rules (productions) and successor sequences (Sect. 8.4.1 and 8.4.2);
- mutation of parameters and parametric expressions (Sect. 8.4.3 and 8.4.4);
- mutation of growth functions and symbol ages (Sect. 8.4.5).

In addition, predefined surfaces, included as part of the turtle interpretation may themselves be mutated, using, for example, techniques described by Sims [23]. The connection between surface mutation and L-system mutation is illustrated in Fig. 8.1. The two processes proceed independently of each other, coming together at the generation of the geometric model.

8.4.1 Rule Mutation

For each production, symbols and successors have the possibility of being mutated. For the moment, we assume a D0L-system $G = \langle V, \omega, P \rangle$ as defined in Sect. 8.2.1. Mutation of a single production can be represented $p_n \Rightarrow p_n^*$, where $p_n \in P$ is the original production and p_n^* the mutated version. The probability of each type of mutation is specified separately. The types of mutations possible are listed below:

(A.1) A symbol from a production successor may be removed; e.g.
$p_n = \{a \rightarrow ab\} \Rightarrow p_n^* = \{a \rightarrow b\}$.
(A.2) A symbol α, where $\alpha \in \chi_{p_n}$ may be added to the production successor;
e.g. $a \rightarrow ab \Rightarrow a \rightarrow abb$, where α is b.
(A.3) A symbol α, where $\alpha \in \chi_{p_n}$ may change to another symbol β, where $\beta \neq \alpha$; e.g. $a \rightarrow ab \Rightarrow a \rightarrow bb$, where successor a has changed to b.

The above rules work on a previously defined set of symbols, typically a subset of the L-systems alphabet, $\psi : \psi \subseteq V$. This subset can be specified by the user. In addition:

(A.4) A new symbol, α, where $\alpha \in \psi$ may be added to the production; e.g. $a \to ab \Rightarrow a \to abc$, where α is c in this example.

In addition, it is sometimes necessary to disallow mutation of certain symbols whose purpose is some kind of control (such as output resolution), or to limit a search space. Special symbols, such as the brackets ('[' and ']') representing turtle state *push* and *pop* operations, need to be matched to prevent a stack overflow or underflow. Here there are two options:

- ignore stack underflows and kill any genotypes that reach an upper limit of stack size when the phenotype is generated;
- execute a bracket *balancing operation* following the final mutation of a given production. That is for any given production ensure the number of '[' and ']' symbols is equal. This can be achieved by adding or deleting bracket symbols as necessary. This is the preferred option.

8.4.2 Production Set Mutation

In addition to individual symbol changes within successors, productions may be created and deleted. Here we assume a stochastic 0L-system, $G_\pi = \langle V, P, \omega, \pi \rangle$ as defined in Sect. 8.2.2:

(B.1) A stochastic production p_n with associated probability $\pi(p_n)$ may split into two stochastic productions, p'_n and p'^*_n, i.e.: $p_n \Rightarrow \{p'_n, p'^*_n\}$, where $\pi(p'_n) + \pi(p'^*_n) = \pi(p_n)$. For example: $a \xrightarrow{0.5} ab \Rightarrow \{a \xrightarrow{0.3} ab, a \xrightarrow{0.2} bb\}$ The successor of p'_n is the same as that of p_n . The successor of p'^*_n is a mutated version of p'_n, created using the mutations A.1–A.4 as specified in the previous section.

(B.2) A new production can be created. Essentially, this is for identity productions of the form $a \to a$, which are implicitly assumed for all members of V with no explicit productions. If all members of V have non-identity productions then this mutation can't take place. If an identity production does exist then the successor is assigned to be a randomly selected element from the set ψ (defined in the previous section). Note that this may result in the identity production if the predecessor is an element of ψ. This rule is superfluous to a degree, since if we assume the existence of identity productions they could be subject to the mutations A.1–A.4 as specified in the previous section. It is introduced as a convenience due to the internal representation of productions in the system.

(B.3) An existing production may be deleted. If the production is stochastic then the productions with the same predecessor gain in probability in equal amounts totalling to the probability of the deleted rule. e.g.

$$\{a \xrightarrow{0.4} ab, a \xrightarrow{0.3} bb, a \xrightarrow{0.3} abc\} \Rightarrow \{a \xrightarrow{0.5} bb, a \xrightarrow{0.5} abc\}.$$

(B.4) The probability of a stochastic production may change. This change is selected from the interval $(-\pi(p_n), 1 - \pi(p_n))$, where $p_n \in P$, is the production selected for this mutation. The addition or difference redistributes probabilities over all the other stochastic productions involving that successor, e.g.

$$\{a \xrightarrow{0.5} ab,\ a \xrightarrow{0.5} bb\} \Rightarrow \{a \xrightarrow{0.2} ab,\ a \xrightarrow{0.8} bb\}.$$

8.4.3 Parametric Mutation

We now assume parametric 0L-systems, $\langle V, \Sigma, P, \omega \rangle$ as defined in Sect. 8.2.3. For the sake of efficiency and ease of implementation, symbols may not gain or lose parameters during mutation. However, new modules[4] may be created with the default number of parameters, or if no default exists, a random number of parameters up to a fixed limit. Productions involving predecessor parametric module may split as follows:

(C.1) Productions involving modules with no conditions may split, and gain conditions. Some examples:

$$a(l) \rightarrow a(2l)\, b(l^2) \Rightarrow \begin{cases} a(l) : l \leq 10 \rightarrow a(2l)\, b(l^2), \\ a(l) : l < 10 \rightarrow a(2l)\, c(l/2) \end{cases}$$

$$d(s_1, x_2) \rightarrow d(x_1 + x_2, x_2 + 1) \Rightarrow \begin{cases} d(x_1, x_2) : x_2 > 8 \rightarrow \\ \qquad d(x_1 + x_2, x_2 + 1), \\ d(x_1, x_2) : x_1 \leq x_2 \,|\, x_2 \geq 8 \rightarrow \\ \qquad d(x_1 - x_2, x_1 + 1). \end{cases}$$

(C.2) For productions involving modules with conditions, the conditions themselves may mutate according to the rules specified in the next section.

8.4.4 Mutation of Expressions

Parameters of modules on the successor side of productions are expressions. They are parsed into a tree structure and executed during the application of productions. For example, the expression $(x_1 + x_2)/x_3$ can be represented:

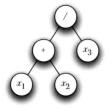

---

[4] A module is a *symbol* and its associated *parameters*, and a current *age* if timed.

The mutation operations outlined here equate to manipulation of such tree structures. Let us assume a predecessor module α from a given production $p_n \in P$, is undergoing parametric mutation. We also assume α has a series of associated parameters $x_1, x_2, \ldots, x_k$. We assume a set of binary operators $\{+, -, \div, \times, \exp\}$ and the unary operator $\{-\}$. Each node on the expression tree can be recursively subject to mutation by the following rules:

(D.1) If the node is a variable it can mutate to another variable. $x_i \Rightarrow x_j$, where $1 \leq j \leq k$ and $i \neq j$.
        The set of possible variables x_i and x_j are drawn from the set of parameters associated with the predecessor symbol under consideration $(i, j \rightarrow [1, n], i \neq j)$.
(D.2) If the node is a constant, it is adjusted by the addition of some normally distributed random amount.
(D.3) If the node is an operator it can mutate to another operator. Operators must be of the same 'arity' (i.e. unary or binary), e.g. $x_1 + 5 \Rightarrow x_1 \times 5$.
(D.4) Nodes may become the arguments to a new expression, e.g. $x_1 + 5 \Rightarrow x_2 \times (x_1 + 5)$. This example is illustrated graphically below:

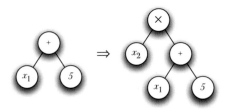

(D.5) An expression may reduce to one of its operands, e.g. $x_1 + 5 \Rightarrow x_1$.
        In the current implementation, only simple arithmetic operators are supported (addition, subtraction, division, multiplication, negation and power). Other functions (such as trigonometric functions) could be added if required. Formal parameters on the left side of productions do not mutate, as this serves no useful purpose.

8.4.5 Growth Function and Module Age Mutation

If a module is timed it must have a birth age (α) and terminal age (β). Both these values must be constants.[5] Both constants can be mutated by the addition of a normally-distributed random value, with mean 0. The variance is proportional to the constant's current value.

The *growth function*, $g : \mathbb{R}^n \rightarrow \mathbb{R}$ (defined in Sect. 8.2.4) for a timed module controls the behaviour of that module over its growth period. It is expressed in the form $g_s(\lambda_s, \tau_s, \beta_s)$, where s is the symbol associated with

---

[5] In non-evolutionary versions of the system, birth ages can be expressions.

the function, λ_s the parameter vector for s, τ_s is the current age and β_s the terminal age. A simple linear growth function would be $g_s = \tau/\beta$, increasing from α/β to 1 from birth to terminal age. However, such a simple function is not typical in nature. More common is some form of exponential function with damping increasing with age such as the Bertalanffy equation:

$$g_s = k_1(1 - e^{-k_2\tau})^{k_3}, \tag{8.8}$$

where k_1, k_2 and k_3 are constants. Many different growth functions have been studied in the biological literature, the exact choice of function dependent on the particular feature or phenomena being modelled[6] (e.g. tree trunk diameter, length or height of the organism, population size, etc.). Unification of the major growth functions has been proposed, but this results in far more complex equations with a large number of constants [27].

Since growth functions are expressions, their internal and external construction is the same as for a module's parameter expressions. Thus the mutations are identical to those described for expressions in the previous section.

8.4.6 Mutation Probabilities

Different kinds of mutations occur with differing probabilities. The term *mutation probability* means the probability a particular kind of mutation will occur at a given time. Mutation probabilities are expressed as floating-point numbers with the interval $[0, 1]$. A probability of 0 for a particular mutation means that it will never occur ('always off'); 1 means it will always occur ('always on'); any value in between sets the overall frequency of occurrence (e.g. 0.1 means, on average, 1 mutation every 10 times the mutation is considered). Some mutation probabilities are dependent on the size of the genotype being mutated – with fixed mutation probabilities, more mutations will occur on larger genotypes than on shorter ones.

An important consideration for successfully evolving structures is correctly adjusting mutation probabilities as the evolution progresses. For example, it is better to set production mutation probabilities to maintain or slightly shrink current production size. If rule mutation is biased towards adding productions and/or modules then genotypes tend to become larger without necessarily producing phenotypes of greater fitness. Large sets of productions take longer to parse and in general, take longer to generate.[7] This does not stop rules evolving complexity by selection. An upper limit is set for any newly mutated rule's resource consumption (i.e. using too much space or time). Rules that

---

[6] It is important to consider when modelling growth using L-systems that a large number of modules may contribute to the total growth of the 'organism'. A biological growth function, such as (8.8) represents the growth of the organism in total, not necessarily the individual components modelled as growing modules.

[7] Although a production as simple as $a \rightarrow aa$ doubles the size of the produced string at each derivation (assuming an axiom of a).

consume above this limit during derivation are automatically stopped by the system and removed.

The user can change mutation probabilities interactively during the evolutionary process. A wide variety of mutation probability controls are provided in the system developed, as this affords a useful aid when one appears to be approaching the desired result and wishes to limit mutations to specific areas. Hierarchical controls also permit changing groups of associated parameters as a whole, while maintaining individual ratios between the associated parts.

8.5 The Interactive Process

To evolve forms interactively we begin with a germinal genotype as described in Sect. 8.1.3. This genotype may be drawn from an existing library of L-systems, defined by the user, or an 'empty' L-system consisting only of identity productions can be used. The subset of the alphabet of the current L-system suitable for mutation, ψ (Sect. 8.4.1), can be specified at this time. A list of external surfaces to be used is also supplied. The parent production set is then mutated according to probabilities specified. The axiom is mutated in a manner similar to that of productions. After mutation, the L-system is parsed and derived to a specified level, or time-period in the case of timed models. The user may interrupt this process at any time. The software will automatically interrupt the process if the computation time or space requirements exceeds a specified limit. In traditional uses of the genetic algorithm, population sizes can be large, as the selection process is automatic. In the case of this type of system, selection is based on the subjective notion of aesthetics. This is one reason why the population size in this case is limited. A much more overpowering reason is the limitation of space on the screen to display phenotypes and the computation time involved in generating large populations.

Eleven to fifteen mutations per generation are performed. This provides a trade off between generation time and variety. With increased computational power (or more patience) larger populations can be created. The screen resolution also limits the number of phenotypes that can be displayed and manipulated simultaneously. The parent phenotype is displayed in the upper left-hand corner of the screen followed by its mutated children. The user adjusts the sliders to control the mutation probabilities and selects a child to act as the new parent. Selecting the current parent is also possible, normally indicating that the user is unsatisfied with any of the current child mutations.

Each phenotype may have a temporal dimension (in the case of timed L-systems) and three spatial dimensions. Thus the system allows real-time rotation, translation and scaling in three-dimensions along with control over playback (play, pause, rewind) for observing the development over time.

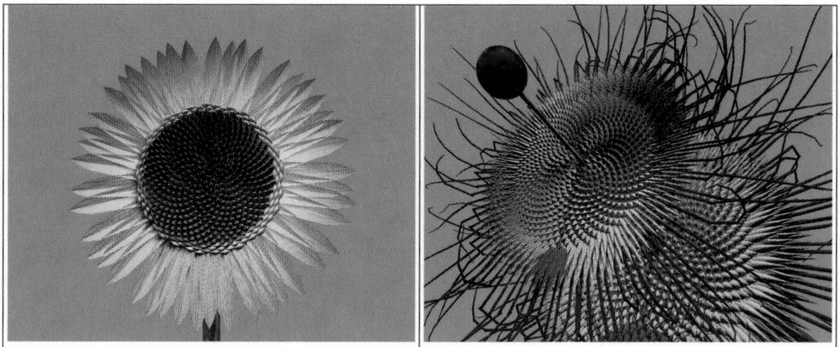

A

B

```
surface stamen;
surface floret;
surface floret2;
surface leaf2;
productions:
A(n) -> +(137.5)[f(n^0.5) C(n)]
        A(n+1);
C(n) : n <= 440 -> floret
     : n > 440 & n <= 565 ->
                     floret2
     : n > 565 & n <= 610 ->
              ^(90) S(0.3) leaf2;
axiom:
        stamen A(0);
```

```
surface stamen;
surface floret;
surface floret2;
surface ball;
productions:
M1(l,s) : l < 2 ->
   [S(s)[f(-1.0*s)stamen]
   F(1.60*s)&(90)A(0)]!(s*0.07)
   F(4.14*s)M1(l+1,s*0.67);
         : l >= 2 -> [ballh];
A(n) : n < 1000 -> +(137.5)
   [f(1.10*n^0.5)^(90-(n/1000*90))
   C(n)]A(n+1)
     : n > 1000 -> ;
C(n) : n >= 10 & n < 600 : ->
        floret2
     : n >= 600 & n < 900 : ->
        floret
     : n >= 900 : -> [
        M2(n,0,1.76,90.40,0.05,3)];
M2(p0,p1,p2,p3,p4,p5): p0<8 : ->
/(p5,p0)^(p5,p0)!(p4)
f(-p2/9)F(p2,p0 + 0.07)^(p3,p0)
M2(p0,p1+1,p2/1.4,p3/2.0,p4*0.8*
   p5,p5*1.8);
axiom:
        M1(0,2.11);
```

Fig. 8.9. (A) Original form (a Sunflower) and (B) the form after many generations of aesthetic evolution using the system described in this chapter. The L-systems generating each model are shown below the image

8.6 Results

The system, in various forms, has been used over a number of years to produce both still and animated works. Figure 8.9 is an example of the results achieved with the system. Starting with the figure shown on the left (A, a common sunflower head), after more than 50 generations of aesthetic selection the resultant form is shown on the right (B). The key productions that generate each form are shown underneath the corresponding image. In the interests of clarity, some control and rendering support symbols have been removed from the productions shown.

The model also includes morphogenic growth, and the resulting animation is smooth and continuous from birth to the fully developed model (Fig. 8.10). The emphasis on the system has been one of a sculptural tool for modelling and evolving growth, form and behaviour (Fig. 8.11).

Fig. 8.10. Sequence showing the temporal development (left to right) of the evolved model shown in Fig. 8.9. Each element in the sequence is approximately 1 second apart

8.6.1 A Case Study: *Bloom*

Thus far, only technical and descriptive details have been covered in this chapter. Technical details are interesting, however the aesthetic, narrative and production methodologies are also a vital part of practically using any creative system [14]. This section briefly looks at the process involved in using the system described in this chapter to create a commissioned artwork, titled *Bloom.*

Bloom is an artwork commissioned for the largest public billboard in Australia, located in the 'Creative Industries Precinct' of the Queensland University of Technology in Brisbane. The billboard image is approximately 45 m × 9.5 m. Figure 8.12 shows the artwork in situ. The basic concept for the work was to use artificial evolution methods to evolve five plant-like forms, reminiscent of the five Platonic solids, but reinterpreted in a different sense of the concept of 'form'.

The original source for the plant forms was derived from native species, local to the area. An L-system corresponding to each original plant was first

Fig. 8.11. A walking creature. The gaits of the legs, body movements and geometry are all created and controlled by timed L-systems.

Fig. 8.12. *Bloom* on display at QUT, Brisbane, Queensland (photo by Peter Lavery)

Fig. 8.13. *Deniflex*: detail of one of the five *Bloom* forms

developed by hand-coding the L-system. These hand-coded L-systems represented the germinal genotypes for evolution. Each genome was subject to the interactive evolutionary process outlined in this chapter. In addition, portions of each of the genomes were spliced and intermixed, resulting in parts of one plant's structure appearing in a different plant. After many generations, five final hybrid forms were evolved and rendered at high-resolution. Each of the five 'new' species[8] is presented in isolation, starkly displayed on a black background, giving a clinical, formalist feel. Each form has a certain softness and synthetic beauty, drawing on the dualistic nature of synthetic biology (Fig. 8.13 shows the detail of one of the forms). This dualism promises im-

---

[8] Of course, this method of mixing species would not necessarily be possible in real biology.

mense new possibilities, yet these possibilities also indicate a cost to our own nature and environment of potentially dangerous proportions.

The billboard is located on a busy freeway in a developed city area. Native species that might have once been prolific in the area have largely been destroyed or removed to make way for human 'progress'. The giant billboard of evolved plants reminds viewers of a displaced nature. Viewers of the artwork may recognise, for example, elements of the Bunya pine and a variety of Banksia forms. These iconic forms remind us of native species, yet their appearance is clearly strange and unwieldy. The work also conjures a sense of irony, as the best views of nature are now synthetic ones, large, bright and easily viewed from the comfort of any of the more than 70,000 cars that drive past the billboard each weekday morning, relentlessly spewing emissions into the atmosphere.

The deliberate use of decay and mutation in these models plays against the 'super-real' feel of most computer graphics, suggesting a conflict between the ideal of perfection in computer graphics and a diminishing biosphere. The concepts of mutation, cross-breeding and genetic manipulation used to create these images signal a growing concern over the consequences of recent scientific research in genetically modified plants and animals. We humans have always manipulated nature to our advantage, but the pace and possibilities now possible with genetic manipulation are unprecedented. As the philosopher John Gray reminds us, while science has enabled us to satisfy our desires, it has done nothing to change them [6]. *Bloom* uses the generative evolutionary computational models of morphogenesis, discussed in this chapter, to remind us of the cost of those desires.

References

1. Abelson, H., DiSessa, A.: Turtle geometry: the computer as a medium for exploring mathematics. The MIT Press series in artificial intelligence. MIT Press, Cambridge, Mass. (1982)
2. Agin, G.: Representation and description of curved objects. Stanford Artificial Intelligence Report: Technical Memo AIM-173, Stanford, California (1972)
3. Bloomenthal, J., Barsky, B.: Modeling the mighty maple. In: Proceedings of the 12th annual conference on Computer graphics and interactive techniques, pp. 305–311. ACM, New York (1985)
4. Dawkins, R.: The Blind Watchmaker. Longman Scientific & Technical, Essex, UK (1986)
5. Dorin, A.: Aesthetic fitness and artificial evolution for the selection of imagery from the mythical infinite library. In: J. Kelemen, P. Sosík (eds.) Advances in Artificial Life. Lecture Notes in Artificial Intelligence 2159, pp. 659–668. Springer-Verlag (2001)
6. Gray, J.: Straw dogs: thoughts on humans and other animals. Granta Books, London (2002)
7. Hanan, J.: Parametric L-Systems and their application to the modelling and visualization of plants. Ph.D. thesis, University of Regina, Saskatchewan (1992)

8. Herman, G., Rozenberg, G.: Developmental Systems and Languages. North-Holland, Amsterdam (1975)
9. Kitano, H.: Designing neural networks using genetic algorithms with graph generation system. Complex Systems **4**, 461–476 (1990)
10. Lindenmayer, A.: Mathematical models for cellular interactions in development, I and II. Journal of Theoretical Biology **18**, 280–315 (1968)
11. McCormack, J.: The application of L-systems and developmental models to computer art, animation, and music synthesis. Ph.D. thesis, Monash University, Clayton (2003)
12. McCormack, J.: Aesthetic evolution of L-systems revisited. In: G. Raidl et al. (ed.) Applications of Evolutionary Computing (EvoWorkshops 2004). Lecture Notes in Computer Science 3005, pp. 477–488. Springer-Verlag, Berlin (2004)
13. McCormack, J.: Generative modelling with timed L-systems. In: J. Gero (ed.) Design Computing and Cognition '04, pp. 157–175. Kluwer Academic Publishers, Dordrecht (2004)
14. McCormack, J.: Impossible nature: the art of Jon McCormack. Australian Centre for the Moving Image, Melbourne (2004)
15. McCormack, J.: Open problems in evolutionary music and art. In: F. Rothlauf et al. (ed.) Applications of Evolutionary Computing (EvoWorkshops 2005) Lecture Notes in Computer Science 3449, pp. 428–436. Springer-Verlag, Berlin (2005)
16. Mech, R., Prusinkiewicz, P., Hanan, J.: Extensions to the graphical interpretation of L-systems based on turtle geometry. Technical report 1997-599-01, University of Calgary, Alberta, Canada (1997)
17. Monod, J.: Chance and necessity – an essay on the natural philosophy of modern biology. Penguin, London (1971)
18. Parish, Y., Müller, P.: Procedural modeling of cities. In: Proceedings of the 28th annual conference on Computer graphics and interactive techniques, pp. 301–308. ACM, New York (2001)
19. Prusinkiewicz, P., Lindenmayer, A.: The algorithmic beauty of plants. Springer-Verlag, New York (1990)
20. Prusinkiewicz, P., Mündermann, L., Karwowski, R., Lane, B.: The use of positional information in the modeling of plants. In: Proceedings of the 28th annual conference on Computer graphics and interactive techniques, pp. 289–300. ACM, New York (2001)
21. Rozenberg, G., Salomaa, A.: The Mathematical Theory of L-systems. Academic Press, New York (1980)
22. Salomaa, A.: Formal Languages. Academic Press, New York (1973)
23. Sims, K.: Artificial evolution for computer graphics. In: Proceedings of the 18th annual conference on Computer graphics and interactive techniques, pp. 319–328. ACM, New York (1991)
24. Takagi, H.: Interactive evolutionary computation: fusion of the capabilities of EC optimization and human evaluation. Proceedings of the IEEE **89**, 1275–1296 (2001)
25. Thom, R.: Structural stability and morphogenesis: an outline of a general theory of models, 1st English edn. W. A. Benjamin, Reading, Mass. (1975)
26. Wainwright, S.: Axis and circumference: the cylindrical shape of plants and animals. Harvard University Press, Cambridge, Mass. (1988)
27. Zeide, B.: Analysis of growth equations. Forest Science **39**, 594–616 (1993)

Part III

Embryogeny

Evolutionary Design in Embryogeny

Daniel Ashlock

University of Guelph, Department of Mathematics and Statistics, 50 Stone Road East, Guelph, Ontario N1G 2W1, Canada dashlock@uoguelph.ca

In biology texts *embryogeny* is defined as "the development or production of an embryo." An embryo is a living creature in its first stage of life, from the fertilized egg cell through the initial development of its morphology and its chemical networks. The study of embryogeny is part of *developmental biology* [1, 2]. The reader may wonder why a book on evolutionary design should have a section on embryogeny. Computational embryogeny is the study of representations for evolutionary computation that mimic biological embryogeny. These representations contain analogs to the complex biological processes that steer a single cell to become a rose, a mouse, or a man. The advantage of using embryogenic representations is their richness of expression. A small seed of information can be expanded, through a developmental process, into a complex and potentially useful object. This richness of expression comes at a substantial price: the developmental process is sufficiently complex to be unpredictable.

Let us consider a pseudo-random number generator in analogy to embryogeny. This is an algorithm that takes an initial number, called a seed, and generates a long sequence of numbers whose value lies in the difficulty of telling that the sequence is not, in fact, random. While the algorithm driving a pseudo-random number generator is typically both short and fast, it exploits subtle aspects of number theory to achieve a high degree of unpredictability. Computational embryogeny seeks to take a small data object and expand it into a large, complex, and potentially useful object such as a robot controller or a detailed image. In order to be useful, the *expression algorithm* that transforms an artificial embryo into a finished structure must run rapidly enough to be used in fitness evaluations in evolutionary computation. It must also not be so complex that direct design of the final object by the usual methods of design in engineering is a compelling alternative. An expression algorithm must thus mimic a pseudo-random number generation in being both short and relatively fast. If the expression algorithm were also predictable, the design problem being solved would be simple. We can deduce that useful expression algorithms have a behavior that is not easy to predict from the artificial embryos they act upon.

Unpredictable expression algorithms that transform a small collection of parameters or a simple initial geometric state into a complex object are natural targets for evolutionary computation. Getting an unpredictable process to exhibit a desired behavior cannot be done with optimization algorithms like gradient search. Either there is no simple notion of gradient or the gradient is extremely rough (unpredictable). The representation of the computational embryogeny cannot itself be completely random, but any sort of heritability in the interaction of the representation and the variation operators chosen by the designer will permit evolution to search for useful structures. It remains to motivate why computational embryogeny is a useful alternative to direct representations.

The famous no free lunch theorem proves that no technique is universally useful. The correct question is therefore "when is computational embryogeny useful?" The three chapters in this section form a set of three case studies of when computational embryogeny is useful and their references point to many more. Using an expression algorithm to expand a small informational seed into a complex structure can be viewed as mapping the space of possible seeds into a subspace of the enormously larger space of structures. If the subspace of structure space selected by the expression algorithm contains no viable structures, then the computational embryogeny fails. If, on the other hand, the subspace selected by the expression algorithm is highly enriched in viable structures, then the computational embryogeny succeeds. One can view computational embryogeny as a technique for radically reducing the size of the search space via the selection of a subspace of structure space. In Chap. 10 of this section, Chris Bowers quotes Altenberg and Williams as noting that a monkey, given a typewriter, is far more likely to produce a verse of Shakespeare than a monkey given a pencil and paper. This is because the typewriter selects a tiny subspace of the space of possible black-on-white paper documents.

Another point of view one can take is that the expression algorithm is a point at which expert knowledge and known constraints can be embedded into the representation. Rather than using penalty functions to encourage search to take place in an admissible portion of structure space, a well designed expression algorithm constructively forces search to take place in a subspace containing only viable structures. Computational embryogeny is thus useful when it is possible to construct an expression algorithm that embeds known constraints and design principles. This makes the use of computational embryogeny a somewhat opportunistic enterprise. It is important to keep this in mind as a way of avoiding a nail hunt in celebration of the acquisition of a new methodological hammer. Computational embryogeny can be remarkably effective for some problems but is inappropriate for others. A recent survey classifying applications of computational embryogeny appears in [4].

The Chapters

When creating a computational embryogeny system the designer may choose to create a highly specific system for a single task or a general framework. Likewise, the designer can make a weak or strong biological analogy. At its weakest any indirect representation using a form of expression algorithm, even one as simple as a greedy algorithm for completing a partial structure, can be considered an instance of computational embryogeny. Incorporating analogies of genetic control networks, cellular developmental paths, and biochemical signaling networks strengthens the biological analogy to the point where the computational embryogeny becomes an interesting model of biological embryogeny.

Chapter 9

Embryogenisis of Artificial Landscapes, is an example of a highly task-specific embryogenesis system that generates artificial landscapes using an expression algorithm derived from Lindenmayer systems [3]. Lindenmayer systems are developmental models originally devised for plants. They appear in the section of this book on evolved art as well. The landscape evolution in Chap. 9 is at the weaker extreme of the biological analogy. An indirect representation is used, and the expression algorithm creates a large image from a small informational seed. There is, however, no structure analogous to the biological signaling that drives the embryogeny of living plants.

Chapter 10

Modularity in a Computational Model of Embryogeny gives a model of embryogeny that is novel in incorporating a type of modularity. Modularity is a key feature of biological systems. Different subsystems in a living organism manage the development of different organs, overall morphology, skeletal development, and the management of basal metabolism. Incorporating modularity into models of embryogenesis is thus key for biological plausibility. Modularity is also a key feature of successful engineering and design techniques, and so its incorporation into computational embryogeny is likely to enhance its usefulness as a design tool. The new model is applied to a pattern formation task and compared with other techniques.

Chapter 11

On Form and Function: The Evolution of Developmental Control outlines and demonstrates a general framework called the *Evolutionary Development System* (EDS) for computational embryogeny. This system makes the strongest biological analogy of the three chapters and provides a software framework for exploring the utility of computational embryogeny. EDS is demonstrated on two types of tasks. In the first, an artificial organism is evolved to fill a cube or a planar region as best it can. The second EDS task is to evolve robot controllers for memorizing a path, effectively using sensors to navigate, and avoiding obstacles.

Daniel Ashlock
Embryogeny Area Leader

References

1. Davidson, E.: Genomic Regulatory Systems: Development and Evolution. Academic Press, New York (2003)
2. Gilbert, S.: Developmental Biology, 7th edn. Sinauer Associates (2003)
3. Lindenmayer, A.: Mathematical models for cellular interaction in development, I and II. Journal of Theoretical Biology **16**, 280–315 (1968)
4. Stanley, K., Miikkulainen, R.: A taxonomy for artificial embryology. Artificial Life **9**, 93–130 (2003)

9

Embryogenesis of Artificial Landscapes

Daniel Ashlock[1], Stephen Gent[2], and Kenneth Bryden[3]

[1] University of Guelph, Department of Mathematics and Statistics, 50 Stone Road East, Guelph, Ontario N1G 2W1, Canada dashlock@uoguelph.ca
[2] Iowa State University, Department of Mechanical Engineering, Ames, Iowa 50011, USA sgent@iastate.edu
[3] Iowa State University, Department of Mechanical Engineering, Ames, Iowa 50011, USA kmbryden@iastate.edu

9.1 Introduction

This chapter examines the artificial embryogeny of landscapes intended for use with virtual reality which consist of collections of polygons encoded using L-systems. *Artificial Embryogeny* is the study of indirect representations. A recent survey that attempts to classify different types of artificial embryogeny appears in [18]. A *representation* is a way of encoding a model or a solution to a problem for use in computation. For example, an array of n real numbers is a representation of the value of a function in n variables. A representation is *indirect* if it gives a set of directions for constructing the thing it specifies rather than encoding the object directly. The process of following the directions given in the indirect representation to obtain the final object is called *expression*. Indirect representations require an *interpreter* to express them and, because of this, are more difficult to understand at the specification or genetic level. There are a number of advantages to indirect representations that more than balance this genetic obscurity in many situations. The most general of these advantages is that the transformation from the indirect specification to the final model or solution can incorporate heuristics and domain knowledge. This permits a search of a genetic space that is far smaller than the space in which the expressed objects reside and has a much higher average quality. Another advantage, showcased in this chapter, is compactness of representation. The indirect representations we evolve in this chapter use a few hundred bytes to specify megabyte-sized collections of polygons.

One kind of indirect representations specifies initial conditions for a dynamical system that generates the desired final state. Dynamical systems based on reaction–diffusion equations were proposed by Turing [20] as a possible expression mechanism for permitting biochemistry to create biological morphology. Beautiful examples of pattern formation, in the form of mammalian coat patterns, using reaction–diffusion-based dynamical systems can be found in [16]. Discrete dynamical systems such as cellular automata [7] can also be used as indirect representations of this type.

   Cellular representations are an indirect representation that use a grammar as the basis of their encoding. The representation consists of a specification of strings of production rules in the grammar that, when applied, yield the final structure. These representations were first applied to artificial neural nets [8,9] but have also been applied to finite state game-playing agents [6]. *Lindenmayer systems* or (L-systems) [13] which were devised as computational models of plants [17] are an older type of grammatical indirect representation that share many features with cellular representations. L-systems have been applied to targets as diverse as music [4], error correcting codes [3], and the morphology of virtual creatures [11]. In this chapter, L-systems will be used to evolve a diverse collection of virtual landscapes. This encoding of virtual landscapes uses a few hundred bytes of data to specify a collection of large, complex virtual images of the same landscape at different resolutions.

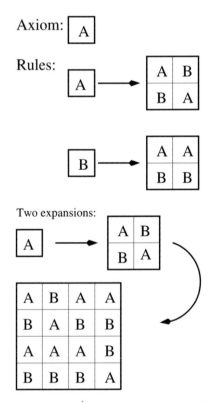

Fig. 9.1. The axiom and rules for a simple two-dimensional L-system, together with two expansion of the axiom

An L-system has two parts. The first is a grammar which specifies an axiom and a collection of replacement rules. The L-system creates a sequence of objects, starting with the axiom, by applying the rules. This application is called an *expansion*. The replacement rules used in this chapter operate on a two-dimensional array of characters. Rules are applied simultaneously to every symbol in a two-dimensional array to create a new, larger array. An example of this type of two-dimensional L-system is shown in Fig. 9.1.

The second part of an L-system is the *interpreter*. The interpreter turns the expression of the L-system into the object it specifies. In this chapter, the interpreter's task is to render the symbols into the polygons of a virtual landscape. The L-systems used here are called *midpoint* L-systems and create midpoint fractals. For midpoint L-systems, interpretation (or expression) is integrated with the expansion process.

The simplest sort of *midpoint fractal* is generated starting with a horizontal line segment. Take the midpoint, displace it vertically by a number selected from a probability distribution, creating two line segments that meet. Repeat the process on each of the resulting line segments possibly using a different probability distribution at each level of the process. Continue until you have the desired appearance, cycling through a sequence of probability distributions $D_1, D_2, D_3, \ldots$ as you go. Such a fractal is shown in Fig. 9.2. You can also

Fig. 9.2. An example of a midpoint fractal generated from a line segment

generate midpoint fractals starting with a square. This is done just as it was for the line segment, except that you displace the center of the square instead of the midpoint of the line segment. Each time you displace the center of a square, you create four new squares. The choice of D_i controls the character of the resulting fractal. For example, if the D_i are Gaussian with a small positive mean and variance that decreases with i, the resulting midpoint fractal will look like part of a mountain range [14].

For midpoint L-systems the D_i are replaced with fixed numbers associated with the symbols in the L-system together with a decay parameter ω. Let $\{S_1, S_2, \ldots, S_k\}$ be the symbols of the L-system (A and B in Fig. 9.1). Let h_i be the number associated with symbol S_i. The interpretation of the midpoint L-system starts with the specification of the heights of four corners of a square. In each expansion the value $\omega^{k-1} \cdot h_i$ is added to the center of the square where symbol S_i is being expanded. The value k is the index of the expansion (first,

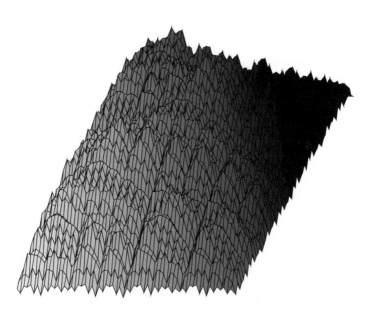

Fig. 9.3. The result of six expansions of the midpoint L-system given in Fig. 9.1
with $h_1 = 0.8$ $h_2 = 0.2$ and $\omega = 0.7$

second, etc.). Thus, on the first expansion, the value added to the center of
the initial square when it is partitioned is the appropriate h_i for the axiom.
In the next expansion, $\omega \cdot h_i$ will be added to the heights of the centers of the
four squares the initial square was divided into. In the third expansion, the
values added will have the form $\omega^2 \cdot h_i$ for the centers of all 16 squares present
in that expansion, and so on. The value ω controls the degree to which the
heights associated with symbols decay with expansion. In the experiments
in this chapter, values for ω between 0 and 1 are used. A rendering of the
example midpoint L-system given in Fig. 9.1 is shown in Fig. 9.3.

An application of evolving L-system grammars appears in [12, 15] where
the L-system provided the connection topology of an artificial neural net.
The parameters of the L-system interpreter were fixed in that study, not
evolved. Evolutionary algorithms that set the parameters of an L-system appear in [1, 2]. In [10] L-systems are used to specify a body for an artificial
agent that is co-evolved together with a control system. The type of L-system
presented in this chapter is unique in co-evolving the parameters used in interpretation together with the L-system grammar. The current application is
intended to generate rugged versions of idealized smooth landscapes for use in
virtual reality applications. Ideally, the algorithm will generate a selection of

different landscapes, none too different from the original idealized landscape but differing in details of appearance.

The remainder of this chapter is structured as follows. Section 9.2 specifies the evolvable representation for midpoint L-systems. Section 9.3 gives the experimental design for the evolutionary algorithm used to locate midpoint L-systems that specify virtual landscapes. Section 9.4 summarizes the results of the evolutionary runs. Section 9.5 discusses the results and suggests possible variations. Section 9.6 suggests some possible next steps for the creation and application of midpoint L-systems.

9.2 The L-system Representation

It is straightforward to put midpoint L-systems into an evolvable form. Specifying a midpoint L-system requires the following parameters:

1. the initial heights of the corners of the first square;
2. the number of symbols used;
3. the 2×2 grids of symbols used to expand each symbol;
4. the midpoint displacement associated with each symbol; and
5. the decay parameter ω.

In this study the decay parameter ω is specified globally and thus has a fixed value in any given experiment. Likewise, the initial square heights have a fixed, global value of zero. The axiom for all midpoint L-systems in this study is the single symbol **A**. (Since expansion rules for symbols are generated at random when initializing populations of L-systems, this choice of a fixed axiom has no cost in expressiveness of the system.)

The representation used to evolve L-systems contains two linear structures. For an n-symbol L-system there is a string of $4n$ symbols that specifies the replacement rules, in adjacent groups of four, for each of the symbols. There is also an array of n real values that specify the midpoint displacements associated with each symbol. An example of a midpoint L-system specified in this fashion is given in Fig. 9.4. The upper-case Roman alphabet symbols (followed by the lower-case if necessary) in alphabetical order expand into the groups of four. (In this case, A–H.)

The variation operators we defined for this representation are as follows. Two crossover operator are used: two-point crossover of the string specifying the expansion rules and two-point crossover of the array of weights. Both crossover operators are used 100% of the time. Two mutation operators are used. The *displacement* mutation picks one of the midpoint displacement values uniformly at random and adds a number uniformly distributed in the range $[-0.05, 0.05]$ to that displacement value. The *symbol* mutation picks one of the symbols in the string that specifies the replacement rules and replaces it with a symbol chosen uniformly at random from those available. Both of these mutations are used, once each, on each new structure.

Expansion rules:
CCCC.AAGF.DGDG.GDGD.FBHC.GEDA.GGGG.CEEB

Displacement values:
(0.790, −0.002, −0.102, 0.010, 0.116, −0.058, 0.036, 0.143)

Fig. 9.4. An example of an eight-symbol midpoint L-system evolved to fit a hill. Dots in the string of symbols that specify the expansion rules are inserted for readability. The four symbols in each expansion rule are places in the 2×2 array in reading order

Midpoint L-systems are evolved to fit a specified landscape. All landscapes are placed on the unit square. The landscapes used in this chapter (shown in Fig. 9.5) are: a hill:

$$H(x, y) = \frac{1}{1 + (6x - 3)^2 + (6y - 3)^2} \tag{9.1}$$

and a crater:

$$C(x, y) = \frac{1}{1 + (r - 2)^2} \tag{9.2}$$

where $r(x, y) = \sqrt{(12x - 6)^2 + (12y - 6)^2}$. Both these shapes are radially symmetric about the center of the unit square making them somewhat challenging for the midpoint L-systems which find shapes that divide naturally into squares easiest to fit.

9.3 Experimental Design

The fitness of a given midpoint L-system is evaluated by expanding the L-system seven times to generate a 128×128 grid of height values. These values are placed on a regular grid covering the unit square. The RMS error of the agreement of the height values with the desired landscape is computed. This error value, to be minimized, is the fitness function used to drive evolution.

The evolutionary algorithm operates on a population of 120 midpoint L-systems. The initial population is created with symbols for expansion rules chosen uniformly at random from those available. The initial values for the midpoint displacements are chosen uniformly at random in the range [0.1, 1] but are permitted to leave the interval later via mutation. The algorithm is a steady-state algorithm [19] using size seven tournament selection. For each landscape and set of parameters tested, we performed a collection of 100 evolutionary runs. Each of these runs consisted of 10,000 mating events. For each mating event, a tournament of size seven is selected, and the two most fit members of the tournament are bred to produce children that replace the two least fit members. Breeding consists of copying the parents, performing

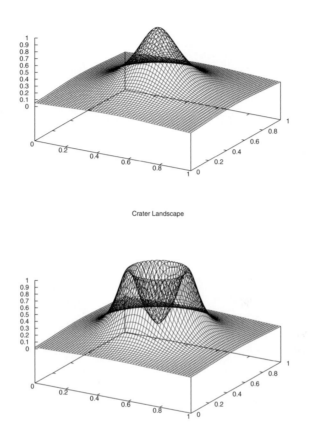

Fig. 9.5. Surface plots of the hill and crater landscapes

both kinds of crossover on the copies, and then performing one displacement mutation and one symbol mutation on each of the resulting new structures.

For the hill landscape, given by (9.1), nine experiments were performed to explore the impact of changing the decay parameter ω and the number of symbols. These experiments used the nine possible combinations available when $\omega \in \{0.8, 0.9, 0.95\}$ and $n \in \{4, 8, 16\}$. Visual inspection of the resulting landscapes led to the choice of $\omega = 0.8$ and $n = 16$ for the subsequent set of experiments performed on the crater landscape. An additional experiment was performed using $\omega = 0.8$ and $n = 32$ for the crater landscape to study the impact of allowing more symbols in the L-system.

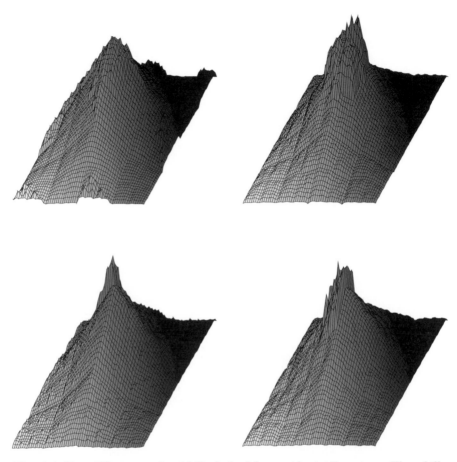

Fig. 9.6. Four different rendered hills derived from midpoint L-systems. These hills are all developed from $\omega = 0.8$, $n = 16$ L-systems. This rendering uses six expansions of the midpoint L-system for each hill. Note the diversity of appearances

9.4 Results

The parameter study of the impact of varying the number of symbols and the decay parameter is summarized in Table 9.1. When ω was 0.8 or 0.9, the average RMS error of the L-system to the target surface improved as the number of symbols increased. This improvement was statistically significant, as documented by the disjoint confidence intervals, for both increases of symbol number for $\omega = 0.8$ and for the change from eight to 16 symbols when $\omega = 0.9$. When $\omega = 0.95$ the RMS error becomes steadily worse as the number of symbols increases with a significant difference between four and eight symbols. This initial exploration shows that the impact of varying ω and n is not independent. Both visual inspection of landscapes rendered from the

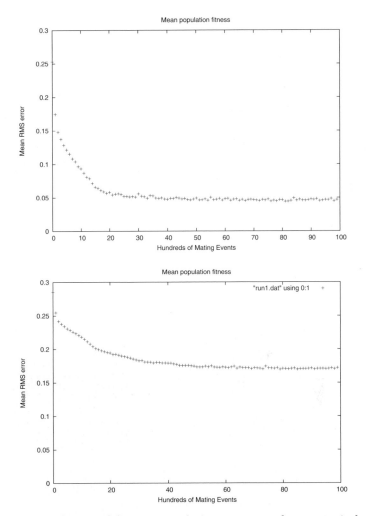

Fig. 9.7. The behavior of fitness as evolution progresses for two typical runs of the midpoint L-system training algorithm. The upper plot shows fitness on the hill landscape given by (9.1), while the lower plot shows fitness on the crater landscape given by (9.2). The much higher RMS error values for the crater run reflect the relative difficulty of fitting the crater

L-systems and the fitness results supported the choice of $\omega = 0.8$ and $n = 16$ for subsequent experiments. Examples of the types of hills evolved with this system are given in Fig. 9.6.

Fitting the crater landscape is a considerably more challenging problem than fitting the hill landscape (as illustrated in Fig. 9.7). The confidence interval for the craters using $\omega = 0.8$ and $n = 16$ symbols, analogous to the ones in Table 9.1, is $(0.1661, 0.1710)$. Many of the evolved craters took the

Table 9.1. 95% confidence intervals on the mean RMS error (fitness) for agreement between the best midpoint L-systems in each run and the target landscape for each of the nine sets of 100 evolutionary runs performed to study the impact of changing the decay parameter ω and number of symbols n

| | Mean RMS error, 95% confidence intervals | | |
| | Value of ω | | |
n	0.8	0.9	0.95
4	(0.0490, 0.0500)	(0.0487, 0.0497)	(0.0487, 0.0479)
8	(0.0466, 0.0475)	(0.0489, 0.0499)	(0.0511, 0.0523)
16	(0.0426, 0.0434)	(0.0459, 0.0469)	(0.0514, 0.0527)

form of a collection of spikes at various positions along the crater's rim. One of the better looking craters, also the most fit, is shown in Fig. 9.8. The 95% confidence interval for mean RMS error for the craters evolved with $n = 32$ symbols is $(0.1397, 0.1478)$.

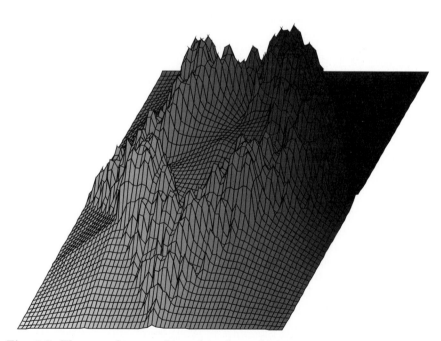

Fig. 9.8. The most fit crater located in the evolutionary runs on the crater landscape. This crater uses $n = 32$ symbols in its midpoint L-system

9.5 Discussion and Conclusions

The evolutionary algorithm used for locating midpoint L-systems worked well for the hill landscape and for some runs for the crater landscape. It is a novel example of the techniques of artificial embryogenesis. The RMS error for the best set of parameters ($n = 16$, $\omega = 0.8$) has hills deviating 4.3%, on average, from the height profile of the idealized hill surface. The midpoint L-system approximations of the crater with $n = 16$ symbols varied roughly 17% from the ideal crater, and even the best evolved crater deviated 14% on average. The parameter study on the hill suggests that increasing n or decreasing ω might yield a better approximation of the ideal landscape. This hypothesis was tested in the runs that evolved craters with $n = 32$ symbols. The confidence interval on the mean RMS error of end-of-run best fits to the crater dropped substantially with the best fitness found having 12.1% average error.

The pattern of fitness values in Table 9.1 is a little startling. Initially we conjectured that lowering ω and raising n would make independent, positive contributions to fitness. This is not the case when ω is near 1. This is probably due to problems with fitting both the flat and curved portions of the landscape accurately. The parameter fitting study demonstrated that the parameters interact, and so a parameter sweep should be performed in any other work done with this system.

The algorithm using both trial surfaces located a variety of different midpoint L-systems that produce landscapes with distinct appearances. This meets the goal of supplying a palette of landscape features for a virtual landscape designer. The polygons specified by a midpoint L-system at the evolved resolution of 128×128 form a data object with a size a little over 800 kilobytes; this grows to megabytes when rendered as an uncompressed image. The midpoint L-system itself, with 16 symbols, requires 48 characters and 16 real number midpoint displacements, for a total of 144 bytes of storage. The midpoint L-systems can thus specify a virtual landscape in far less space than direct storage of the polygon models. Since the midpoint L-systems, once evolved, use no random numbers, they can rapidly be expanded to whatever resolution is required.

9.5.1 Level of Detail

The level of detail (LOD) for a midpoint L-system is determined by the number of expansions performed. The evolved midpoint L-systems in this chapter were all evolved with a fitness evaluation using seven expansions of the L-system which yielded a grid of $128 \times 128 = 16384$ height values.

A benefit of modeling landscapes with a midpoint L-system is that the LOD can be varied by simply running different numbers of expansions of the L-system. The time to expand a midpoint L-system is linear in the final number of polygons, and the expansion is almost trivial to parallelize since

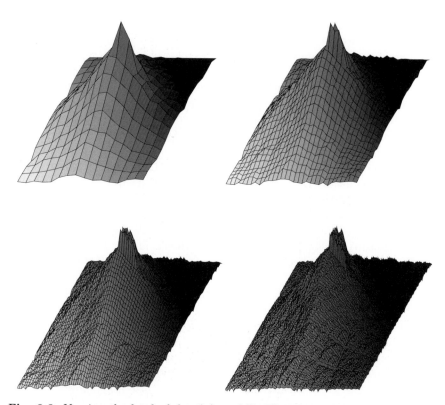

Fig. 9.9. Varying the level of detail for a hill. The same midpoint L-system for the hill landscape is rendered on $N \times N$ grids for N=16, 32, 64, and 128. This corresponds to four, five, six, and seven expansions of the L-system which yield increasing levels of detail

different quadrants of the expansion share information only at their boundaries, and that information does not change during the expression process. Figure 9.9 shows an example of a single midpoint L-system for the hill landscape expanded to four different levels of detail. The highest level of detail shown, seven expansions of the L-system, is the level of detail at which the L-system was evolved. Figure 9.10 shows a crater with five and six expansions.

If the level of detail of the evolved midpoint L-systems is increased beyond the level at which it was evolved, the behavior can become unpredictable. Expansions performed beyond the evolved level of detail with the L-systems located in this study produced remarkably spiky pictures, and so it is probably a good idea to evolve midpoint L-systems to the maximum level of detail required. The most expensive part of the midpoint L-system training software is fitness evaluation. The time for a fitness evaluation is proportional to 2^n

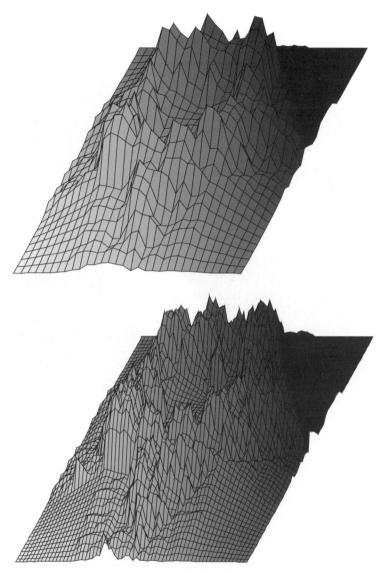

Fig. 9.10. Varying the level of detail in a crater. The same midpoint L-system for the hill landscape is rendered on a 32×32 and a 64×64 grid representing five and six expansions of the L-system

where n is the number of expansions of the L-system used during fitness evaluation.

Fig. 9.11. Varying the value of ω. The above crater was evolved using $\omega = 0.8$ but then rendered for $\omega = 0.7, 0.75, 0.8$, and 0.85. The smaller ω the flatter the landscape

A compromise between the conflicting constraints of training time and final level of detail may be possible. As the fitness plots in Fig. 9.7 demonstrate, the initial population is relatively unfit. Using a small number of expansions initially and increasing the number as evolution progresses could well result in a substantial savings in training time. The schedule for increasing the parameter n during the course of evolution would have to be experimentally determined. We conjecture that effective patterns for increasing n will be different for different landscapes.

9.5.2 Varying the Decay Parameter

The decay parameter controls the relative importance to the landscape's final appearance of the earlier and later applications of the expansion rules. A high value of ω permits later expansions to have more impact. Figure 9.11 shows what happens when ω is varied for a crater evolved with $\omega = 0.8$. This suggests that, if evolved land-forms of the sort demonstrated here are needed to build a complex landscape, then varying ω will permit an inexpensive source of additional diversity. If ω varied functionally with position in the virtual landscape, it could be used as a parameter to set the tone for various regions. The more rugged the area, the higher the ω.

While all the values of ω used in this chapter are in the range $[0.7, 0.95]$, other values yield interesting effects. If $\omega > 1$, then later expansions have *more* of an effect. Evolved landscapes from this chapter rendered in this fashion are remarkably spiky; if rules were evolved with $\omega > 1$, then the results would probably be quite different from those we present.

9.6 Next Steps

The parameter study performed for the hill landscape tested a small number of parameters on a single landscape. The characterization of the system for evolving midpoint L-systems as an evolutionary algorithm is thus barely begun. Larger parameter sweeps may yield unexpected results in the part of the parameter space not yet explored, and many more landscapes should be tested. The rate of application of the crossover and mutation operators was set to the defaults used in the authors' own in-house evolutionary algorithm code; exploration here might also improve performance.

The unexpected interaction of the number n of symbols and the decay parameter ω suggest that letting ω become an evolvable parameter could also help. A step that was not taken was to normalize the height values of the expanded midpoint L-system to have the correct average height. All such normalization was left for evolution. Permitting ω to float during evolution would give evolution a far more effective tool, an overall normalization parameter, for adjusting the height of the virtual landscape. In retrospect, it seems obvious that an evolvable ω would improve performance. "Obvious" and "correct" are not always the same in evolutionary computation, and so experiments with evolvable ω are an early goal of additional research on this system.

9.6.1 Haptic and Visual L-textures

The goal of this study was to create compact descriptions, in the form of midpoint L-systems, for large features of a virtual landscape. This is not the only potential application. If midpoint L-systems were fit to a relatively flat landscape they could be used to supply tactile texture. Virtual reality systems are starting to implement haptic (touch) feedback devices and the height data supplied by a midpoint L-system could be used to create any desired degree of roughness by simply modifying the vertical scale. In effect, the ω parameter could be used to control the rugosity of a haptically perceived surface.

While the systems evolved in this chapter used a square area for fitness evaluation and rendering, it is not difficult to create non-square midpoint L-systems. Midpoint L-systems can be used to "paint" any rectangle. If the

Fig. 9.12. Examples of rendering midpoint L-systems as gray-scale values rather than heights. The L-systems rendered above are hand designed, not evolved

height values are instead displayed as colors or shades of gray, then midpoint
L-systems can be used to generate the more traditional visual textures used in
virtual reality as well as interesting pictures. Examples of gray-scale renderings
of two hand-designed midpoint L-systems appear in Fig. 9.12.

The images shown in Fig. 9.12 suggest that colored midpoint L-systems
could be used to design interesting, symmetric objects. Rather than mimicking
natural land-forms, these L-systems could be used to create three dimensional
crystals, complex patterns for floor tiles or rugs, and other dressing of a virtual
environment. In this case the ability of midpoint L-systems to create complex
forms that use modest amounts of storage and which can be rendered quickly
would be quite valuable. If symmetry and a requirement to stay within a
particular volume were imposed, then an evolutionary algorithm could be
used to generate thousands of crystalline forms or non-repeating floor-tile
designs for an entire virtual city.

9.6.2 Recursive Application

The initial height values for the squares in which our L-systems were expanded
were set to 0 throughout the chapter. The ability to set these initial values
to any height means that a land-form specified by a midpoint L-system can
be expanded into any four-sided polygon within the landscape. This, in turn,
yields the potential for recursive application of midpoint L-systems. At its
simplest, one might render a broad flat hill (one of the hills evolved in this
chapter with $\omega = 0.4$) at a low LOD and then place an evolved crater in
several of the resulting squares. A more complex application would place the
crater L-system in the expansion process at different levels of the expansion
of the hill. This would permit craters of varying size to populated the virtual
hill.

A more complex form of this idea would use existing midpoint L-systems
as tiles of a new one. The evolutionary algorithm presented in this chapter
would generate a palette of land-forms that would then be available to a higher
level system that creates virtual landscapes that satisfy various strategic goals
such as lines of sight or lowland connectivity but which are otherwise diverse.
This representation would decompose the problem of fitting complex landscapes into pieces represented by the lower and higher level L-systems. This
application might be valuable for providing an automatic source of variety in
a video game.

Acknowledgments

This work expands and illuminates a paper presented at the 2005 Congress
on Evolutionary Computation [5]. The authors would like to thank the Iowa
State University Virtual Reality Applications Center, particularly Douglas
McCorkle and Jared Abodeely, who helped evaluate over 1000 renderings of

midpoint L-systems produced during the parameter study. The authors also thank the Department of Mathematics and Statistics at the University of Guelph for its support of the first author.

References

1. Ashlock, D., Bryden, K.: Evolutionary control of L-system interpretation. In: Proceedings of the 2004 Congress on Evolutionary Computation, vol. 2, pp. 2273–2279. IEEE Press, Piscataway, NJ (2004)
2. Ashlock, D., Bryden, K., Gent, S.: Evolutionary control of bracketed L-system interpretation. In: Intelligent Engineering Systems Through Artificial Neural Networks, vol. 14, pp. 271–276 (2004)
3. Ashlock, D., Bryden, K., Gent, S.: Evolving L-systems to locate edit metric codes. In: C. Dagli et al. (ed.) Smart Engineering System Design: Neural Networks, Fuzzy, Evolutionary Programming, and Artificial Life, vol. 15, pp. 201–209. ASME Press (2005)
4. Ashlock, D., Bryden, K., Meinert, K., Bryden, K.: Transforming data into music using fractal algorithms. In: Intelligent Engineering Systems Through Artificial Neural Networks, vol. 13, pp. 665–670 (2003)
5. Ashlock, D., Gent, S., Bryden, K.: Evolution of L-systems for compact virtual landscape generation. In: Proceedings of the 2005 Congress on Evolutionary Computation, vol. 3, pp. 2760–2767. IEEE Press (2005)
6. Ashlock, D., Kim, E.: The impact of cellular representation on finite state agents for Prisoner's Dilemma. In: Proceedings of the 2005 Genetic and Evolutionary Computation Conference, pp. 59–66. ACM Press, New York (2005)
7. Ball, P.: The Self-made Tapestry, Pattern Formation in Nature. Oxford University Press, New York (1999)
8. Gruau, F.: Genetic synthesis of modular neural networks. In: Proceedings of the Fifth International Conference on Genetic Algorithms, pp. 318–325. Morgan Kaufmann, San Francisco (1993)
9. Gruau, F., Whitley, D., Pyeatt, L.: A comparison between between cellular encodings and direct codings for genetic neural networks. In: Genetic Programming 1996, pp. 81–89. MIT Press, Cambridge MA (1996)
10. Hornby, G., Pollack, J.: Body-brain co-evolution using L-systems as a generative encoding. In: L. Spector et al. (ed.) Proceedings of the Genetic and Evolutionary Computation Conference (GECCO-2001), pp. 868–875. Morgan Kaufmann, San Francisco, California, USA (2001)
11. Hornby, G., Pollack, J.: Evolving L-systems to generate virtual creatures. Computers and Graphics 25(6), 1041–1048 (2001)
12. Kitano, H.: Designing neural networks using genetic algorithms with graph generation system. Complex Systems 4, 461–476 (1990)
13. Lindenmayer, A.: Mathematical models for cellular interaction in development, parts I and II. Journal of Theoretical Biology 16, 280–315 (1968)
14. Mandelbrot, B.: The fractal geometry of nature. W. H. Freeman and Company, New York (1983)
15. Miller, G., Todd, P., Hegde, S.: Designing neural networks using genetic algorithms. In: Proceedings of the 3rd International Conference on Genetic Algorithms, pp. 379–384. Morgan Kaufmann, San Mateo, CA (1989)

16. Murray, J.: Mathematical Biology II. Springer-Verlag, New York (2003)
17. Prusinkiewicz, P., Lindenmayer, A., Hanan, J.: The algorithmic beauty of plants. Springer-Verlag, New York (1990)
18. Stanley, K., Miikkulainen, R.: A taxonomy for artificial embryology. Artificial Life **9**, 93–130 (2003)
19. Syswerda, G.: A study of reproduction in generational and steady state Genetic Algorithms. In: Foundations of Genetic Algorithms, pp. 94–101. Morgan Kaufmann (1991)
20. Turing, A.: The chemical basis of morphogenesis. Philosophical Transactions of the Royal Society(B) **237**, 37–72 (1952)

On Form and Function: The Evolution of Developmental Control

Sanjeev Kumar

Sibley School of Mechanical & Aerospace Engineering, Cornell University, Ithaca, New York, 14853, USA `sanjeev.kumar8@gmail.com`

10.1 Introduction

How does the genome control the construction of a complex multi-cellular system with well defined form and structures from a single cell? Artificial life and developmental biology overlap on some quite important topics. The question above reveals two such pivotal topics: construction and control.

Construction of complex adaptive systems, or indeed intricate *forms*, in a robust, self-organizing manner is a notoriously difficult problem that highlights fundamental issues of scalability, modularity, and self-organization. The need to build such complex technology has, over recent years, sparked renewed interest in using approaches inspired by developmental biology [4,8,12,14,19]. For example, the problem domain seeing the most success with the application of the developmental metaphor for construction-related tasks is Evolutionary Design [1]. Examples of typical tasks include the synthesis and use of models of development that incorporate the use of embryos, cells, proteins, diffusion, receptors, differentiation, cell signaling, and other cell and developmental processes to build forms ranging from buildings [11] to geometric shapes [14] to creatures [2,5].

Control is an important *function*. Be it a robot controller or the control of the construction process, devising novel techniques to implement robust control has been an important research motivation for many years and has only recently begun taking inspiration from non-neural developmental biology. For example, in the field of robotics aspects of developmental biology, such as hormone control, have been modeled with much success for distributed robot control [9,22]. In Evolutionary Robotics (ER), robot control using developmental biology inspired approaches has typically been achieved indirectly, for example, through the development of a controller such as a neural network. Several researchers have investigated such an approach [4,8]. The focus of such investigations, however, is on evolving a developmental process which specifies how the topology of a neural network (NN) is to be grown using concepts from developmental biology in which cells are able to divide, differentiate, and form connections with other cells. After a period of growth, the resulting network of cells is interpreted as a neural network used to control robot behavior and navigation. Common to all such works is that once the neural network is constructed the genome is discarded [20].

Yet nature has solved problems of construction, control, scalability, modularity, self-organization, and even self-repair through the evolution of development: the process or set of processes responsible for constructing organisms [24], and the complex Genetic Regulatory Networks (GRNs) that are found in the genome. Furthermore, nature does not discard the genome once the brain is built; quite the contrary, developmental processes and mechanisms continue to operate long after embryonic development has ceased in order to maintain the highly dynamic state of organisms.

In nature the ability to react fast to changing situations and circumstances is crucial to the success of an organism, for example, fleeing a predator. As Marcus [17] points out, neurons react on a faster time-scale than genes, typically on the order of milliseconds, whereas genes are relatively slower. However, this should not detract from the fact that the genome is an immensely complex, real-time control system that builds bodies, brains, and immune systems capable of remarkable abilities.

Development offers an oft-forgotten alternative route to tackling problems of construction and control. Yet despite all its abilities, the process of development remains a relatively untapped, and frequently overlooked source of inspiration, mechanisms, and understandings. Consequently, inspired by biology, the field of computational development is seen as a potential solution to such construction and control related problems [15]. An important point worth mentioning is that as computer scientists and engineers, we are not constrained by the same issues as biology. As a result, genetic regulatory networks need not operate on a slower time-scale than Neural Networks (NNs); they could operate at the same time-scale. Furthermore, in using GRNs as the control system for generating, say, reactive robot behaviors in which current sensor readings are mapped into actions; or as, say, controllers of a process that specifies the construction of form, future research can begin to explore how one might harness their powerful robust, fault-tolerant, regenerative, and dynamical abilities.

Work by Kumar [13] has shown GRNs to be a viable control architecture for reactive robot control for the problem domain of evolutionary robotics, along with other works such as [1,20,23]. Alternative, and related, approaches to robot control and behavior generation that have seen much success are rule systems [21]. Still, problems with rule systems exist: although they can be made to have state, traditionally state was lacking.

In this chapter, I describe previous works that address the dual problems of form and function. In particular, this chapter addresses the ability of development to produce both form (morphology of geometric shapes that have proven useful in, for example, computer graphics such as planes and cubes) and function (robot control, i.e. the ability to specify reactive robot behaviors through the evolution of real-time GRN robot controllers).

The chapter is structured as follows: first a brief description is given of the model of development, called the Evolutionary Developmental System (EDS), and the underlying genetic regulatory network model that lies at the core of the EDS. This is followed by two self-contained sections detailing experiments that investigate the use of development for form and function respectively. The first of these sections documents experiments that not only demonstrate how the EDS and its GRN model are able to construct – in a self-organizing manner – geometric forms (shapes), but also provide detailed evolved developmental mechanisms that specify the self-construction of these forms. Two sets of experiments, which differ only in the target form evolved, are detailed. The first set of experiments explores the construction of a plane, while the second experiment examines the construction of a cube.

The second of these sections demonstrates how an adapted version of the GRN model within the EDS can be evolved for function, i.e. the problem of reactive robot control. The function experiments are broken up into three sets of experiments. The first set of experiments shows controllers that are able to memorize routes through the environment using no sensors. The second set shows controllers that are able to interact with the environment through sensors for both navigation and behavior generation. However, analysis revealed that these controllers, although appearing to display obstacle avoidance behavior, memorized routes through the environment. In order to address this, a third set of experiments is detailed that evolves GRN controllers to perform general purpose obstacle avoidance while being subjected to small amounts of sensory noise.

10.2 The Evolutionary Developmental System (EDS)

The Evolutionary Developmental System is an object oriented computer model of many of the natural processes of development [14]. At the heart of the EDS lies the developmental core. This implements concepts such as embryos, cells, cell cytoplasm, cell wall, proteins, receptors, transcription factors (TFs), genes, and cis-regulatory regions. Genes and proteins form the atomic elements of the system. A cell stores proteins within its cytoplasm and its genome (which comprises rules that collectively define the developmental program) in the nucleus. The overall embryo is the entire collection of cells (and proteins emitted by them) in some final conformation attained after a period of development. A genetic algorithm is wrapped around the developmental core. This provides the system with the ability to evolve genomes for the developmental machinery to execute. The following sections describe the main components of the developmental model: proteins, genes, and cells.

10.2.1 Proteins

In the EDS, the concept of a protein is captured as an object. In total there are 40 proteins (see [14] for more details), each protein having four member variables:

- an ID tag (simply an integer number denoting one of 46 pre-defined proteins the EDS uses to control cellular behavior);
- source concentration (storing the concentration of the protein);
- two sets of coordinates (isospatial [6] and Cartesian);
- a bound variable (storing whether or not a receptor has bound a protein).

(The latter is only used in receptor proteins.)

10.2.2 Genes

In nature, genes can be viewed as comprising two main regions: the cis-regulatory region [3] and the coding region [16]. Cis-regulatory regions are located just before (upstream of) their associated coding regions and effectively serve as switches that integrate signals received (in the form of proteins) from both the extra-cellular environment and the cytoplasm. Coding regions specify a protein to be transcribed upon successful occupation of the cis-regulatory region by assembling transcription machinery.

The EDS uses a novel genetic representation termed the cis-trans architecture (Fig. 10.1), based on empirical genetics data emerging from experimental biology labs [3]. The first portion of the genome contains protein specific values (e.g. protein production, decay, diffusion rates). These are encoded as floating-point numbers.

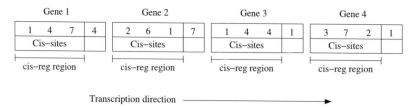

Fig. 10.1. An arbitrary genome created by hand. Genes consist of two objects: a cis-regulatory region and a coding region. Each number denotes a protein

The remaining portion of the genome describes the architecture of the genome to be used for development. It describes which proteins are to play a part in the regulation of different genes. It is this latter portion of the genome that is employed by each cell for development.

Currently, the EDS's underlying genetic model assumes a "one gene, one protein" simplification rule [16] to aid in the analysis of resulting genetic regulatory networks. The genome is represented as an array of gene objects (Fig. 10.1). Each gene object contains two members: a cis-regulatory region and a protein coding region. The cis-regulatory region contains an array of TF target sites. These sites bind TFs in order to regulate the activity of the gene. The gene then integrates these TFs and either switches the gene "on" or "off".

Details of the equations used to calculate activity of a single gene are given in Table 10.1. Integration is performed by summing the products of the concentration and interaction strength (weight) of each TF to find the total activity of all TFs occupying a single gene's cis-regulatory region; see (10.1). This sum provides the input to (10.3), yielding a probability of the gene firing [10, 14].

10.2.3 Cells

Cell objects in the EDS have two state objects: current and new. During development, the system examines the current state of each cell, depositing the results of the protein interactions on the cell's genome in that time step into the new state of the cell. After each developmental cycle the current and new states of each cell are swapped, ready for the next cycle.

The EDS supports a range of different cell behaviors, triggered by the expression of certain genes. The behaviors used for the experiments described in this work are:

Table 10.1. Equations used to calculate the activity of a single gene by summing the weighted product of all transcription factors regulating a single structural gene

$$\text{input}_j = \sum_{i=1}^{d} \text{conc}_i \times \text{weight}_{ij} \qquad (10.1)$$

where input_j = total input of all TFs assembling upon the jth gene's cis regulatory region; i = current TF; d = total number of TF proteins visible to the current gene; conc_i = concentration of the ith TF at the centre of the current cell; and weight_{ij} = interaction strength between TF i and gene j;

$$\text{activity}_j = \frac{\text{input}_j - \text{THRESHOLD\_CONSTANT}}{\text{SHARPNESS\_CONSTANT}} \qquad (10.2)$$

where activity_j = total activity of the jth gene; input_j = total input to the jth gene; and SHARPNESS_CONSTANT = a constant taken from the range 0.1–0.001 and is typically set to 0.01;

$$\text{activation\_probability}_j = \frac{1 + \tanh(\text{activity}_j)}{2} \qquad (10.3)$$

where $\text{activation\_probability}_j$ = activation probability for the jth gene; and activity_j = total activity of the jth gene.

- division (when an existing cell "divides", a new cell object is created and placed in a neighboring position);
- the creation and destruction of cell surface receptors; and
- apoptosis (programmed cell death).

The EDS uses an n-ary tree data structure to store the cells of the embryo, the root of which is the zygote (initial cell). As development proceeds, cell multiplication occurs. The resulting cells are stored as child nodes of parents' nodes in the tree. Proteins are stored within each cell. When a cell needs to examine its local environment to determine which signals it is receiving, it traverses the tree, checks the state of the proteins in each cell against its own and integrates the information.

The decision for a cell to divide in the EDS is governed by the ratio of division activator protein to repressor. The direction (or isospatial axis) the daughter cell is to be placed in is non-random and is specified by the position of the mitotic spindle within the cell, see [14] for more details.

10.3 Experiments

10.3.1 Form: Multi-cellular Development and Differentiation

This section details experiments to investigate the evolution of GRNs for multi-cellular development. GRNs were evolved for the task of constructing multi-cellular 3D geometric form and structure in a robust, self-organizing manner. The target forms are a plane and cube.

10.3.2 System Setup

The experiments in this section used the following parameter settings: 100 runs with a population size of 100 evolving for 250 generations. One-point crossover is applied all the time, while Gaussian mutation was applied at a rate of 0.01 per gene. Tournament selection was used with a tournament size of 33 (although 33 is regarded as high, informal experimentation with the system, not reported here, provided this value). Thirty developmental iterations, 100 generations, six proteins and two cis-sites per gene were used for the multi-cellular cube experiment. Fitness for multicellular cubes was determined using the equation for a cube. The number of cells inside and outside the enclosed cube was determined, resulting in a function to be minimized (10.4):

$$\text{fitness} = \left(\frac{1}{\text{cells\_inside}} \right) + \left(\frac{\text{cells\_inside}}{\text{SCALE}} \right) \tag{10.4}$$

where SCALE refers to a shape dependant constant defining total number of cells in shape.

10.4 Results and Analysis

10.4.1 Multi-cellular 3D Morphology

Plane

Figure 10.2(c) shows a crucial change in development from the five-cell stage (not shown) – two divisions occur in iteration 9. These divisions are due to the zygote and the fifth cell. The zygote manages to increase its level of protein 6 above a threshold and change division direction to 1, while the fifth cell (shown in the lower left corner of Fig. 10.2d) also divides but in direction 0, giving rise to this seven cell embryo. The new cell in the lower left corner is fated to die.

Figure 10.2(d) shows the state at iteration 10 in which the new cell in the lower left corner of the image ceases dividing. Instead, the upper right portion of the image shows division has occurred again in direction 1. The cell in the

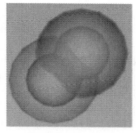

(a) Zygote, iteration 1

(b) Iteration 2 – zygote divides in direction 1

(c) Iteration 9 – seven-cell stage, growth in two directions

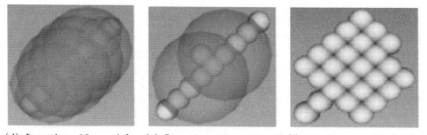

(d) Iteration 10 – eight-cells, growth occurs in upper right corner, this cell is fated to die

(e) Iteration 11 – 10-cell T-shape embryo, with division in directions 0 and 11

(f) Final state of plane – with all proteins removed

[27, 11 | 36] [15, 4 | 20] [15, 36 | 17] [7, 37 | 6] [11, 27 | 33]
[7, 15 | 30] [17, 36 | 2] [22, 5 | 22] [8, 13 | 10] [28, 7 | 0]

Fig. 10.2. The development of the best plane (top) and its evolved genome (below)

lower left and upper right corners of Fig. 10.2(d) are ephemeral additions and are ultimately fated to commit suicide (or apoptosis).

Figure 10.2(e) shows the zygote's first daughter cell has managed to divide in direction 11 resulting in a T-shape. Thereafter, cells eventually start to divide in direction 11 and others in direction 10, thus giving rise to the main body of the plane.

Cube

Figure 10.3 shows snapshots of the development of the best cube, which attained a fitness score of 0.006173. Figure 10.3 also shows the evolved genome. Evolution has only evolved a single gene for directional control of cell division through gene 7, which emits protein 4. Protein 4 rotates the division spindle anti-clockwise by 1 direction (full analysis is beyond the scope of this chapter, see [14]).

In the zygote, for example, two proteins control (more or less) the activation of gene 3: proteins 0 and 4, conferring inhibitory and excitatory stimuli, respectively. In the daughter cell, levels of both proteins 0 and 4 are low due to division, and so do not provide sufficient inhibition or activation (not shown, see [14]).

Instead, it falls not only to other proteins, such as proteins 24 and 37 (not shown), to provide inhibition, but also to cell signaling, which initially delivers large inhibitory stimuli through the receptor 13-proteins 4 and 31 signal transduction pathways from the first division in iteration 3.

Over time, as receptor 13 decays so does the inhibitory stimulus received through that pathway. Leaving the job of inhibiting gene 3 in the daughter cell to an alternative pathway. Note a full analysis is beyond the scope of this chapter; see [14].

It must be noted that both cells by virtue of expressing a different subset of genes also have a different subset of active receptors. The zygote begins development with an assortment of receptors, while the daughter cell (and later progeny) inherit their state including receptors from their parent, and then begin to express different genes and consequently different receptors.

In addition, the behaviour of certain receptors (for example, 9 and 10) reflects the fact that different receptors may interact with exactly the same proteins, but the outcome of the interaction may be different. This type of differential receptor–protein interaction is important in development.

The gene expression plots of Fig. 10.4 reveal important differential gene expression patterns between the two cells, i.e. the cells have differentiated. Noticeably, genes 3 and 9 are expressed, albeit sparingly, in the zygote, but not at all in the daughter cell. Other important differences in gene expression between the two cells are the expression of genes 7 and 8, which are both increasingly activated, in the zygote, over time, but are seldom activated in the daughter cell.

(a) Zygote | (b) Iteration 4 – three-cell stage | (c) Iteration 5 – four-cell stage with new cell placed in the lower front of the embryo

 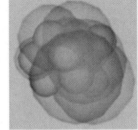 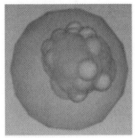

(d) Iteration 10 – five iterations later and the embryo remains at the four-cell stage | (e) Iteration 15 – 11-cell stage with many long-range proteins removed for clarity | (f) Iteration 20 – the core of the cube begins to take form

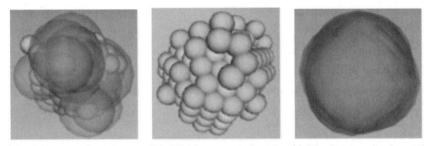

(g) Iteration 25 – approximate cube structure is established | (h) Final state of cube with all proteins removed from view for clarity | (i) Final state of cube with all proteins present

[32, 3 | 28] [0, 35 | 20] [4, 30 | 28] [26, 28 | 0] [26, 36 | 30]
[33, 32 | 31] [25, 8 | 4] [27, 23 | 21] [22, 1 | 27] [37, 8 | 14]

Fig. 10.3. The development of the best cube (top) and its evolved genome (below)

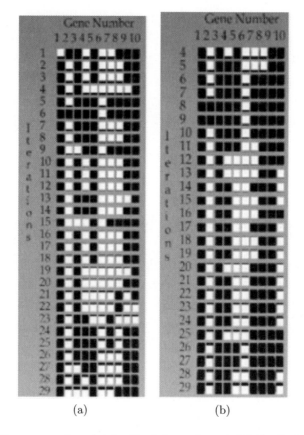

Fig. 10.4. Gene expression plot for the **(a)** zygote and **(b)** the first daughter cell of the best cube. Vertical axis denotes time, horizontal axis denotes gene ID, white and black squares denote activated and inactivated genes, respectively

10.4.2 Function Experiments: Reactive Robot Control

Having successfully explored the evolution of form using developmental biology-inspired methods, via the EDS, this section explores the evolution of function (i.e. reactive robot control) using developmental methods. This work had two main objectives:

- show that GRNs are a plausible technology for real-time reactive robot control; and
- generate reactive behaviors such as traversing an obstacle laden world.

However, during the course of the experiments, analysis of evolved solutions generated an additional objective:

- to evolve general purpose obstacle avoidance irrespective of the robot's position in the environment.

Due to the inherent noise in the GRN's gene activation mechanism and the addition of Gaussian noise to the sensory inputs, reactive robot control would be achieved in the face of noise at both levels. The number of proteins was changed between the experiments due to problem requirements such as the use of sonar proteins, for example.

Noise

The GRNs used in this work employed noise at two different levels:

- gene activation (rule firing); and
- sensory input (in the third experiment).

At the gene activation level, the firing of genes (rules) was performed probabilistically, thus introducing noise at the level of rule firing making the task of evolving robot controllers more difficult. The incorporation of noise at this level is justified on the basis that it enables the system to evolve robust rule sets. Noise at the level of sensory input (only employed in experiment 3) was provided by adding a small degree of Gaussian noise (σ in the small range [0.01–0.08]) to the sonar values before being input to the GRN controller.

10.4.3 Robots

This work employed two different robots – a Pioneer 2DX (see Fig. 10.5a), and an Amigobot (see Fig. 10.5b) – in simulation using the University of Southern California's robot simulator, player-stage [7]. This simulator was selected for many reasons of which the most salient was the wide range of robots supported, and the inherent noise in the system. This element of stochasticity aides the transition of the controller over, what has now become known as, the "reality gap" from simulation to real hardware [8].

(a)   (b)

Fig. 10.5. Robot images

Three sets of experiments were performed. In the first and second set of experiments, both robots were permitted to rotate either left or right and move forwards, but were prevented from moving backwards so as to encourage clear paths to emerge while avoiding obstacles. In the third set of experiments, however, the robots were allowed to reverse.

Fitness Function

All experiments involved using Euclidean distance as part of the fitness function. For the first set of experiments, the task set for the robots was to maximize Euclidean distance – to move as far as possible – within a set period of time, approximately 12 seconds with no sensory information. In the second experiment, the robot was allowed to use sensors.

The third set of experiments used Euclidean distance as one of the terms, the full fitness function is shown in (10.5). This equation is an adapted version of the obstacle avoidance fitness function used by [18], it defines an equation to be maximized. It is, however, lacking an additional term, which [18] included: a term to measure whether or not the robot is moving in a straight line. This was included in Mondada's work to prevent evolution awarding high fitness values to controllers that cause the robot to traverse a circle as fast as it can. In place of this term, the Euclidean distance was used.

$$\text{Avoidance} = \sum (e + s + \text{SonarValue}^{\text{MAX}}) \tag{10.5}$$

where e is the Euclidean distance, s the speed of the robot, and $\text{SonarValue}^{\text{MAX}}$ the reading from the sonar with the most activity.

Experiment 1: Memorizing a Path

The first experiment requires the memorization of a path through an environment with obstacles using no sensors or start seeds (i.e. no maternal factors – solutions are evolved completely from scratch). Although this experiment does not class as "reactive" robot control, a solution to this problem does re-quire internal state; thus the experiment was deemed necessary in order to ascertain the level of state embodied within GRNs. The GRN could only use the first four proteins shown in Table 10.2. Note that this experiment was only performed with the Pioneer 2DX in simulation. This task implicitly requires internal state in order to solve the problem. This set of experiment used four proteins, a population size of 100 evolving for 100 generations, with 10 genes, two cis-sites, and a simulation lifetime of 12 seconds per fitness function trial.

Experiment 1: Results

As can be seen from Fig. 10.6, the EDS evolved genetic regulatory network controllers that were successfully able to memorize a path through the environment without any sensors. The figures illustrate two different paths discovered

Table 10.2. List of proteins and their purpose

Protein ID	Purpose
0	Move forwards
1	Move backwards
2	Rotate counter-clockwise
3	Rotate clockwise
4	No purpose
5	Rear left sensor
6	Left sensor
7	Front-left sensor 1
8	Front-left sensor 2
9	Front-right sensor 1
10	Front-right sensor 2
11	Right sensor
12	Rear-right sensor

by two different controllers evolved from scratch. Figure 10.6(a) provides an example of the robot traversing a circular trajectory with minor deviations. To encourage a more intricate path to be found, the wall in the upper left corner of Fig. 10.6(a) was brought down, thus blocking the robot's path in the upper left direction, see Fig. 10.6(b). With the wall now blocking the robot's path Fig. 10.6(b) shows a different evolved GRN resulting in a more intricate path with no environment interaction. In this example, the robot is able to navigate quickly through the environment negotiating left and right turns past obstacles. This example reflects internal state in GRNs.

Using no sensors, a population size of 100 individuals and 100 generations, good solutions capable of discovering clear paths through the environment emerged at around the 54th generation.

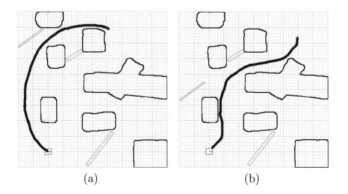

(a)   (b)

Fig. 10.6. No sensors. The robot has no sensory feedback whatsoever. It learned a route through the obstacle course. Moving objects in its path would cause the controller to break

Experiment 2: System–Environment Interaction

While memorizing paths through an environment using no sensory feedback may seem to be of limited use, it does demonstrate that GRNs are capable of encoding such paths and that these kinds of solutions are evolvable. In the second experiment, both the Pioneer 2DX and the Amigobot were used, again in simulation, but with different types of sensors. In the case of the Pioneer, SICK LMS 200 laser-range finder sensors were employed, while the Amigobot used its eight sonar sensors. The laser values were compiled into two values: one which corresponded to a left side sensor and the other, which corresponded to a right side sensor. The Amigobot has sonar emitters and receivers distributed around the robot with six arranged around the sides and front, while two are located at the rear of the robot. Since the robot was not permitted to reverse, the two sonars at the rear of the robot were not used (only in experiments 1 and 2). Consequently, the remaining six sonars were split in two and provided left and right sonar sensing. The settings used were: seven proteins, 15 genes, six cis-sites and a population size of only 25 individuals evolving for 50 generations.

Experiment 2: Results

Figures 10.7 and 10.8 show that despite noisy gene transcription (noisy rule firing) GRNs are evolved that are able to control both a Pioneer 2DX robot equipped with SICK LMS lasers, and an Amigobot equipped with sonars through the same environment more convincingly.

Additionally, coupling the system and environment through sensors enables faster evolution of successful solutions, for example, all experiments using sensors resulted in good solutions, able to cope with noisy gene transcription, emerging within just four generations using 25 individuals. Compare this to the performance of GRNs evolved with no sensors in experiment 1.

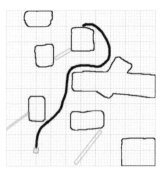

Fig. 10.7. Amigobot sonar. Only sonar sensors are used, which immediately provide sufficient information for the robot to take intricate paths through the obstacle course. This solution was found much more quickly than without sensors

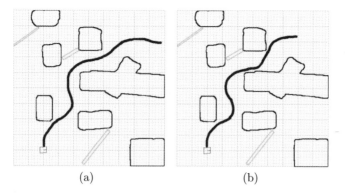

Fig. 10.8. Laser avoidance. Sensory feedback in the form of laser sensors allowed the robot to navigate intricate paths through the obstacle course. These solutions were found much more quickly than without sensors

Experiment 3: General Purpose Obstacle Avoidance (GPOA)

Analysis of the results in experiment two revealed that evolution had managed to cheat: the controllers had learned routes through the world just as in the first experiment. Except this time, solutions were found much sooner since evolution was able to exploit sensory information from the laser and sonars. With this in mind a third objective was set: to evolve a general purpose obstacle avoidance GRN controller.

In this experiment, in order to ensure the controllers were using sensory information and not simply using additional proteins to memorize the route, only sensor information – in the form of protein concentrations – was used in the GRN. In other words, the GRNs could only use sensor proteins (proteins 5 through 12, see Table 10.2) to trigger gene expression. This was achieved by forcing cis-site inputs to genes to be sonar proteins. The total number of cis-sites per gene was increased for this experiment to six. Thus, out of a total of eight sonar proteins only six were used per gene, note the particular sonar proteins used were set by evolution. A total of 13 proteins were used for this set of experiments: four proteins for movement, eight for sonar readings, and one protein with no purpose (evolution can use this protein as it sees fit).

Noise was added at the level of gene activation in the form of probabilistic gene activation. In addition, Gaussian noise was added to all sonar values with a small standard deviation, $\sigma = 0.02$, (note experiments with $\sigma = 0.2$ were also performed successfully) arbitrarily selected. A single fitness evaluation consisted of controlling the robot for twelve seconds using the currently evolved GRN. The parameter settings for this experiment are: seven genes, six cis-sites, 13 proteins, a simulation lifetime of 12 seconds per fitness function trial, and 25 individuals evolving for 25 generations. As Fig. 10.9(a) shows, a new world was created with corridors, open spaces, and obstacles in the form of walls.

Experiment 3: Results

A solution displaying general purpose obstacle avoidance behavior was found in generation five and gradually optimized over the subsequent twenty generations, see Fig. 10.9(b). As the robot moved around the world it tended to display a general anti-clockwise turning motion while avoiding obstacles.

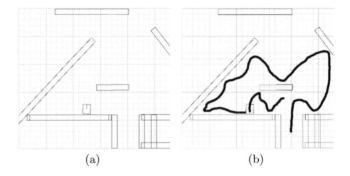

(a)   (b)

Fig. 10.9. General purpose obstacle avoidance controller

This particular solution's approach to negotiating obstacles was as follows: on approaching an obstacle head-on, and fast, the controller caused the robot to reverse. On a slower approach the robot got within a certain distance of the object and either reversed or very slowly bumped into the object, upon which reverse is triggered immediately. An obstacle sensed behind the robot, however, always resulted in immediate forward movement away from the object.

It is worthwhile noting that despite being evolved in a static environment, where each fitness assessment consisted of a single trial, the Amigobot can be moved to different areas of the environment and still maintains general purpose obstacle avoidance behavior. Additionally, informal experiments have shown that evolved GRN controllers are quite robust with respect to the level of noise added, for example, this GPOA controller evolved using sonar noise with a standard deviation $\sigma = 0.02$ (very small), yet as the noise is increased (to a maximum of $\sigma = 0.2$) the controller manages to cope with no adverse effects.

10.5 Discussion

This chapter has demonstrated the application of the developmental metaphor for the dual tasks of generating form and function. Form was explored through the evolution of robust controllers for the construction of 3D, multi-cellular form and structure in a self-organizing manner, akin to the development of a single cell into a complex multicellular system. While function, in the form of reactive robot control, was evolved through genetic regulatory network robot controllers, which replace the function of a neural network.

This chapter has summarized successful experiments to evolve processes of development controlled by genetic regulatory networks. This was achieved through the use of a bespoke software testbed, the Evolutionary Developmental System (EDS), which represents an object-oriented model of biological development.

The successful evolution of genetic regulatory networks that are able to specify the construction of 3D, multi-cellular forms that exhibit phenomena such as cell differentiation and subroutining was shown.

In addition, the successful application of the developmental metaphor to the field of evolutionary robotics was demonstrated by exploring the ability of GRNs to specify reactive robot behaviors. Through the successful evolution of GRN robot controllers that provide general purpose obstacle avoidance, the following was shown:

- the developmental biology metaphor can be productively applied to the problem domain of evolutionary robotics;
- GRNs are a plausible technology for real-time reactive robot control;
- GRNs have internal state and are capable of generating reactive behaviors such as traversing an obstacle laden world; and
- GRN controllers can be evolved for general purpose obstacle avoidance that are robust to small amounts of varying noise at two important levels: rule firing (or gene expression) and sensor noise.

In biology, the developmental and genetic machinery combine to provide a method by which nature can construct, in real-time, robust, complex forms capable of complex functions.

In order to ensure GRNs remain a viable option for the field of evolutionary robotics, further research is required into the ability of GRNs to specify, multiple low-level reactive robot behaviors in a modular manner. The ability of GRNs to construct modular networks while being robust to damage makes GRNs very well suited to the task of generating form and function. The preliminary results shown here offer encouraging potential for future research into developmental-based controllers, and indeed developmental biology inspired approaches to evolutionary design, evolutionary robotics and robotics in general.

Acknowledgments

Many thanks to Peter Bentley, Ken De Jong, Michel Kerszberg, Sean Luke, Paul Wiegand, Lewis Wolpert, and Binita Kumar for helpful advice and criticism. This research was funded by Science Applications International Corporation.

References

1. Bentley, P.: Adaptive fractal gene regulatory networks for robot control. In: J. Miller (ed.) Genetic and Evolutionary Computation Conference Workshop Proceedings – Workshop on Regeneration and Learning in Developmental Systems (2004)
2. Bongard, J.: Evolving modular genetic regulatory networks. In: Proceedings of the 2002 IEEE Congress on Evolutionary Computation (CEC 2002), pp. 1872–1877 (2002)
3. Davidson, E.: Genomic Regulatory Systems: Development and Evolution. Academic Press (2001)
4. Dellaert, F.: Towards a biologically defensible model of development. Master's thesis, Case Western Reserve University (1995)
5. Eggenberger, P.: Evolving morphologies of simulated 3D organisms based on differential gene expression. In: Proceedings of the Fourth European Conference on Artificial Life, pp. 205–213 (1997)
6. Frazer, J.: An Evolutionary Architecture. Architectural Association Publications (1995)
7. Gerkey, B., Vaughan, R., Howard, A.: The player/stage project: tools for multi-robot and distributed sensor systems. In: Proceedings of the 11th International Conference on Advanced Robotics (2003)
8. Jakobi, N.: Harnessing morphogenesis. In: Proceedings of the International Conference on Information Processing in Cells and Tissues (1995)
9. Jiang, T., Wideltz, R., Shen, W., Will, P., Wu, D., Lin, C., Jung, J., Chuong, C.: Integument pattern formation involves genetic and epigenetic controls operated at different levels: feather arrays simulated by a digital hormone model. International Journal on Developmental Biology (2004)
10. Kerszberg, M., Changeux, J.: A simple molecular model of neurulation. BioEssays **20**, 758–770 (1998)
11. Kicinger, R., Arciszewski, T., De Jong, K.: Evolutionary design of steel structures in tall buildings. Journal of Computing in Civil Engineering **19**(3), 223–238 (2005)
12. Kitano, H.: Designing neural networks using genetic algorithms with graph generation system. Complex Systems **4**(4), 461–476 (1990)
13. Kumar, S.: A developmental biology inspired approach to robot control. In: Proceedings of the Artificial Life 9 Conference (2004)
14. Kumar, S.: Investigating computational models of development for the construction of shape and form. Ph.D. thesis, University College London (2004)
15. Kumar, S., Bentley, P.: On Growth, Form and Computers. Elsevier Academic Press (2004)
16. Lewin, B.: Genes VI. Oxford University Press (1999)
17. Marcus, G.: The Birth of the Mind. Basic Books (2004)
18. Mondada, F., Floreano, D.: Evolution of neural control structures: Some experiments on mobile robots. Robotics and Autonomous Systems **16**(2–4), 183–195 (1995)
19. Prusinkiewicz, P., Lindenmeyer, A., Hanan, J.: Developmental models of herbaceous plants for computer imagery purposes. In: Proceedings of the 15th Annual Conference on Computer Graphics and Interactive Techniques (1988)

20. Quick, T., Dautenhahn, K., Nehaniv, C., Roberts, G.: Evolving embodied genetic regulatory network-driven control systems. In: Proceedings of the 7th European Conference on Artificial Life (ECAL 2003) (2003)
21. Schultz, A., Grefenstette, J.: Evolving robot behaviors. In: Proceedings of the 1994 Artificial Life Conference (1994)
22. Shen, W., Salemi, B., Will, P.: Hormone-inspired adaptive communication and distributed control for CONRO self-reconfigurable robots. IEEE Transactions on Robotics and Automation **18**(5) (2002)
23. Taylor, T.: A genetic regulatory network-inspired real-time controller for a group of underwater robots. In: Proceedings of Intelligent Autonomous Systems 8, pp. 403–412 (2004)
24. Wolpert, L., Beddington, R., Brockes, J., Meyerowitz, E.: Principles of Development. Oxford University Press (2002)

11

Modularity in a Computational Model of Embryogeny

Chris P. Bowers

School of Computer Science, University of Birmingham, UK
C.P.Bowers@cs.bham.ac.uk

11.1 Introduction

Natural evolution is about searching for adaptations in order to survive in a constantly changing environment. Of course, in nature these adaptations can occur at many levels from the single gene to entire populations of individuals. Adaptation can consist of rapid successful genetic changes, common in many forms of bacteria, to changes in phenotypic behaviour, such as the ability of humans to develop new skills and technologies.

Nature has evolved techniques, such as multi-cellular organisms and the complex unfolding of genetic information, in order to provide new, and more successful, ways of adapting. One could argue that the evolutionary search process is not the really interesting part of evolution. After all, it is just another form of search comparable in many ways to other search processes. What is really interesting about evolution is the systems and behaviours that have been observed to have built up around it in both natural and simulated evolutionary systems. On the other hand, simulations of evolutionary processes are typically very constrained, in how they adapt, by the representations that are used.

We are becoming increasingly constrained by the approaches we take to engineering and developing new technologies as our technological requirements become more demanding [25]. Nature can be of great inspiration to those scientists and engineers who strive to create a new generation of technologies that overcome the limitations of existing ones. Such approaches are often termed "nature inspired" and the benefits are widely displayed. Biological systems are capable of adaptation, self-organisation, self-repair and levels of complexity against which our current engineering endeavours pale in comparison.

One approach is to try to integrate some of the adaptations that nature has produced into a simulated evolutionary framework. This requires careful, and often extremely difficult, untangling of these adaptations from the biological and physical framework in which they developed. However, striving to achieve these same capabilities through simulated evolution may be an extremely fruitful direction of research.

One such adaptation is embryogeny which embodies the natural process by which the genetic representation within the single initial cell (zygote) controls the development of that cell to form a functioning multi-cellular organism. This process allows an extraordinarily complex organism (phenotype) to be produced from a relatively simple set of genetic instructions (genotype). This can be achieved because the zygote cell divides to produce additional cells that continue to proliferate to eventually produce every cell within the organism. All of these cells originated from the zygote cell and all contain the same genetic information. Therefore, differences in the behaviour of cells within an organism, as a result of this genetic information, must be derived from the state of the cell and its environment. However, the state of the cell and its environment can also depend upon the actions of cells. This forms a complex interaction between the phenotype and genotype and results in an extremely complex mapping between the two representations.

11.2 Why a Computational Model of Embryogeny?

A commonly used argument to justify the simulation of an embryogeny-like process is to point to the issue of scalability in relation to phenotype size [6, 24, 30]. In a typical evolutionary algorithm the representation used to transform the genotype into a phenotype is either a direct (one-to-one) mapping or an indirect (many-to-one) mapping. This may lead to an association of size between the genotype and phenotype. In order for simulated evolution to tackle increasingly complex engineering and research problems, the search space of a direct encoding would also become increasingly large and difficult to navigate. This has been a major influence upon the movement to more indirect encoding approaches [5,6,15,30]. Since modelling embryological processes introduces a complex (indirect) mapping between the genotype and phenotype, this association of size no longer exists. However, it is important to note that this effectively creates two different spaces between which neighbourhood relationships are not necessarily preserved.

A second argument is that nature is very good at maintaining a system that is undergoing perpetual change with cells constantly dying and being replaced. These natural systems demonstrate high levels of robustness and self-repair and work has begun to try to build these characteristics into man-made systems [23, 24].

With the rapid advancement of computational resources, investigation utilising simulation of such processes has become a rapidly growing field. It is a subject that has been widely recognised as being of fundamental importance by researchers in various strands of evolutionary computation and evolutionary biology. However, even with ever improving resources, progress has proven to be slow in coming and significant results rare and difficult to obtain.

A further side effect of a computational model of embryogeny, due to the complex nature and potential redundancy within the mapping, is the existence of many individuals which, whilst genetically diverse, perform identically when evaluated. These individuals are considered to be "neutral" relative to each other. Such neutrality has important implications for computational search processes because it changes the nature of the search space. Local optima become much rarer with most individuals belonging to networks of individuals of equal fitness (neutral networks) which can be vastly distributed across the entire search space.

What happens to a population, which inhabits a neutral network, is largely dependent upon how the evolutionary process has been defined. Typically, the result is a random drifting (neutral drift) of the population until a path to a fitter individual is found and then adaptive evolution recommences. This enables the population to appear converged on local optima at the phenotypic level whilst neutral drift allows for genetic divergence.

During neutral evolution, by definition, there can be no directed selection pressure. However selection is still ongoing, albeit random. Wilke [34] argued that some of these randomly selected parents will be more likely to produce viable offspring than others. The more densely connected the neutral network an individual occupies the more likely that its offspring will be neutral rather than deleterious. This will place a bias on evolution toward more densely connected neutral networks.

However, the effect of neutrality on a search process is still an open question. Miglino et al. [22] argued that neutral changes could be functional in the sense that they enable *pre-adaptations* which subsequently become the basis for an adaptive change. Hogeweg [16] showed, that using a model of morphogenesis, the behaviour of mutated individuals during neutral evolution, which was termed the *mutational shadow* of the neutral path, was noticeably different than during adaptive evolution.

There have been many arguments for how neutrality may result in an improved evolvability. Ebner et al. [13] suggested that the redundancy causing the neutrality also increases the likelihood that evolution can find a smooth and amiable search path and that neutral networks help to prevent early convergence.

Introducing such complex mappings to the genotype-phenotype map is often viewed as an added complication that makes simulating and understanding evolution even more difficult. However, it is important to recognise that this mapping process can effectively be used to direct the evolutionary search path and without it, there may be no possibility of truly understanding evolutionary systems. Wagner and Altenberg [33] used the analogy of a monkey and a verse of Shakespeare to describe the importance of a mapping process. If a monkey randomly presses keys on a typewriter it is far more likely to produce a verse of Shakespeare than if it was given a pencil and paper. This is due to mechanical constraints of the typewriter to produce only viable text characters. In the same fashion, genetic variation can have a constrained effect on the phenotype via the genotype-phenotype map.

The hypothesis upon which this work is based is that an evolutionary search will bias towards individuals which are more densely connected within a neutral network. Since each neutral network will have in common a set of genes which generate the shared fitness of that network, a modularity within the mapping between genotype and phenotype is created around these genes.

11.3 A New Computational Model of Embryogeny

Biological systems are subjected to, and are highly entrenched in, the physics of the environment in which they develop. This makes accurate simulation of biological systems incredibly difficult since many of the underlying features are heavily entangled with processes dependent upon the precise physics of the environment.

The model presented here is an attempt to simulate the dynamics achieved within biological systems between the genetic and phenotypic representations. The model strikes a balance between being both computationally feasible and expressing the growth dynamics observed in real biological systems. The purpose of this model is to provide a tool for investigating the consequences of such a distinction between genotype and phenotype. In this sense, the biological references are, to a large extent, only inspirational.

11.3.1 Cell Position

Models of cell growth often use grid-based or iso-spatial coordinate systems [7, 20, 24]. This makes the simulation of chemical diffusion and choice of data structure much easier. However, it introduces problems for cell division and physical cell interactions due to the positional limitations enforced by the coordinate systems on the cells. For example, cell division becomes awkward if all neighbouring elements of a dividing cell are already occupied. The most common solution is to simply over-write an existing cell if a new cell is to be placed in that location [9, 24].

In our new model, a different approach is taken which removes these limitations and retains the ability to model chemical diffusion whilst allowing more rational physical interactions between cells. Each cell exists in a real valued position with coordinates stored as a two-dimensional vector. This allows a much greater variety of structural forms than models using grid or iso-spatial coordinates. The position of a cell is measured from its geometrical centre and can be placed anywhere within a unit square, $[0, 1]^2$, based upon a standard Cartesian coordinate system.

11.3.2 Physical Body

Cells are structurally discrete with a definitive boundary between the internal state of the cell and the outside environment. The cell membrane is a semipermeable layer, which allows some molecules to pass through whilst preventing others. Transport mechanisms exist which allow the exchange of molecules that cannot permeate through the cell membrane, under certain conditions.

In this model, chemicals from external sources are freely allowed to enter the cell. Restrictions only apply to chemicals that exist within the cell. The rate at which chemicals from an internal source can leave a cell is controlled by a diffusion rate.

The choice of physical cell shape for many previous cell growth models was based on a grid or iso-spatial approach, where the cell is modelled as a polygon such as a square or hexagon. Since this model uses real-valued positions there is no simple constraint in terms of the form or shape of a cell. For simplicity, cells in this work are considered to be circular in shape, requiring only a single variable for their definition, the radius.

11.3.3 Cell State and Differentiation

The cell state is defined by a set of real-valued protein concentrations and diffusion rates. The concentration represents the amount of a given chemical that exists within a cell. The diffusion rate defines the amount of that chemical which is free to diffuse out of the cell and into the environment. In both this model and in natural systems, the cell state is highly dependent upon proteins, since these are almost the only products of the genetic representation stored within the cell. The genes act as protein factories and it is the proteins that they produce which define the state of the cell and thus determines the specific cell behaviour and function.

Cell differentiation is the process by which a cell becomes specialised in form or function. Essentially a cell is considered to have differentiated if its form or function has changed in some manner. Cell differentiation is an important process in enabling multi-cellular organisms to consist of a large variety of highly functionally specialised cells. The ability to redifferentiate is also a key aspect of multi-cellular growth.

This has been shown to be critical to the regenerative capabilities of some creatures such as the newt which can regenerate limbs which have been subject to damage [11], since the cells have enough plasticity to allow them to redifferentiate into stem-like forms which then go on to re-grow the damaged component. This capability is, of course, of vast interest to those aspiring to enhance these biological traits or implement similar capabilities into manmade structures such as electronic hardware [23, 24].

The cell type is a result of the cell being in a specific cell state. Since the model represents the cell state as a set of real-valued chemical concentrations there are no obviously identifiable discrete cell types. In order to identify a cell as belonging to a specific cell type, a definition of that cell type must be given as a set, or range, of values to which the cell state must correspond in order to be identified as a cell of that given type. The simplest way to achieve this is to allocate a single protein concentration to a cell type such that an increase in that protein concentration indicates a closer affinity to the corresponding cell type. There are no externally enforced restrictions on cell types with any cell able to redifferentiate freely.

11.3.4 Division

Cell division consists of, firstly, the duplication of the genetic material held within the cell and then physical partitioning of the cell into two. This whole process is triggered as a consequence of the cell state. In most cases, cell mass is doubled and vital components duplicated before division. In this case, daughter and parent cells can be considered almost identical. There are exceptions, such as cleavage, where division occurs without increase in mass. In some cells, especially those that form in layers, there is a polarity in the internal structure of the cell. In these cases, it is common for the division process to be orientated relative to this polarisation [1].

In this work, division is considered to be a replication of a cell. When a cell divides it creates an almost identical copy of itself such that the only difference between parent and daughter cell is position. The daughter cell is placed alongside the parent cell and a vector stored within the parent cell, the cell spindle, determines the direction of this placement. Each cell stores its own independent spindle that it is free to re-orientate and enables the cell to store directional information.

Cell division in biological systems must be regulated to ensure that growth ceases at some point. This is dependent upon many factors associated with the cell development process including energy constraints and physical forces. This ability for a simulated cell growth process to regulate its own size has been identified as a very difficult problem [24]. In order to try to avoid such a nontrivial issue, without detriment to the computational model of embryogeny, it is appropriate to directly control the regulation of phenotype size by restricting both the cell radius and environment size such that the growth process is terminated once the phenotypic environment is filled with cells. This is easily achieved since the boundaries to the growth environment are a fixed unit square, and the cell radius is defined as the ratio between the size of a cell and this growth environment. Since all the cells have the same radius the hexagonal closest circle packing ratio, $\pi/\sqrt{12}$, can be utilised since it gives the optimal ratio between the area of the circles and the space in which they are packed. Given the most prolific rate of division will result in 2^g cells, where g is the number of iterations of cell division (growth steps), (11.1) approximately relates the cell radius, r, to the number of iterations of cell division required to fill the environment.

$$g \approx \log_2 \left(\frac{\sqrt{3} \left(1 + 2r\right)^2}{2\pi r^2} \right) : g \in \mathbb{N}. \tag{11.1}$$

It can only be an approximation since the curved surface of the cell does not allow perfect tessellation but it does provide a sufficient upper bound on cell numbers. The number of growth steps must clearly be a non-negative integer value.

Although this removes provision for self-regulation of overall phenotypic size there is still the ability to regulate the size of internal structures within the phenotype relative to each other.

11.3.5 Cell Death

Cells can die from direct external influences that cause damage to the cell structure (necrosis), but can also undergo a self-inflicted death for a purposeful means. Programmed cell death refers to the case where it forms an essential part of the multi-cellular growth process. Examples include the separation from a single mass of the fingers and toes in the embryo state and the death of a cell as an immune system response.

In this model, cells are capable of programmed cell death, at which point they are removed from the phenotype environment, such that they could be considered never to have existed. Obviously, in natural systems, the dead cells remain and must be physically dealt with accordingly.

11.3.6 Cell Signalling

One can argue that, in order for cells to form a multi-cellular structure with behaviour consistent with a single organism, there must exist the capability for cells to communicate. Cell signalling provides the ability to communicate and there are two primary methods through which it is conducted.

Direct cell–cell communication describes the case in which a cell directs the form of communication to a specific cell. Since the extra-cellular environment has no means to support this, direct contact between cells is required. Information is then either passed directly through the cell wall to an internal signal receptor or via signal receptors on the surface of the cell.

Indirect cell–cell communication prevents direct control over which cells receive the signal, but allows longer-range communication. Cells secrete cell-signalling molecules that diffuse in the extra-cellular environment. Cells can receive these diffused signalling molecules in the same way as direct cell–cell communication either at the cell surface or within the body of the cell.

Of course, as in all areas of biology, there are exceptions to these cases. For example, neuronal cells are specialised for direct cell–cell interactions over relatively larger distances by using neuritic protrusions from the cells that use a mixture of chemical and electrical signals.

For the purposes of this model, direct cell signalling is considered to be a further adaptation of the cell whilst indirect cell signalling is a fundamental feature of the underlying physics. For this reason only indirect communication, through the diffusion of chemicals, is represented.

11.3.7 Chemical Diffusion

The capability of cells to differentiate based on positional information highlights the capability of multi-cellular organisms to organise their internal shape, form and structure. This process is known as morphogenesis and is highly dependent on the presence of diffusible chemicals, called morphogens. These morphogens control the topological structuring of cells based on chemical concentration patterns.

Diffusion is the spontaneous spreading of a physical property and is usually observed as a lowering in concentration often demonstrated using an ink drop in a glass of water. Typically diffusion is modelled using a partial differential equation which describes the changes in the density, ρ, of some physical property, with diffusion rate c, over time t:

$$\frac{\delta \rho}{\delta t} = c \nabla^2 \rho. \tag{11.2}$$

In the case of cell growth models based on a grid layout, this equation can be adapted to a discrete grid based environment. If each element of the grid has a neighbourhood, N, and a discrete Cartesian density at time t of $\rho_{x,y}(t)$, then:

$$\rho_{x,y}(t) = (1 - c)\rho_{x,y}(t - 1) + \frac{c}{|N|} \sum_{i,j \in N} \rho_{i,j}(t - 1). \tag{11.3}$$

Another approach is to approximate the long term effects of diffusion from a point source using some localised function. Such *instantaneous diffusion* can be calculated for any given point from a number of sources using a set of localised functions. This assumes that the long term behaviour of the diffusion process will be stable with a constant rate of release and absorption of morphogen. Kumar [20] chose to use a Gaussian function centred over the protein source, s, which assumes the morphogen has diffused equally and consistently in all directions:

$$\rho = s e^{-d^2/2c^2} \tag{11.4}$$

where c is the width of the Gaussian and replicates the behaviour of the diffusion rate and s is the height of the Gaussian which corresponds to the strength of the chemical source or *release rate*. The distance, d, is measured between the point at which the diffusion is being calculated, ρ, and the location of the chemical source.

Although it makes vast assumptions about the diffusion characteristics of a chemical, instantaneous diffusion is by far the simplest and quickest approach to take. It removes the need for iterative calculations that other diffusion models require and there is no need to worry about boundary conditions. Also, calculating diffusion using an iterative process requires a decision about the rate at which chemicals diffuse in relation to the rate of cell growth. In this sense, instantaneous modelling of chemical diffusion suggests that cell development is occurring on a much longer time scale than chemical diffusion. So any dynamics between the diffusion process and cell development are lost, as is the ability to model reaction–diffusion systems. What remains is the ability for diffusion to represent global information for cells such as direction, magnitude and distance. This provides all the information that a real diffusive system would provide, given a much faster rate of diffusion than rate of cell growth and where morphogens are considered to be chemically inert to each other. This also means that any interaction between chemicals is strictly through processes within the cell. Therefore, the dynamics of the system are solely a result of cell states such that any emergent behaviour is a result of the embryogeny model alone.

In natural systems there is evidence that the growing organism can construct orthogonal gradients in order to provide more detailed positional information. This ability seems to be particularly important in the early growth of the brain [31]. The model supports diffusion from predefined external sources such as from a boundary edge or source positioned outside of the boundary to support such positional information.

11.3.8 Intercellular Forces

The choice of real valued positioning of cells solves the problem of cell overwriting by removing the constraints of a grid layout, but this also creates the potential for cells to overlap. In real organisms cells cannot overlap, but they can exert forces upon each other.

By modelling cells as particles, which exert forces based upon a spring model, overlap can be reduced since overlapping cells result in spring compression that exerts an opposing force to increase the distance between the cells. The trajectory of these particles due to these forces can be modelled using Verlet integration [32]:

$$x_{(t+\Delta t)} = 2x_{(t)} - x_{(t-\Delta t)} + a_{(t)}\Delta t^2 \tag{11.5}$$

where $x_{(t)}$ is the position at time t, Δt is the time step and a the acceleration vector resulting from the application of forces upon the cell. The great advantage of the Verlet integration method is its improved stability and efficiency when compared with other methods for modelling the trajectories of particles [19].

One situation, which can easily occur, is the constant division of cells with division spindles in the same direction, resulting in a one-dimensional chain of cells. Since the forces applied by the Verlet algorithm would all be acting in the same direction there is no possibility for cells to move out of the direction of the chain. To overcome this problem, a small amount of random noise is applied to the direction of division which enables the Verlet algorithm to apply forces in directions slightly out of line with the other cells in order to push the cell out of this chain. Generating random numbers with a random seed derived from the cell state enables the deterministic nature of the process to be retained.

11.3.9 Genes and the Genetic Regulatory Network

Sequencing the genes from an organism alone will tell you very little about how that organism functions and how it came to be. In order to achieve a more functional understanding, the nature of the interaction between genes is required [26]. Genetic Regulatory Networks (GRNs) are a fundamental feature of natural developmental processes, and we are just beginning to unravel how these networks of genes interact with evolution. In essence, a genetic regulatory network is a complex protein factory. It produces proteins that can interact with and alter the surrounding environment, but it is also dependent upon proteins for its function.

Enabling the representation of GRNs requires that genes can support regulatory dependencies upon each other. In this work, genes are modelled using the concept of an Operon which describes how genes can interact and control the expression of functional groupings of genes. Since the only function of a gene, if expressed, is to specify some given protein, and the transcription and thus expression of these genes is dependent upon proteins, then proteins are how the genetic regulatory network must interact within itself and the external environment. In this model two methods of gene regulation are supported.

The first is through the chain of expression created by the genes. This chain allows a string of genes to regulate each other in sequence until a gene that ignores the expression from the previous gene breaks the sequence. This form of regulation is dependent upon the gene's position in the genotype.

The second is through protein regulation, whereby one gene controls the characteristics of a protein that affects the performance of another gene. With this form of regulation there is less dependence upon position of the genes within the genotype.

Each gene consists of the following elements:

- An expression strength which is obtained from the previous gene in the genome and unless altered by the gene is passed on to the next gene in the genome.
- A protein dependency which can be used by the gene for evaluative or functional purposes.
- An output which may either be in the form of some functional operation on the cell or to control the expression of the next gene in the genome by altering its expression strength.

Table 11.1. Genes and their functions

Gene ID	Dependent	Function	Express
		Terminal	
0	-	-	$e = 0.0$
1	-	-	$e = 1.0$
		Expressive	
2	-	-	$\rho(p)$
3	-	-	$1.0 - \rho(p)$
4	-	-	$r(p)$
5	-	-	$1.0 - r(p)$
		Evaluative	
6	if $e > \rho(p)$	-	$e = 1.0$
7	if $e < \rho(p)$	-	$e = 1.0$
8	if $e > \rho(p)$	-	$e = 0.0$
9	if $e < \rho(p)$	-	$e = 0.0$
		Functional	
10	-	$r(p) = e$	-
11	-	$r(p) = 1.0 - e$	-
12	if $e = 1.0$	spindle towards protein	-
13	if $e = 1.0$	spindle opposite protein	-
14	if $e = 1.0$	divide	-
15	if $e = 1.0$	die	-

From this basic set of features, several classes of gene type can be defined.

- *Terminal* genes have no dependents or function. They simply override the expression chain to ensure that the next gene is either expressed or not.
- *Expressive* genes determine the expression of the next gene based on the concentration of the protein dependency.
- *Evaluative* genes are similar to expressive genes except that they are not only dependent upon the protein dependency concentration but also on the expression strength through some form of evaluation.
- *Functional* genes are dependent upon expression strength. If they are expressed then they perform some function utilising a protein dependency. These functions can directly alter the cell state or cause the cell to divide or die.

Table 11.1 lists 16 gene types used in the model, each having a unique identifier or *gene ID*. For each gene, the manner of the dependency on the dependency protein and any effects on the cell state, must be defined. In addition, any ability to control the current strength of the expression chain, e, must also be defined.

The protein dependency, p, can be used to determine the diffused concentration of the dependency protein, $\rho(p)$ or the release rate of the dependency protein from with the cell, $r(p)$.

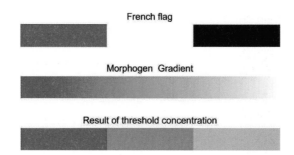

Fig. 11.1. The French flag model of spatial differentiation

A genome consists of a finite list of genes, 20 in this case. Each gene is represented by two integer values, the first value represents the *gene ID* and the second identifies the genes *protein dependency, p.* Each gene in the genome can be of any one of the 16 gene types with any combination of protein.

11.4 The Pattern Formation Problem

One key advance in understanding the processes of differentiation and morphogenesis was to highlight its relationship with position. Wolpert [35] noted that cells can use positional information in order to determine their cell type. He used the analogy of a French flag to explain how the diffusion of a chemical from one side of the flag could determine in what section of the French flag a cell resided using a threshold as outlined in Fig. 11.1.

The use of Wolpert's French flag analogy has become widespread when discussing the self-organisation of cells into some form of structure. For this reason many previous attempts to model cell growth within a computational system have often used the French flag as a target structure for self-organising pattern formation for both reasons of continuity and appropriateness [8,9,24].

The simulated evolution of a complex mapping such as that introduced by a computational model of embryogeny faces a number of issues. Firstly, neighbourhood structure is not preserved which may result in an extremely noisy and discontinuous search space. Secondly, such a complex mapping process has the potential for much redundancy which means that neutrality is prevalent. It is important to ensure that the population is still able to move through these neutral areas when a selection pressure is non-existent.

Bearing in mind these issues, the following evolutionary algorithm is defined: selection is performed by choosing the best 50% of the population. Each of these selected individuals is subjected to a random two point crossover operation with another selected individual. This is performed by replacing all the values between the two randomly selected points from the first selected parent with the values between those same two points from the second. The product of this crossover is then subjected to a further random single point

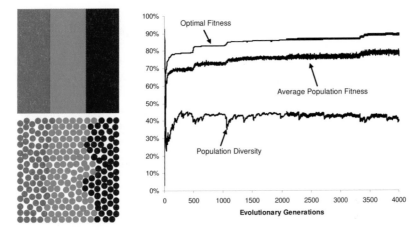

Fig. 11.2. Averaged evolutionary behaviour for the French flag pattern problem

mutation. Mutation is conducted by replacing a value, chosen at random, with a randomly generated value.

The resulting offspring then replace the worst 50% of the population. This ensures a consistent generational selection pressure during adaptive evolution since the population is effectively fitness ranked. The selection method also supports a sufficient level of random drift during neutral evolution by ensuring that at least 50% of the population is replaced with a set of newly generated offspring.

Figure 11.2 shows a typical individual, and the template against which it was evaluated. The averaged behaviour of this evolutionary approach is also shown in the same figure with the average population fitness following just below the current best fitness suggesting that the population is fairly converged in terms of fitness. Fitness is measured as the percentage of cells which are identified as being in the correct cell state given their position. The genetic diversity is also plotted and is measured between two genotypes as the percentage of genes which differ. This is averaged over every every possible combination of individuals in the population. Genetic diversity, although noisy, is generally maintained during evolution.

11.5 Modularity in a Computational Model of Embryogeny

11.5.1 Definitions of Modularity

The concept of "modularity" or "module" is often used in a very fluid way. In software engineering, modularity is used to refer to software components which are developed or used in a largely independent manner whilst interacting to produce some overall function such as object orientated design [27] and functional programming [17]. Cognitive scientists often refer to the modularity of mind [14], arguing that the mind is composed of a set of independent, domain specific, processing modules, whilst evolutionary psychologists believe this modularity occurred due to selection pressures at transitional points in the evolution of the mind. The hardware production industry uses modularity to describe how sets of product characteristics are bundled together. For example, the automotive industry is considered to be becoming more modular by enabling common automotive components to be easily transferred between various models of vehicles in a bid to reduce production and maintenance costs [28].

Such ubiquitous and often vague usage of the term often makes its use very easy to misinterpret and difficult to defend. For this reason it is often advantageous to refer to the more fundamental definition of a module as a set of sub-components of a system which are greatly inter-dependent whilst remaining largely independent of the rest of the system.

11.5.2 Modularity in Evolution and Development

Modularity is an important aspect of natural evolution. It is clear from the physiology of animals that they are inherently modular. This is perhaps most obviously apparent in the symmetry of appendages such as legs and arms and in the organs of the body which have very specific but necessary functions, largely independent of each other, such as the heart, lungs and liver. It is a widely held belief amongst evolutionary biologists that this apparent modular architecture of animals is generated at every hierarchy of the developmental process right down to the interactions between genes or even molecules [29].

Modularity in evolution occurs in many forms and one of the greatest challenges is to clearly identify prospective modularity from within such incredibly complex systems. This is becoming increasingly important as we gather a greater understanding of biological systems. Much of the discussion raised by evolutionary biologists on this topic is highly relevant to the work presented here since it consists of observing modularity in various structures and determining how these modules interact and influence each other [3, 4, 29]. Also, in the same fashion, insights into aspects of modularity in cognitive and neurological systems can be as equally useful [12, 18, 21].

The importance of modularity can be no better highlighted than by its numerous and extensive practical benefits. Modularity can reduce disruptive interference, provide a mechanism for robustness and enable complex systems to be built from simple subsystems. These are just a few of the arguments for why a modular approach is useful.

11.5.3 Modularity in the Genotype-Phenotype Mapping

Measuring modularity in the genotype and phenotype independently makes little sense in a system where the two entities are so highly inter-dependent. Here, the occurrence of modularity within the mapping process between genotype and phenotype is investigated.

The term "pleiotropy" is used to refer to the situation when a single gene influences multiple phenotypic traits. This has important ramifications for the mapping between genotype and phenotype. Altenberg [4] considers modularity to exist when the genotype–phenotype map can be decomposed into the product of independent genotype–phenotype maps of smaller dimension. He adds:

> The extreme example of modularity would be the idealized model of a genome in which each locus maps to one phenotypic trait. For the converse, the extreme example of non-modularity would be a genotype–phenotype map with uniform pleiotropy in which every gene has an effect on every phenotypic variable. Real organisms, one could argue, have genotype–phenotype maps that range somewhere in between these extremes.

If such an association were to exist, then identifying definitive relationships between elements of the genotype with certain elements of the phenotype would allow the identification of such modules and indicate that modularity exists.

Identifying modules is not always easy. For the French flag pattern there are three obvious choices, the three differently coloured stripes, and previous work has shown such modularity to be evident [10]. However, this form of modularity measure is limited since it would discount other potential sources of modularity, within a solution, which are not intuitively obvious. A better way to identify the existence of modularity in the embryogeny mapping would be to measure the alignment of structure in the genotype with structure in the phenotype without needing to define what those structures might be. This requires representations of structure from the genotype and phenotype which are comparable. There is already a clearly defined topology for the genotype in the form of the genetic regulatory network. A similar network of gene–gene relationships can be derived from the phenotype.

For each cell within a phenotype, the dependency upon the internal cell state for each gene in the genome can be measured. To do this, each gene in turn is ignored when mapping the genotype to the phenotype. Differences for each cell are measured against the original phenotype. All cells that have an altered state due to ignoring a gene have a shared dependency upon that gene. Figure 11.3 shows such dependencies for each of the 20 genes in a sample genome where cells dependent upon a gene are marked with a cross. In this example case, a large number of the genes have absolutely no impact on the phenotype when they are removed, evidenced by the lack of crosses in the

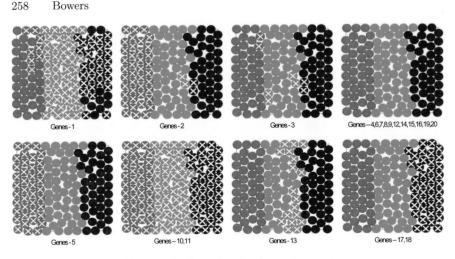

Fig. 11.3. Phenotypic dependencies for each gene in a genome

phenotype. However, a number of the genes are clearly very specific in the cells that they affect.

The relationship between genes can be determined through their interactions with the phenotype. This can be achieved by measuring the similarity between the phenotypic dependencies of any given pair of genes. If this is done for every combination of genes in the genotype, then a network based upon these gene–gene relationships can be formed where nodes represent the genes and edges represent the similarity of the respective gene's impact upon the phenotype. The similarity between the phenotypic dependencies of two genes is measured as the percentage of the total cells which are affected by both genes.

The alignment between genotypic and phenotypic structures can be determined by comparing the genetic network derived from phenotypic interactions to the network defined by the genetic structure itself, the GRN. For each gene there is a corresponding node in each of the two comparative networks, and each of these nodes has a set of edges. Since these edges are weighted then the similarity between a given node in the two comparative networks can be measured by using the Euclidean distance between the set of edges belonging to those nodes. A smaller Euclidean distance represents a greater similarity. This is then averaged over each node to form a measure of similarity for the entire network.

Figure 11.4 shows the result of measuring the similarity between the GRN and the genetic network derived from the phenotype. This is measured over evolutionary time for the best individual in the population. The results are averaged over 10 evolutionary runs and shown with the standard deviation. There is a clear progressive alignment between these two networks during evolution.

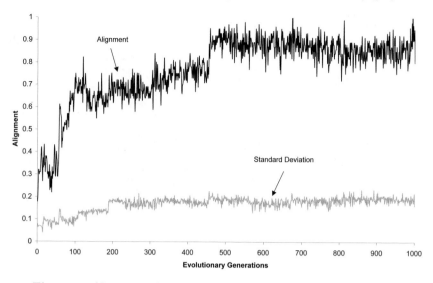

Fig. 11.4. Alignment of structures derived from genotype and phenotype

Throughout this work, results have been obtained using a specified evolutionary algorithm. However, if this trait of modularity is truly dependent upon the interaction between simulated evolution and a computational model of embryogeny then there should be a discernible difference in the measure of module alignment between the individuals obtained through simulated evolution and other search techniques. Here, the module alignment is compared for simulated evolution, a hill climbing local search and random search.

For random search, each iteration of the search algorithm consists of the creation, at random, of a new population of individuals. The best individual found so far is retained.

A local search is conducted using a hill-climbing algorithm. From the current point of search, a subset of the neighbourhood is generated using single point mutations from the current position. If any neighbouring individual in the generated sample has an improved or equal fitness to the current point of search it becomes the new point of search.

To ensure the comparisons between the three search algorithms are as fair as possible, the number of evaluations required at each generation is consistent across each search algorithm. However, simply comparing this measure of alignment over the number of iterations of the search process is not meaningful since, especially in the case of random search, the performance of these algorithms can vary greatly. A means of comparison is required that is less dependent upon the performance of the search process over time. The time efficiency of the search algorithms can essentially be ignored by plotting the measure of alignment directly against the fitness of individuals found by the various search processes. Figure 11.5 shows the alignment, between genotypic

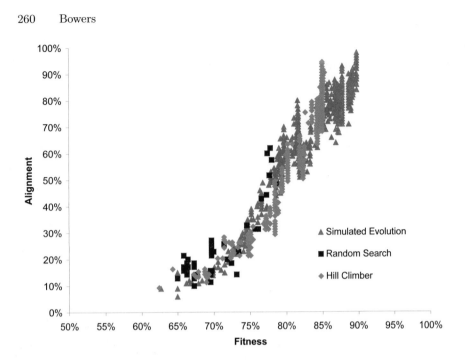

Fig. 11.5. Alignment of genotypic and phenotypic structures against their relative fitness

and phenotypic structures, plotted against the fitness of the individual for all three forms of search.

Although the results for random search do not reach the same levels of fitness as those of local and evolutionary search, the results suggest that all the search processes investigated here result in an increasing alignment between genotypic and phenotypic structures with increasing fitness and that they follow a very similar trend. Therefore, each of the search algorithms is capable of producing individuals with modular alignment and that this level of alignment is dependent upon the fitness of an individual. Since a random search is sampling uniformly across the entire search space, thus is not dependent upon a search trajectory through the search space, it can be established that the modular behaviour must be independent of the search process. This would suggest that an inherent property of the embryogeny mapping process is that fitter individuals will have a greater inclination to modular structures.

11.6 Conclusion

Results of this work suggest that the observed modular behaviour is an inherent property of the computational model of embryogeny and, as such, is likely to be obtained from any form of search algorithm. This does not necessarily discredit an evolutionary search as a good choice of search process. In comparison to local and random search, simulated evolution is certainly the more efficient optimiser. This could be simply due to the noisy nature of the search landscape, and the presence of neutrality, to which an evolutionary approach is the better suited of the three search algorithms.

However, there are some interesting observations that can be made about the nature of the alignment between genotypic and phenotypic structures and what this means for a search process. Altenberg's hypothesis of Constructional Selection [2,3] states:

> Selection during the origin of genes provides a filter on the construction of the genotype-phenotype map that naturally produces evolvability.

Altenberg suggests that a search process that incrementally changes the degrees of freedom of a genetic representation will do so to more closely match the degrees of freedom of the phenotypic domain. This is because such individuals are more evolvable, since a change in a variable in such a genotype is more likely to result in some measurable response, in the associated variables, in the phenotype.

Another way of viewing this argument is to say that an evolutionary search process will bias towards individuals that, once mapped from genotypic space to phenotypic space, will form a relatively smooth search landscape that is easier to traverse. The increasing alignment between genotypic and phenotypic structure shown in the results of Fig. 11.5 are symptomatic of such an alignment of the degrees of freedom.

Therefore, one could argue that an algorithm which utilises iterative adaptations to existing solutions would be more suited to navigating the search space embodied by the model of computational embryogeny introduced in this work.

References

1. Alberts, B., Bray, D., Lewis, J., Ra., M., Roberts, K., Watson, J.: Molecular Biology of the Cell, 3rd edn. Garland Publishing (1994)
2. Altenberg, L.: The evolution of evolvability in genetic programming. In: J. Kinnear (ed.) Advances in Genetic Programming, pp. 47–74. MIT Press (1994)
3. Altenberg, L.: Genome growth and the evolution of the genotype-phenotype map. In: W. Banzhaf, F. Eeckman (eds.) Evolution and Biocomputation: Computational Models of Evolution, pp. 205–259. Springer (1995)
4. Altenberg, L.: Modularity in evolution: Some low-level questions. In: W. Callebaut, D. Rasskin-Gutman (eds.) Modularity: Understanding the Development and Evolution of Complex Natural Systems, pp. 99–131. MIT Press (2004)
5. Banzhaf, W., Miller, J.: The challenge of complexity. In: A. Menon (ed.) Frontiers of Evolutionary Computation, pp. 243–260. Kluwer Academic (2004)

6. Bentley, P., Gordon, T., Kim, J., Kumar, S.: New trends in evolutionary computation. In: Proceedings of the 2001 Congress on Evolutionary Computation, pp. 162–169. IEEE Press (2001)

7. Bentley, P., Kumar, S.: Three ways to grow designs: a comparison of embryogenies for an evolutionary design problem. In: Proceedings of the 1999 Genetic and Evolutionary Computation Conference, pp. 35–43. Morgan Kaufmann (1999)

8. Beurier, G., Michel, F., Ferber, J.: Towards an evolution model of multiagent organisms. In: Multi-Agents for Modeling Complex Systems (MA4CS'05): Satellite Workshop of the European Conference on Complex Systems (ECCS'05) (2005)

9. Bowers, C.: Evolving robust solutions with a computational embryogeny. In: J. Rossiter, T. Martin (eds.) Proceedings of the UK Workshop on Computational Intelligence (UKCI-03), pp. 181–188 (2003)

10. Bowers, C.: Formation of modules in a computational model of embryogeny. In: Proceeding of 2005 Congress on Evolutionary Computation, pp. 537–542. IEEE Press (2005)

11. Brockes, J., Kumar, A.: Plasticity and reprogramming of differentiated cells in amphibian regeneration. Nature Reviews Molecular Cell Biology **3**, 566–574 (2002)

12. Bullinaria, J.: To modularize or not to modularize. In: J. Bullinaria (ed.) Proceedings of the 2002 UK Workshop on Computational Intelligence (UKCI-02), pp. 3–10 (2002)

13. Ebner, M., Shackleton, M., Shipman, R.: How neutral networks influence evolvability. Complexity **7**(2), 19–33 (2001)

14. Fodor, J.: Modularity of Mind. MIT Press (1983)

15. Gordon, T., Bentley, P.: Development brings scalability to hardware evolution. In: Proceedings of the 2005 NASA/DoD Conference on Evolvable Hardware, pp. 272–279. IEEE Computer Society (2005)

16. Hogeweg, P.: Shapes in the shadows: Evolutionary dynamics of morphogenesis. Artificial Life **6**, 85–101 (2000)

17. Hughes, J.: Why functional programming matters. The Computer Journal **32**(2), 98–107 (1989)

18. Jacobs, R.: Computational studies of the development of functionally specialized neural modules. Trends in Cognitive Sciences **3**(1), 31–38 (1999)

19. Jakobsen, T.: Advanced character physics. In: Proceedings of the 2001 Game Developer's Conference (2001)

20. Kumar, S.: Investigating computational models of development for the construction of shape and form. Ph.D. thesis, Department of Computer Science, University College London, UK (2004)

21. Lam, C., Shin, F.: Formation and dynamics of modules in a dual-tasking multilayer feed-forward neural network. Physical Review E **58**(3), 3673–3677 (1998)

22. Miglino, O., Nolfi, S., Parisi, D.: Discontinuity in evolution: How different levels of organization imply pre-adaption. In: R. Belew, M. Mitchell (eds.) Adaptive Individuals in Evolving Populations: Models and Algorithms, pp. 399–415. Addison-Wesley (1996)

23. Miller, J.: Evolving developmental programs for adaptation, morphogenesis, and self-repair. Lecture Notes in Artificial Intelligence **2801**, 256–265 (2003)

24. Miller, J.: Evolving a self-repairing, self-regulating, French flag organism. In: Proceedings of the Genetic and Evolutionary Computation Conference (GECCO 2004), pp. 129–139. Springer (2004)

25. Miller, J., Thomson, P.: Beyond the complexity ceiling: evolution, emergence and regeneration. In: Proceedings of the Workshop on Regeneration and Learning in Developmental Systems – Genetic and Evolutionary Computation Conference (GECCO 2004) (2004)
26. Monk, N.: Unravelling nature's networks. Biochemical Society Transactions **31**, 1457–1461 (2003)
27. Pressman, R.: Software Engineering: A Practitioner's Approach, 5th edn. McGraw-Hill (2000)
28. Ro, Y., Liker, J., Fixson, S.: Modularity as a strategy for supply chain coordination: The case of U.S. auto. IEEE Transactions on Engineering Management **54**(1), 172–189 (2007)
29. Schlosser, G., Wagner, G.: Modularity in Development and Evolution. University of Chicago Press (2004)
30. Stanley, K., Miikkulainen, R.: A taxonomy for artificial embryogeny. Artificial Life (2003)
31. Trisler, D.: Cell recognition and pattern formation in the developing nervous system. Journal of Experimental Biology **153**(1) (1990)
32. Verlet, L.: Computer experiments on classical fluids I – thermodynamical properties of Lennard-Jones molecules. Physical Review **159**(1) (1967)
33. Wagner, G., Altenberg, L.: Complex adaptations and the evolution of evolvability. Evolution **50**(3), 967–976 (1996)
34. Wilke, C.: Adaptive evolution on neutral networks. Bulletin of Mathematical Biology **63**(4), 715–730 (2001)
35. Wolpert, L.: Positional information and the spatial pattern of cellular differentiation. Theoretical Biology **25**(1), 1–47 (1969)

Part IV

Engineering

Evolutionary Design in Engineering

Kalyanmoy Deb

Department of Mechanical Engineering, Indian Institute of Technology Kanpur, PIN 208016, India deb@iitk.ac.in

Engineering design is an age-old yet important topic, taught in most engineering schools around the world and practiced in all engineering disciplines. In most scenarios of engineering design, depending on whether there is a need for performing the particular design task, a number of well laid-out steps are followed: (i) *conceptual* design in which one or more concepts are formulated and evaluated, (ii) *embodiment* design in which more concrete shapes and material of components are decided, and (iii) *detail* design in which all dimensions, parameters, tolerancing, etc. of the components are made usually with an optimization framework and with a view to manufacturing aspects of the design. These steps are usually repeated a few times before finalizing the design and making a blueprint for manufacturing the product. Although optimization is an integral part of the engineering design process at every step mentioned above, often such a task is overshadowed by the experience of designers and the race for achieving a record turn-over design time.

Evolutionary design, as we describe in this book, is a task which can be used in any of the above design tasks to come up with optimal or innovative design solutions by using evolutionary computing principles. Evolutionary computing algorithms provide a flexible platform for coming up with new and innovative solutions. In the conceptual design stage, driven by interactive or a fuzzy evaluation scheme, different concepts can be evolved. In the embodiment and detail design stages, a finer optimization problem can be formulated and solved using an evolutionary optimization (EO) procedure.

There are, however, a number of reasons for the lukewarm interest in using optimization procedures routinely in an engineering design task:

- The objective functions and constraints associated with the optimization problem in engineering design tasks are usually non-linear, non-convex, discrete, and non-differentiable, thereby making it difficult to use gradient methods and to use optimality theorems which are developed mainly for well-behaved problems.

- There can be more than one optimal solution where an algorithm can get stuck, thereby failing to find a global optimal solution. In multi-modal problems, more than one global optimal solution exists, thereby requiring a global-optimization method to be applied many times, so as to get an idea of different global optimal solutions.
- Objective functions can be many and conflicting with each other, resulting in multiple trade-off optimal solutions. Again, a repeated application of a classical method is needed in such multi-objective optimization problems.
- Objective functions and constraints can be noisy and involve manufacturing or modeling errors, thereby causing the classical methods to get caught in the details and not find desired optimal solutions.
- Objective functions and constraints can be computationally daunting to evaluate, thereby causing an inefficient search due to the serial algorithmic nature of classical methods.
- The decision variables are usually mixed: some of them can be real-valued, some others can be discrete (integer, for example), yet some others can take one or a handful of options (such as the choice of a material), and some others can be a permutation or a code representing some chain of events et cetera. The solution can be a graph or a tree representing the architecture of a system built in a hierarchical manner from sub-systems et cetera. Classical methods need to have fix-ups to handle such mixed-variable problems.
- The number of decision variables, objectives and constraints may be large, causing "curse of dimensionality" problems.
- Some parameters and decision variables may not be considered as a deterministic quantity. Due to manufacturing uncertainties and the use of finite-precision manufacturing processes, these parameters must ideally be considered as stochastic parameters with a known distribution.
- Knowing the uncertainties involved in parameters, decision variables and the objective/constraint evaluation, an upper limit on the proportion of unacceptable products may be imposed, leading to reliability-based design optimization.

Classical optimization methods were developed mainly by following theoretical optimality conditions and are often too idealized to be applied to such vagaries of engineering design optimization tasks.

However, for the past two decades or so, evolutionary optimization (EO) procedures are becoming increasingly popular in engineering design tasks due to their flexibility, self-adaptive properties, and parallel search abilities. But all these advantages do not come free; the user of an EO must use appropriate and customized EO operators to the problem at hand, which requires a good understanding of how EO algorithms work, a thorough understanding of optimization basics, and good knowledge about the problem being solved. In this section on Engineering Design, we have three chapters which bring out different aspects of EO applications pertinent to engineering design.

Hu et al.'s chapter discusses a genetic programming based EO approach with a bond graph based strategy to evolve new designs in three different engineering applications. In some cases, human-competitive designs are found (designs better than what a well-skilled human would develop). Since the suggested methodology can, in practice, evolve different topologies, the methodology is appropriate for a conceptual design task.

Deb's chapter discusses a number of matters related to engineering optimization: (i) single versus multiple objective optimization, (ii) static versus dynamic optimization, and (iii) deterministic versus stochastic optimization through uncertainties in decision variables. On a hydro-thermal power dispatch problem involving time-varying quantities, uncertainties in power generation units, non-differentiable objectives, and non-linear objectives and constraints, the suggested EO based strategies find theoretically correct Pareto-optimal solutions and suggest ways to choose a single optimal solution for implementation in a dynamically changed optimization problem.

Finally, Ling et al.'s chapter suggests an EO based multi-modal optimization procedure which is capable of finding multiple optimal solutions in a single simulation run. On a holographic grating optimization problem, the proposed EO has found four different optical configurations. The authors also discuss a way to choose a particular solution from the four obtained solutions by keeping in mind the practicalities of implementing the solutions.

The examples portrayed in this book on engineering design are just three application studies taken from a plethora of different engineering problems which are being routinely solved using EO methodologies. The number of such applications and their areas are so vast that it would be a foolish act to even make a list of them. It is our humble understanding that the case studies discussed here will spur the reader on to become interested in the EO literature (conference proceedings, journals, books, etc.). However, before closing these introductory remarks, the author would like to present his five-step view on a systematic application of an EO methodology to a real-world engineering design optimization problem:

1. **Prepare the Case:** Before applying an EO methodology, the problem must be attempted using one or more standard classical methodologies. It may be found that the classical method cannot solve the problem at hand, or that it takes enormous computational time, or that it finds a solution which is not acceptable, or that it does not even find any solution, or that the solution is sensitive to the supplied initial solution, or others. Such a study will build up a case in favor of using an EO methodology for solving the problem at hand.

2. **Develop an EO Methodology:** After studying the problem thoroughly, an EO methodology appropriate to the problem can be developed. This is where some experience with the EO literature and the EO simulation studies are needed. The choice of an appropriate EO can be motivated by the nature of decision variables, objective functions and constraints

the problem has. For example, if all variables are real-valued, it is better to choose a real-parameter EO, such as an evolution strategy, differential evolution or a real-parameter genetic algorithm with real-valued recombination and mutation operators. Some other important issues in developing a methodology are as follows:

Handling Constraints: In the presence of constraints, an important step is to choose an appropriate constraint-handling strategy. It would not be desirable to use penalty-based methodologies in the first instance. If constraints can be handled by using a repair mechanism, this would be an ideal choice. For example, a linear or quadratic equality constraint can be handled by simply finding the root(s) of the equation and replacing the variable values with the root(s).

Customization: While designing the EO algorithm, as much customization as possible must be utilized in creating the initial population and developing EO operators. In most problems, it is possible to develop seed solutions in terms of building blocks or partial good sub-solutions. Sometimes, the creation of an initial population satisfying one or more constraints is easier to achieve. In many instances, the knowledge of one or more existing solutions is already available and these solutions can be used as seed solutions to create an initial EO population by perturbing (or mutating) parts of these seed solutions. Recombination and mutation operators can also be customized to suit the creation of feasible or good solutions from feasible parent solutions. One generic strategy would be to consider recombination as a primary creation operator of new solutions by combining better aspects of a number of parent population members and the mutation to repair these solutions to make them as feasible as possible.

Hybridization: It is always a good idea to use a hybrid strategy involving EO and a local search method. Although there are several ways to couple the two approaches, a common approach is to start the search with an EO and then finish with a local search method.

3. **Perform a Parametric Study:** It is also imperative to perform a parametric study of EO parameters by clearly demonstrating the range of parameter values where the resulting algorithm would work well. The use of design of experiments concepts or other statistical methodologies would be useful here. To reduce the efforts of a parametric study, as few parameters as possible must be used. Some of the unavoidable parameters in an EO procedure are population size, crossover probability, mutation probability, tournament size, and niching parameters. Any other parameters, if they need to be included, should be as few as possible and may well be fixed to some previously suggested values.

In no case should the outcome of the developed EO procedure depend on the random number generator or its seed value. If this happens even after the parametric study, a major change in the representation scheme and/or operators is called for.

4. **Apply to Obtain Optimal Solutions:** After a suitable EO (hybrid or autonomous) is developed through parametric studies, suitable GUI-based software can be developed to find optimal solution(s). If the code is to be supplied to a user, the recommended EO parameter values must be provided.

5. **Verify and Perform Post-optimality Analysis:** After a set of solutions has been found, the onus still remains on the part of the developer/user to analyze the solutions to explain why the obtained solutions are good solutions for the problem. This step is necessary to establish that the obtained solutions from the developed EO methodology are not a fluke. Thus, this may require multiple applications of the developed EO procedure using different (biased or unbiased, whichever is used in the procedure) initial populations. This may require a verification of obtained solutions by means of other optimization algorithms or through the solution of a variant of the original optimization problem leading to a sensitivity analysis. In case of any discrepancies between the results of these different studies, further investigations must be made by repeating the above steps till a satisfactory explanation of the results are obtained. This is a step often by-passed in many optimization studies. Often, an analysis of obtained solutions may bring out important insights about properties which make a solution optimal. Such information may also provide important clues about creating a better initial population, recombination, and mutation operators so as to launch another, yet a more efficient, EO application.

The suggestions made above not only allow a systematic development and application of an EO procedure, but also generate important problem knowledge which is valuable in understanding the problem and the trade-offs among various problem parameters which constitute the region of optimality. None of these tasks is easy, and by no means universal to all problems. Every design optimization problem in practice is different in terms of different complexities associated with variables, objectives, and constraints. Systematic applications of the above procedures may one day unveil salient knowledge about 'What procedure is apt to which problems?' – an ever-pertinent question asked most often in the field of optimization.

Kalyanmoy Deb
Engineering Area Leader

Engineering Optimization Using Evolutionary Algorithms: A Case Study on Hydro-thermal Power Scheduling

Kalyanmoy Deb[*]

Kanpur Genetic Algorithms Laboratory (KanGAL), Indian Institute of Technology Kanpur, PIN 208016, India deb@iitk.ac.in

12.1 Introduction

Many engineering design and developmental activities finally resort to an optimization task which must be solved to get an efficient solution. These optimization problems involve a variety of complexities:

- Objectives and constraints can be non-linear, non-differentiable and discrete.
- Objectives and constraints can be non-stationary.
- Objectives and constraints can be sensitive to parameter uncertainties near the optimum.
- The number of objectives and constraints can be large.
- Objectives and constraints can be expensive to compute.
- Decision or design variables can be of mixed type involving continuous, discrete, Boolean, and permutations.

Although classical optimization algorithms have been around for more than five decades, they face difficulties in handling most of the above problems alone. When faced with a particular difficulty, an existing algorithm is modified to suit it to apply to the problem. One of the main reasons for this difficulty is that most of the classical methodologies use a point-by-point approach and are derivative-based. With one point to search a complex search space, algorithms seem to be inflexible in improving from a prematurely-stuck solution.

Evolutionary algorithms (EAs), suggested around the 1960s and applied to engineering design optimization problems from around the 1980s, are population based procedures which are increasingly being found to be more suited to different vagaries of practical problems. In this chapter, we consider a

---

[*] Currently occupying the Finnish Distinguished Professor position at Helsinki School of Economics, Finland (Kalyanmoy.Deb@hse.fi).

case study of the hydro-thermal power dispatch problem and illustrate how a population-based EA can handle different complexities and produce useful solutions for practice. Particularly, we address the first three complexities mentioned above in this chapter.

In the remainder of the chapter, we first introduce the hydro-thermal power scheduling problem and through this problem discuss various complexities which can arise in solving practical optimization problems.

12.2 Hydro-thermal Power Scheduling Problem

In hydro-thermal power generation systems, both hydroelectric and thermal generating units are utilized to meet the total power demand. A proper scheduling of the power units is an important task in a power system design. The optimum power scheduling problem involves the allocation of power to all concerned units, so that the total fuel cost of thermal generation is minimized, while satisfying all constraints in the hydraulic and power system networks [19]. To solve such a single-objective hydro-thermal scheduling problem, many different conventional methods such as Newton's method [21], the Lagrange multiplier method [14], dynamic programming [20] and soft computing methodologies such as genetic algorithms [12], evolutionary programming [16], simulated annealing [18], etc. have been tried. However, the thermal power generation process produces harmful emission which must also be minimized for environmental safety. Unfortunately, an optimal economic power generation is not optimal for its emission properties and vice versa. Due to the conflicting nature of minimizing power generation cost and emission characteristics, a multi-objective treatment of the problem seems to be the most suitable way to handle this problem [1]. Such an optimization task finds a set of Pareto-optimal solutions with different trade-off conditions between the two different objectives in a single simulation run.

12.2.1 Optimization Problem Formulation

The original formulation of the problem was given in Basu [1]. The hydro-thermal power generation system is optimized for a total scheduling period of T. However, the system is assumed to remain fixed for a period of t_T so that there are a total of $M = T/t_T$ changes in the problem during the total scheduling period. In this off-line optimization problem, we assume that the demand in all M time intervals is known a priori and an optimization needs to be made to find the overall schedule before starting the operation.

Let us also assume that the system consists of N_s number of thermal ($\mathbf{P}_s$) and N_h number of hydroelectric ($\mathbf{P}_h$) generating units sharing the total power demand. The fuel cost function of each thermal unit considering valve-point effects is expressed as the sum of a quadratic and a sinusoidal function and the total fuel cost in terms of real power output for the whole scheduling period can be expressed as follows:

$$f_1(\mathbf{P}_h, \mathbf{P}_s) = \sum_{m=1}^{M} \sum_{s=1}^{N_s} t_m [a_s + b_s P_{sm} + c_s P_{sm}^2 + |d_s \sin\{e_s(P_s^{min} - P_{sm})\}|].$$

(12.1)

Here, parameters a_s, b_s, c_s, d_s, and e_s are related to the power generation units and their values are given in the appendix. The parameter P_s^{min} is the lower bound of the s-th thermal power generation unit (specified in the appendix). This objective is non-differentiable and periodic to the decision variables. Notice that the fuel cost is involved only with thermal power generation units and hydroelectric units do not contribute in the cost objective. Minimization of this objective is of interest to the power companies. A conflicting objective appears from environmental consideration of minimizing the emission of nitrogen oxides for the whole scheduling period T from all thermal generation units:

$$f_2(\mathbf{P}_h, \mathbf{P}_s) = \sum_{m=1}^{M} \sum_{s=1}^{N_s} t_m [\alpha_s + \beta_s P_{sm} + \gamma_s P_{sm}^2 + \eta_s \exp(\delta_s P_{sm})].$$

(12.2)

Here again, fixed values of all parameters except the decision variables are given in the appendix. Like the cost objective, the emission objective does not involve hydroelectric power generation units.

The optimization problem has several constraints involving power balance for both thermal and hydroelectric units and water availability for hydroelectric units. In the power balance constraint, the demand term is time dependent, which makes the problem dynamic:

$$\sum_{s=1}^{N_s} P_{sm} + \sum_{h=1}^{N_h} P_{hm} - P_{Dm} - P_{Lm} = 0, \quad m = 1, 2, \ldots, M,$$

(12.3)

where the transmission loss P_{Lm} term at the m-th interval is given as follows:

$$P_{Lm} = \sum_{i=1}^{N_h + N_s} \sum_{j=1}^{N_h + N_s} P_{im} B_{ij} P_{jm}.$$

(12.4)

This constraint involves both thermal and hydroelectric power generation units. In this stationary optimization problem, it is assumed that the power demand values P_{Dm} at each time period ($m = 1, 2, \ldots, M$) is known a priori.

The water availability constraint can be written as follows:

$$\sum_{m=1}^{M} t_m (a_{0h} + a_{1h} P_{hm} + a_{2h} P_{hm}^2) - W_h = 0, \quad h = 1, 2, \ldots, N_h,$$

(12.5)

where W_h is the water head of the h-th hydroelectric unit and is given in the appendix for the problem chosen in this study.

Finally, the variable bounds are expressed as follows:

$$P_s^{min} \leq P_{sm} \leq P_s^{max}, \quad s = 1, 2, \ldots, N_s, \; m = 1, 2, \ldots, M, \quad (12.6)$$
$$P_h^{min} \leq P_{hm} \leq P_h^{max}, \quad h = 1, 2, \ldots, N_h, \; m = 1, 2, \ldots, M. \quad (12.7)$$

Thus, the two-objective problem involves $(M(N_s + N_h))$ variables, two objectives, $(M + N_h)$ quadratic equality constraints and $(2M(N_s + N_h))$ variable bounds. The specific stationary case considered here involves four $(M = 4)$ changes in demand over $T = 48$ hours having a time window of statis of $t_T = 12$ hours. We use this value to compare our results with another study [1] which also used the same parameter values. The corresponding problem has six (two hydroelectric $(N_h = 2)$ and four thermal $(N_s = 4)$) power units. For the above data, the optimization problem has 24 variables, two objectives, six equality constraints, and 48 variable bounds. All the above parameter values for this case are given in the appendix. Four fixed power demand values of 900, 1100, 1000 and 1300 MW are considered for the four time periods, respectively.

12.3 Solution Procedure for the Stationary Problem Using an EA

A solution is simply represented as a vector of 24 real-parameter variables. An EA procedure for solving an optimization problem helps to reduce the complexity of the problem in a number of ways. We discuss these here.

12.3.1 Handling Multiple Objectives

Multi-objective optimization problems give rise to a set of trade-off optimal solutions. Each such solution is a potential candidate for implementation, but exhibits a trade-off between two objectives. Classical methods require preference information a priori and then optimize a preferred single objective version of the problem [11]. If different trade-off solutions are needed to investigate the effect of different preference values before choosing a final solution, such a classical approach is required to be applied again and again. A recent study has shown that such independent optimizations may be computationally expensive in complex problems [15].

However, EAs are ideal for such problem solving tasks. This is because the EA population can be used to store different trade-off optimal solutions obtained in a single simulation run. A number of efficient methodologies exist for this purpose [3], here we use a commonly-used procedure (NSGA-II [5]), which uses a non-domination sorting of population members to emphasize non-dominated solutions systematically, an elite preserving procedure for faster convergence, and a diversity-preserving mechanism for maintaining a widely distributed set of solutions in the objective space. More about the NSGA-II procedure can be found in the original study [5].

12.3.2 Handling Variable Bounds

The current optimization problem involves a large number (48) of variable bounds, which must be treated as constraints using classical methods. However, in an EA, they all can be easily taken care of directly in the initialization procedure and in the subsequent generation of new solutions. Initial solutions are created randomly within these bounds and subsequent crossover and mutation operators are modified to create solutions only within these limits [3,7], thereby automatically satisfying all variable bounds.

12.3.3 Handling Quadratic Equality Constraints

Since all constraints in this problem are quadratic, we can use them to *repair* an infeasible solution. First, we discuss the procedure for the water availability constraints (12.5), as they involve hydroelectric power generation variables alone. Each equality constraint (for a hydroelectric unit h) can be used to replace one of the M hydroelectric power values $P_{h\mu}$) by finding the roots of the corresponding quadratic equation, thereby satisfying the equality constraints exactly. For a fixed h, the equality constraint can be written for μ-th time period as follows:

$$P_{h\mu}^2 + \frac{a_{1h}}{a_{2h}} P_{h\mu} + \frac{1}{t_\mu a_{2h}} \left(-W_h + a_{0h}T + \sum_{\substack{m=1 \\ m \neq \mu}}^{M} t_m a_{1h} P_{hm} + \sum_{\substack{m=1 \\ m \neq \mu}}^{M} t_m a_{2h} P_{hm}^2 \right) = 0.$$

(12.8)

Since a_{1h}/a_{2h} is always positive, only one root of this equation is positive and it represents the desired hydroelectric power value. Other hydroelectric power values at different time steps are calculated by maintaining the original ratio between them and $P_{h\mu}$. For a solution specifying $(N_h + N_s)$ variables, a systematic procedure is used for this purpose.

Step 1: Start with $m = 1$ (time step) and $h = 1$ (hydroelectric unit).

Step 2: Equation (12.5) is written as a quadratic root-finding problem for variable P_{hm} by fixing other hydroelectric power values at different time steps as they are in the original solution.

Step 3: Find two roots and consider the positive root, if any, and replace original P_{hm} with the positive root. Update other P_{hk} values (for $k \neq m$ as discussed above. If all units satisfy their variable bounds given in (12.7), we accept this solution. We increment h by one and move to the next hydroelectric unit and go to Step 2. We continue till all N_h units are considered and all constraints are satisfied. Then, we go to Step 4.

If none of the roots satisfy the variable bounds, we increment m by one and go to Step 2 for trying with the h-th hydroelectric unit for the next time. We continue till all M variables are tried to find at least one case in which the constraint and the corresponding variable bounds are satisfied. Then, we go to Step 4.

Step 4: If any of the N_h equality constraints are not satisfied by the above procedure, we compute the overall constraint violation (which is the absolute value of the left side of the original equality constraint) and declare the solution infeasible and do not proceed with any further constraint evaluation. Else the solution is declared feasible with zero penalty.

If the above repair mechanism for all N_h water availability constraints is successful, we proceed to repair the solution for thermal power units using power balance equality constraints, else the solution is declared infeasible and no further computation of power balance constraints nor the computation of objective functions are performed. But if the solution is repaired successfully, we proceed to repair the thermal power generation variables by using the power balance equality constraints. Equation (12.3) involves M quadratic equality constraints, each for one time period. We follow a similar procedure as above and repair a particular thermal unit P_{sm} for each time slot. Thus, the above procedures attempt to make a solution feasible. If successful, it computes the objective function values, else it computes a penalty corresponding to the amount of violation of the first constraint it cannot repair successfully and refrain from computing objective values.

The above constraint handling procedure is suitable to be used with the NSGA-II framework, in which a penalty-parameter-less procedure [2] is employed to handle feasible and infeasible solutions. In a tournament selection involving two solutions taken from the NSGA-II population, three scenarios are possible. If one is feasible and other is not, we simply choose the feasible one, If both solutions are feasible, the one dominating the other is chosen. Finally, if both solutions are infeasible, the one with smaller constraint violation is chosen. Thus, it is interesting to note that if a solution is infeasible, objective function computation is not necessary with the above procedure, thereby requiring no penalty parameters. Since this constraint handling procedure involves more than one solution, such a penalty-parameter-less strategy is possible to implement with a population-based optimization procedure. For single-objective optimization, the second condition favors the solution having better objective value [2].

12.4 Simulation Results on Stationary Hydro-thermal Power Scheduling

NSGA-II is combined with the above-discussed constraint handling method for solving the hydro-thermal scheduling problem. NSGA-II parameters used in this study are as follows: Population size is 100, SBX crossover probability is 0.9, polynomial mutation probability is $1/n$ (where n is the number of variables), and distribution indices for crossover and mutation are 10 and 20, respectively. More about these operators can be found in [3,4].

To investigate the optimality of the obtained NSGA-II frontier, each objective is also optimized independently by a single-objective EA (with identical

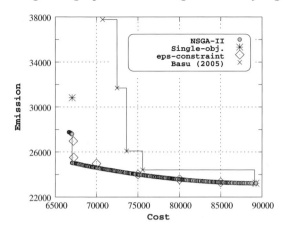

Fig. 12.1. Pareto-optimal front obtained by NSGA-II, verified by single-objective methods, and by a previous study

EA selection, crossover and mutation operators, constraint-handling strategy, and parameter setting) and two individually best objective solutions are plotted in Fig. 12.1. One of these individual-best solutions (the minimum-cost (f_1)) solution gets dominated by a NSGA-II solution and the minimum-emission solution is matched by a NSGA-II solution. To validate the optimality of some other NSGA-II solutions, we also employ the same single-objective EA to solve several ϵ-constraint problems [11] by fixing f_1 value at different values:

$$\text{Minimize } f_2(\mathbf{P}_h, \mathbf{P}_s),$$
$$\text{Subject to } f_1(\mathbf{P}_h, \mathbf{P}_s) \leq \epsilon, \qquad (12.9)$$
$$(\mathbf{P}_h, \mathbf{P}_s) \in \mathcal{S},$$

where $\mathcal{S}$ is the feasible search region satisfying all constraints and variable bounds. The obtained solutions are shown in Fig. 12.1 and it is observed that all these solutions more or less match with those obtained by NSGA-II. It is unlikely that so many different independent optimizations will result in one trade-off frontier, unless it is close to the optimal frontier. These multiple optimization procedures give us confidence about the optimality of the obtained NSGA-II frontier. We believe that in the absence of an EA's proof of convergence for any arbitrary problem, such a procedure of finding solutions and verifying them with various optimization techniques is a reliable approach for practical optimization.

Basu [1] used a simulated annealing (SA) procedure to solve the same problem. That study used a naive penalty function approach, in which if any SA solution if found infeasible, it is simply penalized. For different weight vectors scalarizing both objectives, the study presented a set of optimized solutions. A comparison of these results is made with our NSGA-II approach in Fig. 12.1. It is observed that the front obtained by NSGA-II *dominate* that

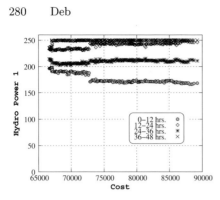

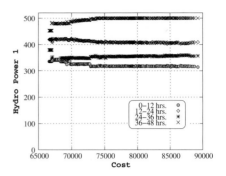

Fig. 12.2. Hydroelectric unit P_{h1} versus f_1 for the original problem

Fig. 12.3. Hydroelectric unit P_{h2} versus f_1 for the original problem

obtained the previous study. This is mainly due to the fact that the previous study used a naive constraint handling method despite the constraints being quadratic. Since we used a simple repair mechanism in our approach, most solutions created during the optimization process are feasible and the algorithm is able to find a better front.

12.4.1 Extending Variable Boundaries

Figures 12.2 and 12.3 show the variation of two hydroelectric power units for different trade-off solutions obtained by NSGA-II. These values are plotted against the corresponding cost objective value. The figure shows that for all solutions P_{h1} in the fourth time period (36–48 hrs.) needs to be set at its upper limit of 250 MW and for most solutions P_{h2} needs to be set at its upper limit of 500 MW. These suggest that there is scope of improving cost and emission values if the upper bounds considered in the original study are allowed to extend for these two variables. In the conceptual stage of design of such a power system when no decision about parameters are fixed, such information is useful. An another study addresses this issue in greater details [22]. Based on this observation and for investigation purposes, we increase the upper limits of these two variables to 350 and 600 MW, respectively. We have also observed that one of the thermal generator T_{s2} almost reaches its upper limit of 175 MW in the fourth time period and hence we also increase its upper limit to 200 MW. Figure 12.4 shows the obtained front using the extended variable boundaries. Individual optima and a number of ϵ-constraint single-objective minimizations suggest that the obtained front is close to the true Pareto-optimal front. To compare, the previous NSGA-II frontier is also shown as a solid line. It is clear that the extension of boundaries allowed a better front to be achieved. The whole new frontier dominates the previous frontier. In both objectives, better individual optimal solutions are obtained.

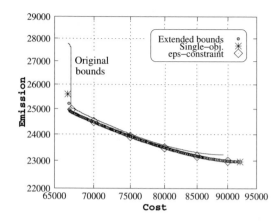

Fig. 12.4. NSGA-II frontier verified by single-objective optimizations for the extended-boundary problem. Original frontier is also shown as a solid line

12.4.2 Detecting Robust Regions of Optimality

The single objective optimization problems solved near the minimum-cost solution (in Fig. 12.1) indicates that this region is difficult to optimize. The cost objective involves a periodic term with the thermal generating units P_{sm}. This objective along with quadratic equality constraint makes the optimization task difficult. To investigate the sensitivities of solutions in this region, we perform a *robust* optimization [8] in which the variables are perturbed within ± 5 MW around each variable using a uniform distribution and 50 different scenarios are considered. Each solution is then evaluated for 50 such perturbations and a normalized change in each function value $(\Delta f_i / f_i)$ is noted. If the maximum normalized change in function between two objectives is more than a threshold value (η), the solution is declared infeasible. Thus, we add a new constraint to the original two-objective optimization problem and optimize to find the robust frontier. Such a constraint is non-differentiable and may cause difficulty, if used with a classical optimization method.

We use two different values of η and plot the obtained robust frontiers in Fig. 12.5 along with the original NSGA-II frontier (from Fig. 12.1). It can be clearly seen that as the threshold η is reduced, thereby restraining the normalized change to a small value, the robust front deviates more from the original non-robust frontier in the area of minimum-cost solution. The figure shows that solutions costing 75,000 or less is sensitive to the variable uncertainties. However, the minimum-emission region remains relatively unperturbed, meaning the near minimum-emission region is already robust. A similar analysis on the modified problem with extended boundaries is performed and the results are shown in Fig. 12.6. It can be clearly seen from the figure that the trade-off frontier for the extended problem is fairly robust except a small portion near the minimum-cost region.

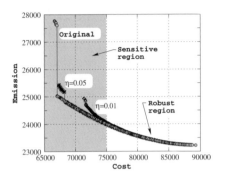
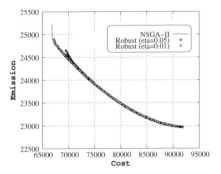

Fig. 12.5. Robust frontier for the original problem

Fig. 12.6. Robust frontier for the modified problem with extended boundaries

Such a robust optimization strategy is of utmost importance in practice, as in the event of uncertainties in achieving decision variables and problems, engineers and designers are in most situations interested in finding a robust solution which is relatively insensitive to such uncertainties. The above study also reveals that the minimum-cost solution is relatively more sensitive to such uncertainties than the minimum-emission solution for the chosen power dispatch problem.

12.4.3 Unveiling Common Principles of Operation

To understand the operating condition of the hydro-thermal power system, we analyze the obtained solutions in the trade-off frontier (Fig. 12.4) obtained using extended boundaries. This process of deciphering important design or operating principles in a problem is discussed in detail in another study [10] and is called the *innovization* – innovation through optimization – task. Figures 12.7 through 12.12 show the variation of output levels with cost objective. Several interesting behaviors (innovizations) can be observed from these plots:

1. Output powers for both hydroelectric units are more or less constant for all trade-off optimal solutions for all time slots. The average values are shown below:

	0–12 Hrs.	12–24 Hrs.	24–36 Hrs.	36–48 Hrs.
Demand P_D (MW)	900.00	1100.00	1000.00	1300.00
P_{h1} (MW)	155.45	232.70	189.29	293.39
P_{h2} (MW)	311.68	406.30	364.37	498.69

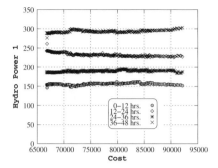

Fig. 12.7. Hydroelectric unit P_{h1} versus f_1

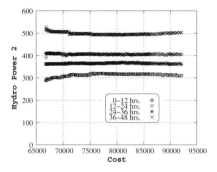

Fig. 12.8. Hydroelectric unit P_{h2} versus f_1

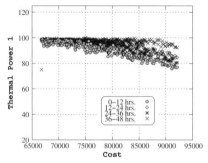

Fig. 12.9. Thermal unit P_{s1} versus f_1

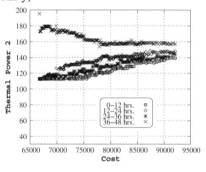

Fig. 12.10. Thermal unit P_{s2} versus f_1

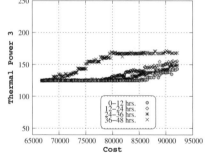

Fig. 12.11. Thermal unit P_{s3} versus f_1

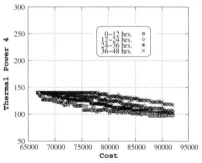

Fig. 12.12. Thermal unit P_{s4} versus f_1

2. These hydroelectric power output values are related to the variation in power demand, as can be seen from the above table. If the demand is more in a time period, the optimal strategy is to increase the hydroelectric power generation for that time period and vice versa.

3. The combined hydroelectric power generation (with only two units) takes care of more than 50% of the total demand in all time periods.

4. The power output of both hydroelectric units for a particular time period reflects the amount of water availability. For the second hydroelectric unit the water availability is more, hence the power generation is also more.
5. For all four thermal units, the power generation must be increased with the demand in that interval.
6. For better cost solutions, the thermal power generation of units 1 and 4 must be increased. This is because a_s and b_s values for these units are lower compared to other two thermal units, thereby causing a comparatively smaller increase in the fuel cost with an increase in power generation in these two units. However, the power generation of units 2 and 3 must be decreased (with an exception at the fourth time period for unit 2) due to the opposite reason to that above.
7. For smaller cost solutions, the thermal power generation is more or less the same for all time periods, except the fourth time period in unit 2. This is because due to handling a large demand in this time period, the emission-effective way of reducing the cost is to increase the thermal power generation in unit 2 (due to the large negative β_s value (see appendix) associated with this unit).
8. Although a large range of power generation is allowed, the optimal values of these power generation units are concentrated within a small region in the search space.

The above properties of the optimized trade-off solutions are interesting and are, by any means, not trivial. Finding the optimized trade-off solutions and then analyzing the solutions for discovering common properties (the innovization process [10]) is a viable procedure for deciphering such important information. Having solved the stationary problem to our satisfaction and comparing the optimized solutions with the existing study, we are now confident and venture to solve the same problem as a non-stationary problem.

12.5 Dynamic Power Scheduling Problem

The hydro-thermal power dispatch problem is truly dynamic in nature due to the changing nature of power demand. An optimal power scheduling for a particular power demand profile is not necessarily optimal for another power demand. In such cases, new optimal solutions must be found as and when there is a change in the power demand. Many search and optimization problems in practice change with time and therefore must be treated as on-line optimization problems. The change in the problem with time t can be either in its objective functions or in its constraint functions or in its variable boundaries or in a combination of the above:

$$
\begin{aligned}
\text{Minimize } &\mathbf{f}(\mathbf{x}, t) = (f_1(\mathbf{x}, t), f_2(\mathbf{x}, t), \ldots, f_M(\mathbf{x}, t)), \\
\text{Subject to } &g_j(\mathbf{x}, t) \geq 0, \quad \forall j, \\
&h_k(\mathbf{x}, t) = 0, \quad \forall k, \\
&x_i^{(L)}(t) \leq x_i(t) \leq x_i^{(U)}(t), \quad \forall i.
\end{aligned}
\tag{12.10}
$$

Such an optimization problem ideally must be solved at every time instant t or whenever there is a change in any of the above functions with t. In such optimization problems, the time parameter can be mapped with the iteration counter τ of the optimization algorithm. One difficulty which may arise in solving the above on-line optimization task is that the underlying optimization algorithm may not get too many iterations to find the optimal solutions before there is a change in the problem. If the change is too frequent, the best hope of an optimization task is to *track* the optimal solutions as closely as possible within the time span allowed to iterate. However, for steady changes in a problem (which is usually the case in practice), there lies an interesting trade-off which we discuss next.

Let us assume that the change in the optimization problem is gradual in t. Let us also assume that each optimization iteration requires a finite time G to execute and that τ_T iterations are needed (or allowed) to track the optimal frontier. In this chapter, we assume that problem does not change (or is assumed to be constant) within a time interval t_T and $\tau_T G < t_T$. Here, initial $\tau_T G$ time is taken up by the optimization algorithm to track the new trade-off frontier and to make a decision for implementing a particular solution from the frontier. Here, we choose $\alpha = \tau_T G/t_T$ to be a small value (say 0.25), such that after the optimal frontier is tracked, $(1 - \alpha)t_T$ time is spent on using the outcome for the time period. Figure 12.13 illustrates this dynamic procedure. Thus, if we allow a large value of t_T (allowing a large number of

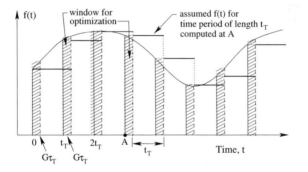

Fig. 12.13. The on-line optimization procedure adopted in this study. For simplicity, only one objective function is shown here

optimization iterations τ_T), a large change in the problem is expected, but the change occurs only after a large number of iterations of the optimization algorithm. Thus, despite the large change in the problem, the optimization algorithm may have enough iterations to track the trade-off optimal solutions. However, this scenario will produce a grossly approximate solution to the real problem, which changes with t continuously. On the other hand, if we choose a small τ_T, the change in the problem is frequent (which approximates the real scenario more closely), but a lesser number of iterations are allowed to track new optimal solutions for a problem which has also undergone a small change. Obviously, there lies a lower limit to τ_T below which, albeit a small change, the number of iterations are not enough for an algorithm to track the new optimal solutions adequately. Such a limiting τ_T will depend on the nature of the dynamic problem and the chosen algorithm, but importantly allows the best scenario (and closest approximation to the original problem) which a particular algorithm can achieve. Here, we investigate this aspect in the context of dynamic multi-objective optimization problem and find such a limiting τ_T for two variants of NSGA-II on a test problem and the real-world hydro-thermal power dispatch optimization problem discussed above. However, the procedure adopted in this study is generic and can be applied to other dynamic optimization problems as well.

The procedure can be applied to its extreme as well. If we allow the problem to change as frequently as the time needed to complete one iteration of the optimization algorithm (that is, $t_T = \tau_T = 1$, yielding $G = 1$), we have a true on-line optimization procedure in which the problem changes continuously with generation counter.

12.5.1 Proposed Modifications to NSGA-II

NSGA-II was developed for solving stationary multi-objective optimization problems and may not be suitable on its own to be applied to dynamic problems. We make minor changes to the original NSGA-II.

First, we introduce a test to identify whether there is a change in the problem. The mathematical formulation of the problem remains the same, but due to change in power demand the evaluation of constraints will give rise to a change in power generation values, thereby causing a change in the objective values. For this purpose, we randomly pick a few solutions from the parent population (10% population members considered here) and re-evaluate them. If there is a change in any of the objectives and constraint functions, we establish that there is a change in the problem. In the event of a change, all parent solutions are re-evaluated before merging the parent and child population into a bigger pool. This process allows both offspring and parent solutions to be evaluated using the changed objectives and constraints.

In the first version (DNSGA-II-A) of the proposed dynamic NSGA-II, we introduce new solutions whenever there is a change in the problem. A $\zeta\%$ of the new population is replaced with randomly created solutions. This helps to introduce new (random) solutions whenever there is a change in the problem. This method may perform better in problems with a large change in the objectives and constraints. In the second version (DNSGA-II-B), instead of introducing random solutions, $\zeta\%$ of the population is replaced with mutated versions of existing solutions (chosen randomly). This way, the new solutions introduced in the population are related to the existing population. This method may work well in problems with a small change in problem.

The dynamic version of the problem involves more frequent changes in the demand P_{Dm}. The objective and constraint functions are the same as those in the stationary problem. We keep the overall time window of $T = 48$ hours, but increase the frequency of changes (that is, increase M from four to 192, so that the time window t_T for each demand level varies from 12 hours to 48/192 hours or 15 minutes). It will then be an interesting task to find the smallest time window of statis which a specific multi-objective optimization algorithm can solve successfully. Any change more frequent than this smallest time limit would be too fast a change in the problem for the algorithm to track the optimal trade-off front. For this purpose, we use the same overall four-step changing pattern in demand as we considered in the stationary case, but make a step-wise linear interpolation between any two consecutive changes to ascertain the intermediate level of demand. Fig. 12.14 explains the interpolation procedure for $M = 16$. The same procedure is followed to calculate the demands for other M values.

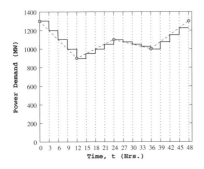

Fig. 12.14. Change in power demand with time for $M = 16$ (3-hourly change)

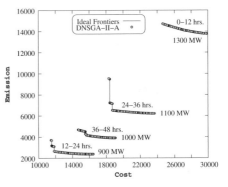

Fig. 12.15. Four fronts, each change in demand, obtained using DNSGA-II-A with $\zeta = 0.2$

Difficulty in Handling Water Availability Constraint

Equation (12.5) requires hydroelectric power generation units from different time intervals to be used together to satisfy the equation. In a dynamic optimization problem, this is a difficulty, as this means that information about all hydroelectric units is needed right in the first generation. This constraint equates the total required water head to be identical to the available value for each hydroelectric system.

It is interesting to note from Figs 12.7 and 12.8 that the optimized hydroelectric power unit is almost proportional to the demand. A proportionate law can be used to allocate P_{h1} and P_{h2} values with time and in such a manner as to satisfy the water availability constraints. We are currently pursuing this proportionate concept and will report results in the due course of time. In this study, we use a simple principle of allocating an identical water head W_h/M for each time interval.

12.5.2 Simulation Results Using Dynamic NSGA-II

To make the time-span of the scheduling problem in tune with the NSGA-II generations, we assume that the demand value P_{Dm} changes with generation counter (as in Fig. 12.14) and the mapping is such that the NSGA-II run reaches its maximum generation number at the end of the scheduling time-span ($T = 48$ hours). Since the proposed procedures have a way to detect if there is a change in the problem, such an optimization algorithm can be directly used for on-line optimization applications.

We apply the two dynamic NSGA-II procedures discussed above (DNSGA-II-A and DNSGA-II-B) to solve the dynamic optimization problem. The parameters used are the same as in the off-line optimization case presented before. First, we consider the problem with a 12-hour statis. To compare the dynamic NSGA-II procedures, we first treat each problem as a static optimization problem and apply the original NSGA-II procedure [5] for a large number (500) of generations so that no further improvement is likely. We call these 'ideal' fronts and compute the hypervolume measure using a reference point which is the nadir point of the ideal front. Thereafter, we apply each dynamic NSGA-II and find an optimized non-dominated front. Then for each front, we compute the hypervolume using the same reference point and then compute the ratio of this hypervolume value with that of the ideal front. This way, the maximum value of the ratio of hypervolume for an algorithm is one and as the ratio becomes smaller than one, the performance of the algorithm gets poorer.

In the dynamic NSGA-II procedures, we run for $960/M$ (M is the number of changes in the problem) generations for each change in the problem. Thus, if a large number changes are to be accommodated, the number of generations needed for each optimization run gets reduced. First, we consider the problem in which we consider a change after every 12 hours ($M = 4$). Figure 12.15 shows the four Pareto-optimal fronts obtained using DNSGA-II-A with 20% addition of random solutions every time there is a change in the problem. The DNSGA-II-A procedure is able to find a set of solutions very close to the ideal frontiers in all four time periods. The figure makes one aspect clear. As the demand is more, the power production demands larger cost and emission values.

Increasing Number of Changes in the Problem

Next, we increase the number of changes from eight to 192, but present results for $M = 16, 48, 96$ and 192. Figures 12.16 – 12.19 show the hypervolume ratio for different number of changes in the problem for the DNSGA-II-A procedure with different proportion of addition of random solutions, ζ. The figures also

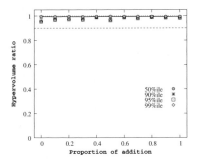

Fig. 12.16. 3-hourly change ($M = 16$) with DNSGA-II-A

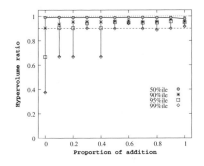

Fig. 12.17. 1-hourly change ($M = 48$) with DNSGA-II-A

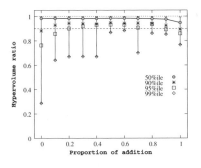

Fig. 12.18. 30-minute change ($M = 96$) with DNSGA-II-A

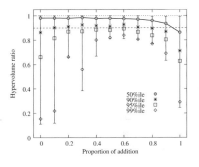

Fig. 12.19. 15-minute change ($M = 192$) with DNSGA-II-A

mark the 50th, 90th, 95th and 99th percentile of hypervolume ratio, meaning the cut-off hypervolume ratio which is obtained by the best 50, 90, 95, and 99 percent of M frontiers in a problem with M changes. The figures show that as M increases, the performance of the algorithm gets poorer due to the fact that a smaller number of generations ($960/M$) was allowed to meet the time constraint. If a 90% hypervolume ratio is assumed to be the minimum required hypervolume ratio for a reasonable performance of an algorithm and if we consider 95 percentile performance is adequate, the figures show that we can allow a maximum of 96 changes in the problem, meaning a change in the problem in every 48/96 or 0.5 hours (or 30 minutes). With 96 changes in the problem, about 20 – 70% random solutions can be added whenever there is a change in the problem to start the next optimization, to achieve a successful run. This result is interesting and suggests the robustness of the DNSGA-II procedure for this problem. It is also clear that no addition of any random solution ($\zeta = 0$) is not a good strategy.

Next, we consider DNSGA-II-B procedure in which mutated solutions are added instead of random solutions. Mutations are performed with double the mutation probability used in NSGA-II and with $\eta_m = 2$. Figures 12.20 – 12.23 show the performance plots. Here, the effect is somewhat different. In general, with an increase in addition of mutated solutions, the performance is better. Once again, with 96 changes in the problem within 48 hours of operation seem to be the largest number of changes allowed for the algorithm to perform reasonably well. However, mutated solutions more than $\zeta = 40\%$ of the population seem to perform well. Once again, DNSGA-II-B procedure is also found to work well with a wide variety of ζ values ($40 - 100\%$).

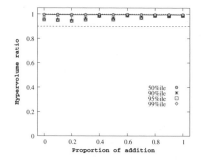

Fig. 12.20. 3-hourly change ($M = 16$) with DNSGA-II-B

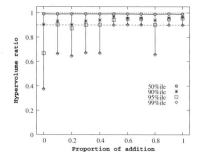

Fig. 12.21. 1-hourly change ($M = 48$) with DNSGA-II-B

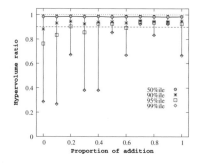

Fig. 12.22. 30-minute change ($M = 96$) with DNSGA-II-B

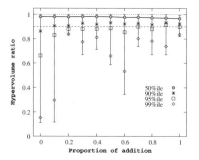

Fig. 12.23. 15-minute change ($M = 192$) with DNSGA-II-B

12.6 Decision-making in a Dynamic Multi-objective Optimization

One of the issues which is not discussed enough in the EMO literature is the decision-making aspect after a set of trade-off solutions is found. In a static multi-objective optimization problem, the overall task is to come up with one solution which would be finally implemented in practice. Thus, choosing a solution from the Pareto-optimal set is as important as finding a set of optimized trade-off solutions. Some studies in this direction have just begun for stationary problems [6, 9, 13, 17] and more such studies are called for. However, in a dynamic multi-objective optimization problem, there is an additional problem with the decision-making task. A solution is to be chosen and implemented as quickly as the trade-off frontier is found, and in most situations before the next change in the problem has taken place. This definitely calls for an automatic procedure for decision-making with some pre-specified utility function or some other procedure. This way, in an on-line implementation, as soon as a dynamic EMO finds a frontier, a quick analysis of the solutions can be made to find a particular solution which optimizes or satisfies the decision-making procedure maximally.

In this chapter, we choose a utility measure which is related to the relative importance given to both cost and emission objectives. First, we consider a dynamically operating decision-making procedure in which we are interested in choosing a solution having almost an equal importance to both cost and emission. As soon as the frontier is found for the forthcoming time period, we compute the pseudo-weight w_1 (for cost objective) for every solution $\mathbf{x}$ in the frontier using the following term:

$$w_1(\mathbf{x}) = \frac{(f_1^{\max} - f_1(\mathbf{x}))/(f_1^{\max} - f_1^{\min})}{(f_1^{\max} - f_1(\mathbf{x}))/(f_1^{\max} - f_1^{\min}) + (f_2^{\max} - f_2(\mathbf{x}))/(f_2^{\max} - f_2^{\min})}. \quad (12.11)$$

Thereafter, we choose the solution ($\mathbf{x}$) with $w_1(\mathbf{x})$ closest to 0.5.

To demonstrate the utility of this dynamic decision-making procedure, we consider the hydro-thermal problem with 48 time periods (meaning a change of the problem at every hour). Figure 12.24 shows the obtained frontiers in solid lines and the corresponding preferred (operating) solution with a circle. It can be observed that due to the preferred importance of 50–50% to cost and emission, the solution comes nearly in the middle of each frontier. The figure also marks the time period on some of the frontiers to give an idea how the frontiers change with time. Since hydroelectric units are assumed to be constant over time in this study, to satisfy two water availability constraints, $T_{h1} = 219.76$ MW and $T_{h2} = 398.11$ MW must be used. However, four thermal power units must produce power to meet the remaining demand and these values for the 50–50% cost-emission solution for all 48 time periods are shown in Fig. 12.25. It is interesting to note that how the thermal units must produce different power values to meet the remaining overall changing demand (shown by a solid line). The changing pattern in overall thermal power generation

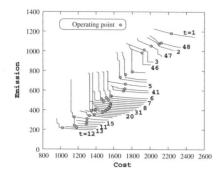

Fig. 12.24. Operating solution for 50–50% cost-emission case on the trade-off frontiers for the 48 time-period problem

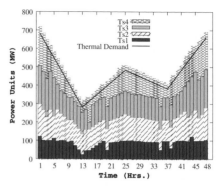

Fig. 12.25. Variation of thermal power production for 50–50% cost-emission case for the 48 time-period problem

varies similar to that in the remaining demand in power. The figure also shows a slight over-generation of power to meet the loss term given in (12.4).

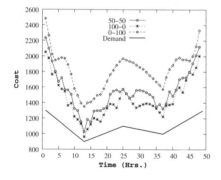

Fig. 12.26. Variation of cost of operation with time for the 48 time-period problem

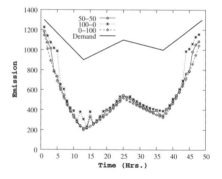

Fig. 12.27. Variation of emission value with time for the 48 time-period problem

Table 12.1. Cost-emission trade-off

Case	Cost	Emission
50-50%	74239.07	25314.44
100-0%	69354.73	27689.08
0-100%	87196.50	23916.09

Next, we compare the above operating schedule of power generation with two other extreme cases: (i) 100–0% importance to cost and emission and (ii) 0–100% importance to cost and emission. Figures 12.26 and 12.27 show the variation of cost and emission, respectively, for these two cases and the 50–50% case discussed above. First, it is interesting to note that the optimal cost and emission values fluctuate the way the power demand varies. Second, the case with 100% importance to cost requires minimum-cost, but causes large emission values and the case with 100% importance to emission causes minimum-emission values, but with large costs. A comparison of overall cost and emission values for the entire 48-hour operation for these three cases are summarized in Table 12.1. The results agree with above argument. Third, the change in emission values is not as much as in cost values, as was also observed in the computation of the robust frontiers in the stationary case (Fig. 12.5).

12.7 Conclusions

In this chapter, we have demonstrated the power of population-based evolutionary algorithms for handling a complex hydro-thermal dispatch optimization problem involving non-linear objective and constraint functions, uncertainties in decision variables, the dynamic nature of the problem due to the change in power demand, and multiple objectives. Following important conclusions can be drawn from this study:

1. The study shows how multi-objective EA results can be found and verified with a number of other single-objective optimization tasks to build confidence in obtained solutions.
2. The study shows how multiple trade-off solutions can be used to decipher important insights about the problem which give a clear idea of the properties needed for a solution to become an optimal solution. This is possible only by finding and analyzing a set of optimal solutions, instead of a single optimal solution.
3. The study also shows how robust solutions (which are relatively insensitive to parameter uncertainties) can be found and good regions of robustness in a problem can be identified.
4. The study also shows how a dynamic optimization task can be executed using an EA and optimal solutions can be tracked, whenever there is a change in the problem.
5. Finally, the study suggests an automatic decision-making procedure for choosing a preferred solution from the multiple trade-off frontier, as and when the problem is changed.

All these tasks have tremendous practical importance and this study is a testimony to some of the possibilities of evolutionary optimization procedures in making a computational optimization task a big step closer to practice.

Acknowledgments

The author would like to thank Ms. Barnali Kar and Mr. Uday Bhaskara Rao for discussions about the hydro-thermal power dispatch optimization problem.

References

1. Basu, M.: A simulated annealing-based goal-attainment method for economic emission load dispatch of fixed head hydrothermal power systems. Electric Power and Energy Systems **27**(2), 147–153 (2005)
2. Deb, K.: An efficient constraint handling method for genetic algorithms. Computer Methods in Applied Mechanics and Engineering **186**(2–4), 311–338 (2000)
3. Deb, K.: Multi-objective optimization using evolutionary algorithms. Chichester, UK: Wiley (2001)
4. Deb, K., Agrawal, R.: Simulated binary crossover for continuous search space. Complex Systems **9**(2), 115–148 (1995)
5. Deb, K., Agrawal, S., Pratap, A., Meyarivan, T.: A fast and elitist multi-objective genetic algorithm: NSGA-II. IEEE Transactions on Evolutionary Computation **6**(2), 182–197 (2002)
6. Deb, K., Chaudhuri, S.: I-MODE: an interactive multi-objective optimization and decision-making using evolutionary methods. In: Proceedings of Fourth International Conference on Evolutionary Multi-Criteria Optimization (EMO 2007), pp. 788–802 (2007)
7. Deb, K., Goyal, M.: A combined genetic adaptive search (GeneAS) for engineering design. Computer Science and Informatics **26**(4), 30–45 (1996)
8. Deb, K., Gupta, H.: Searching for robust Pareto-optimal solutions in multi-objective optimization. In: Proceedings of the Third Evolutionary Multi-Criteria Optimization (EMO-05) Conference (Also Lecture Notes on Computer Science 3410), pp. 150–164 (2005)
9. Deb, K., Kumar, A.: Interactive evolutionary multi-objective optimization and decision-making using reference direction method. In: Proceedings of the Genetic and Evolutionary Computation Conference (GECCO-2007), pp. 781–788. New York: The Association for Computing Machinery (ACM) (2007)
10. Deb, K., Srinivasan, A.: Innovization: innovating design principles through optimization. In: Proceedings of the Genetic and Evolutionary Computation Conference (GECCO-2006), pp. 1629–1636. New York: The Association for Computing Machinery (ACM) (2006)
11. Miettinen, K.: Nonlinear Multiobjective Optimization. Kluwer, Boston (1999)
12. Orero, S., Irving, M.: A genetic algorithm modeling framework and solution technique for short term optimal hydrothermal scheduling. IEEE Transactions on Power Systems **13**(2) (1998)
13. Phelps, S., Koksalan, M.: An interactive evolutionary metaheuristic for multi-objective combinatorial optimization. Management Science **49**(12), 1726–1738 (2003)
14. Rashid, A., Nor, K.: An efficient method for optimal scheduling of fixed head hydro and thermal plants. IEEE Transactions Power Systems **6**(2) (1991)

15. Shukla, P., Deb, K.: Comparing classical generating methods with an evolutionary multi-objective optimization method. In: Proceedings of the Third International Conference on Evolutionary Multi-Criterion Optimization (EMO-2005), pp. 311–325 (2005). Lecture Notes on Computer Science 3410
16. Sinha, N., Chakraborty, R., Chattopadhyay, P.: Fast evolutionary programming techniques for short-term hydrothermal scheduling. Electric Power Systems Research **66**, 97–103 (2003)
17. Thiele, L., Miettinen, K., Korhonen, P., Molina, J.: A preference-based interactive evolutionary algorithm for multiobjective optimization. Tech. Rep. Working Paper Number W-412, Helsinki School of Economics, Helsingin Kauppakorkeakoulu, Finland (2007)
18. Wong, K., Wong, Y.: Short-term hydrothermal scheduling I – simulated annealing approach. IEE Proc-C Gener., Trans., Distrib **141**(5) (1994)
19. Wood, A., Woolenberg, B.: Power Generation, Operation and Control. John-Wiley & Sons (1986)
20. Yang, J., Chen, N.: Short-term hydrothermal co-ordination using multipass dynamic programming. IEEE Transactions Power Systems **4**(3) (1989)
21. Zaghlool, M., Trutt, F.: Efficient methods for optimal scheduling of fixed head hydrothermal power systems. IEEE Transactions Power Systems **3**(1) (1988)
22. Zeleny, M.: Multiple Criteria Decision Making. McGraw-Hill, New York (1982)

Appendix A: Parameters for Hydro-thermal Power Dispatch Problem

Table 12.2. Hydroelectric system data

Unit	a_{0h} (acre-ft/h)	a_{1h} (acre-ft/MWh)	a_{2h} (acre-ft/$(MW)^2$h)	W_h (acre-ft)	P_h^{min} (MW)	P_h^{max} (MW)
1	260	8.5	0.00986	125000	0	250
2	250	9.8	0.01140	286000	0	500

Table 12.3. Cost related thermal system data

Unit	a_s ($/h)	b_s ($/MWh)	c_s ($/$(MW)^2$h)	d_s ($/h)	e_s (1/MW)	P_s^{min} (MW)	P_s^{max} (MW)
3	60.0	1.8	0.0030	140	0.040	20	125
4	100.0	2.1	0.0012	160	0.038	30	175
5	120.0	2.1	0.0010	180	0.037	40	250
6	40.0	1.8	0.0015	200	0.035	50	300

Table 12.4. Emission related thermal system data

Unit	$\alpha_s(Lb/h)$	$\beta_s(Lb/MWh)$	$\gamma_s(Lb/(MW)^2h)$	$\eta_s(Lb/h)$	$\delta_s(1/MW)$
3	50	-0.555	0.0150	0.5773	0.02446
4	60	-1.355	0.0105	0.4968	0.02270
5	45	-0.600	0.0080	0.4860	0.01948
6	30	-0.555	0.0120	0.5035	0.02075

$$B = \begin{bmatrix} 0.000049 & 0.000014 & 0.000015 & 0.000015 & 0.000020 & 0.000017 \\ 0.000014 & 0.000045 & 0.000016 & 0.000020 & 0.000018 & 0.000015 \\ 0.000015 & 0.000016 & 0.000039 & 0.000010 & 0.000012 & 0.000012 \\ 0.000015 & 0.000020 & 0.000010 & 0.000040 & 0.000014 & 0.000010 \\ 0.000020 & 0.000018 & 0.000012 & 0.000014 & 0.000035 & 0.000011 \\ 0.000017 & 0.000015 & 0.000012 & 0.000010 & 0.000011 & 0.000036 \end{bmatrix} \text{ per MW.}$$

13

Multimodal Function Optimization of Varied-Line-Spacing Holographic Grating

Qing Ling[1], Gang Wu[2], and Qiuping Wang[3]

[1] Department of Automation, University of Science and Technology of China,
  Hefei, China `qingling@mail.ustc.edu.cn`
[2] Department of Automation, University of Science and Technology of China,
  Hefei, China `wug@ustc.edu.cn`
[3] National Synchrotron Radiation Laboratory, University of Science and
  Technology of China, Hefei, China `qiuping@ustc.edu.cn`

13.1 Introduction

In practical optimal design problems, objective functions often lead to multi-modal domains. There are generally two basic requirements in the multimodal function optimization problem:

1. to find the global optimum, i.e. global search ability, and
2. to locate several local optima that might be good alternatives to the global optimum, i.e. multi-optimum search ability.

The first one comes from the requirement of optimality, while the second one reflects the needs of practical engineering design. Multiple solutions permit designers to choose the best one in terms of ease of manufacture, ease of maintenance, reliability, etc. which cannot be simply represented by the objective function [2].

Evolutionary algorithms (EAs), such as genetic algorithms (GAs), evolution strategies (ESs), and differential evolution (DE), are powerful tools for global optimization. To locate the global optimum efficiently, we need to improve the diversity of the population to prevent the premature convergence phenomenon. Self-adaptation strategies are helpful in preserving the diversity and achieving global search ability [13,31].

Intuitively, EAs are also proper choices for multi-optimum search problems because of their intrinsic parallel search mechanism. But in multi-optimum search, it is not enough to preserve the diversity of the population only. We need to promote the formation of stable subpopulations in the neighborhood of optimal solutions, both global and local [32]. In this way, multiple solutions are identified at the end of the optimization, rather than only one global optimum. In EAs, niching techniques have been introduced to achieve this goal.

The concept of niche comes from ecosystems. A niche can be viewed as a subspace in the environment that can support different types of life. For each niche, resources are finite and must be shared among the population of that niche. By analogy, in EAs, a niche is commonly referred to as a neighboring area of an optimum of the parameter space. The corresponding fitness function of the optimum represents resources of that niche [32]. Niching techniques reallocate resources among individuals in a niche, thus preventing the population from converging to few optima, and promoting the formation of stable subpopulations in niches.

Niching techniques have achieved successful applications in many engineering design areas. Liu proposed a hybrid niching genetic algorithm in the optimal design of a ultra-broadband amplifier [21] and fiber Raman amplifier [22]. Crutchley searched the quiescent operating point of nonlinear circuits with the aid of a niching EA [4]. Other applications include electrode shape optimization [34], electromagnetics optimization [33], and the planning of multiple paths of robots [15].

Theoretical analysis of niching techniques is very difficult because of the complicated fitness landscape of multimodal functions. Therefore, the analysis is limited to simple trap functions and simplified algorithms [3, 25]. A powerful tool for theoretical analysis is the Markov chain model. Nijssen studied convergence velocity and reliability with the Markov chain [29]. Another example is [7], where relationships between convergence properties, algorithm parameters, and the fitness landscape were investigated.

In this chapter, we mainly focus on the multi-optimum search problem under the framework of multimodal function optimization. Various niching techniques, such as sharing, crowding, clearing, sequential niche sharing, dynamic niche clustering, etc. are discussed. Among the main niching techniques, the restricted evolution evolution strategy (REES) is selected to demonstrate the usefulness of multi-optimum search in practical engineering design.

The rest of this chapter is arranged as follows. Section 13.2 reviews the main niching techniques in multimodal function optimization. REES is described in detail in Sect. 13.3. In Sect. 13.4, the proposed algorithm is compared with standard crowding and deterministic crowding in several multimodal functions. REES is applied to the design of the varied-line-spacing holographic grating (VLSHG) for the National Synchrotron Radiation Laboratory (NSRL) in Sect. 13.5. Section 13.6 concludes the chapter.

13.2 Niching Techniques

A comparison of some existing niching techniques is made in a recent study by Singh and Deb [35]. The key point in niching techniques is how to decide the range of a specified niche, i.e. the niche radius problem. The choice of the niche radius reflects an estimation of the shape of the fitness landscape. There are three classes of methods to deal with the niche radius problem: explicit, implicit, and adaptive methods.

13.2.1 Explicit Niche Radius

Explicit niching methods set a predetermined niche radius before the optimization process. The choice of niche radius influences the performance of multi-optimum search greatly. But the proper value of niche radius is strongly related to the fitness landscape, which is generally difficult to analyze in practical optimization problems.

Sharing

The most important explicit niching method is fitness sharing [5, 11, 12]. The basic idea is to punish individuals for occupying the same regions in the search space by scaling the fitness of each individual. Thus, by penalizing individuals that cluster together, the sharing scheme prevents individuals from converging to a single optimum. The shared fitness $f'(P_i)$ of individual P_i is calculated as follows from the original fitness $f(P_i)$:

$$f'(P_i) = \frac{f(P_i)}{\sum_{j=1}^{j=PopNum} sh(D(P_i, P_j))} \tag{13.1}$$

where $PopNum$ is the population size of the EA, D is the distance metric, and sh is the sharing function. Typically, a triangular sharing function is used:

$$sh(D(P_i, P_j)) = \begin{cases} 1 - (D(P_i, P_j)/\sigma)^\alpha & D(P_i, P_j) < \sigma \\ 0 & D(P_i, P_j) \geq \sigma. \end{cases} \tag{13.2}$$

The constant σ is the predetermined niche radius, and α is the scaling factor to regulate the shape of the sharing function.

The main limitation of the sharing method lies in the choice of niche radius σ. Firstly, setting of σ requires a priori knowledge of the fitness landscape. But for practical optimization problems, we generally do not have enough information about the shape of the objective function and the distance between optimal solutions. Secondly, σ is the same for all individuals. This supposes that all optima must be nearly equidistant in the domain. Sareni [32] proved that sharing failed to maintain desired optimal solutions if they are not equidistant or if the estimated distance between two optima is incorrect. Thirdly, the sharing scheme is very expensive regarding computational complexity. Thus clustering analysis [39] and dynamic niche sharing [27] have been proposed to reduce the computation overhead.

Clearing

Clearing is a similar niching technique to sharing [30]. In a given niche radius σ, the basic clearing algorithm preserves the fitness of the dominant individual while it resets the fitness of all other individuals to zero. Thus, the clearing procedure fully attributes the whole resources of a niche to a single winner – the winner takes all rather than sharing resources with other individuals in the same niche as is done in the sharing method.

Sequential Niche

The sequential niche method executes the EA program with a modified fitness function repeatedly [2]. First, we set the modified fitness function to the raw fitness function, run the EA program, and record the best individual. Second, we update the modified fitness function to give a depression in the region near the best individual. Then the new best individual of the new modified fitness function is recorded by using EA again. By sequentially using EA and sequentially modifying the fitness function, multiple solutions are finally identified. In the sequential niche method, the modification of the fitness function is related to the estimation of niche radius, as in sharing and clearing.

Species Conserving

The species conserving genetic algorithm [18] is based on the concept of dividing the population into several species according to their similarity. Each of these species is built around a dominating individual called the species seed. Species seeds found in the current generation are conserved by moving them into the next generation. Similarity of species is also decided through a predetermined niche radius σ, for which an experimental formula is provided.

Restricted Competition Selection

As in the clearing method, restricted competition selection sets the loser's fitness to zero in a given niche radius. At the end of each generation, M individuals are selected into an elite set. Therefore, the restricted competition selection algorithm is a combination of the clearing method and the elite strategy. By introducing the elite set with size M, we may expect that at least M local optimal or near-optimal solutions are obtained at the end [17].

13.2.2 Implicit Niche Radius

Implicit niche radius methods do not require a predefined niche radius. The information on niches is embodied in the population itself. There are two classes of implicit niching methods: crowding and multi-population.

Crowding methods include standard crowding, deterministic crowding and restricted tournament selection. The main disadvantages of crowding methods are selection error and genetic drift. Selection error means that individuals fail to compete with elements in the same niche. For example, an unsatisfactory individual may not be chosen to compare with other neighboring individuals during the optimization process, and thus survives at the end. Though inevitable for a finite population, selection error will lead to much depressed optimization performance if too many unsatisfactory individuals survive in evolution. On the contrary, genetic drift means that good candidate solutions are chosen by error to compare with dominating individuals, thus individuals are inclined to converge to a few eminent solutions. If we fail to protect the formation of stable subpopulations in the neighborhood of optimal solutions, genetic drift will occur.

Multi-population EAs divide the whole population into some subpopulations, and permit information exchange between subpopulations. The main difficulty in multi-population EAs is how to control information exchange, both maintaining a steady exploitation and achieving an efficient exploration.

Standard Crowding

Standard crowding updates the population through replacing similar parents [36]. For each child individual C, randomly select CF (crowding factor) individuals in the parent population, choose the nearest parent P under some distance metric. If C is better than P, use C to replace P, else preserve P. When CF is small, each individual will only have a small chance to compare with other individuals in the same niche and to improve the quality of solution. Therefore, for small CF, standard crowding introduces great selection error. On the other hand, individuals nearing a dominating solutions may be replaced with high probability, even though they are potential local optima. Thus the standard crowding method will result in noticeable genetic drift.

Deterministic Crowding

In deterministic crowding [24], two parents P_1 and P_2 generate two children C_1 and C_2. For some distance metric D, if $D(P_1, C_1) + D(P_2, C_2)$ is smaller than $D(P_1, C_2) + D(P_2, C_1)$, then introduce competition between P_1 and C_1, P_2 and C_2, else compete between P_1 and C_2, P_2 and C_1. The better solutions survive in the competitions. Deterministic crowding is a special standard crowding strategy where CF equals 2.

Restricted Tournament Selection

Restricted tournament selection is a variant of tournament selection [14]. For each child C, choose WS (window size) individuals from the parent population and select the nearest parent P for competition according to some distance metric. If P is better than C, preserve it, else preserve C. This tournament prevents an individual from competing with individuals of different niches.

Multi-population

In multi-population EAs, the population is partitioned into a set of islands where isolated EAs are executed. Sparse exchanges of individuals are performed among these islands with the goal of introducing some diversity into the subpopulations. Common operators to link subpopulations are exchange and migration [23].

The multinational EA borrows the concept of international relationships. Each subpopulation is regarded as a nation. Operations to link nations are migration and merging [38].

Forking and Dynamic Division

The Forking Operator is an extension of multi-population EAs [37]. For the population of EAs, if certain convergence conditions are satisfied, fork the population into a parent population and a child population. In the parent population, evolution continues in one part of the search space. In the child population, evolution continues inside the other part of the search space.

Dynamic division is similar to the forking method. In the dynamic division algorithm, subpopulations are divided under given criteria, until the convergence of all subpopulations is achieved [6].

Cellular

Compared with the multi-population method, the concept of neighborhood is intensively used in cellular EAs. This means that an individual may only interact with its nearby neighbors. The induced slow migration of solutions through the population provides a kind of exploration, while exploitation takes place inside each neighborhood by genetic operations [1, 28].

13.2.3 Adaptive Niche Radius

To tackle the niche radius problem, several adaptive algorithms have been proposed to adjust the niche radius dynamically. Adaptive niching methods estimate landscape characteristics from population data. The main problem in adaptive niching methods is how to set adaptive niching rules. Complicated rules are difficult to realize while simple rules are unable to extract enough information to form stable subpopulations.

Dynamic Niche Clustering

Gan combined clustering and fitness sharing and proposed a dynamic niche clustering algorithm [8–10]. For each generation, a clustering operation calculates the adaptive niche radius with a dynamically estimated number of clusters. Within a cluster, fitness sharing is implemented with the niche radius provided by clustering. Dynamic niche clustering does not require any prior knowledge of the objective function. But the estimation of number of clusters will affect the quantity and quality of identified optimal solutions greatly.

Restricted Evolution

Restricted evolution (RE) is an adaptive version of restricted competition selection [16]. In each generation, the evolution range of each individual is dynamically expanded or shrunk with a scaling coefficient. The maximum, minimum, and initial evolution ranges are predetermined, as well as the scaling coefficient. Strictly speaking, restricted evolution is just a half-adaptive method. But it is favored for practical optimization problems because of its relatively simple control rules [20].

13.3 Restricted Evolution Evolution Strategy

The procedures of the restricted evolution evolution strategy (REES) are introduced in this section. A consequent local search operator is applied based on identified solutions, as a modification to improve fine search ability.

13.3.1 Procedures of REES

The main ideas in REES are:

1. separating the search space to several subspaces;
2. adapting the radii of subspaces to accommodate the fitness landscape;
3. introducing competition between subspaces to locate better optima;
4. using an ES as the local search operator within each subspace.

The procedures of REES are stated as follows [16]:

1. Initialization of evolution range: a_i is the evolution range for the ith design variable. If the ith design variable is v_i, child generation of the ES is generated within $[v_i - a_i, v_i + a_i]$. $a_{min,j}$, $a_{max,j}$ and $a_{init,j}$ are minimum, maximum and initial value of a_i, respectively.
2. Initialization of elite set: select μ solutions as the initial elite set, assuring that each solution is not in the evolution range of any other solutions. Given two solutions P_1 and P_2, P_2 is in the evolution range of P_1 means that the Euclidean distance of each design variable of P_1 and P_2 is smaller than the evolution range of the corresponding design variable of P_1.

3. Restricted evolution with ES: for each individual in the elite set, run a $(1+\lambda)$ ES [26] in its evolution range to find μ solutions.
4. Adaptation of evolution range: for each individual in the elite set, if the solution of $(1+\lambda)$ ES improves its fitness value, then increase all evolution ranges of its design variables with a factor P_{RE}; else decrease all evolution range of its design variables with a factor $1/P_{RE}$, where $P_{RE} > 1$.
5. Validation of new solutions: delete new solutions which are in the other solutions' evolution range and with worse fitness value. The number of deleted solutions is denoted by ξ.
6. Competition of solutions: generate $\xi + \rho$ new solutions in the whole search space randomly, assuring that they are outside of evolution ranges of existing elite solutions. During the random generation process, they should not invade evolution ranges of other solutions too. Then select the best μ individuals to construct the new elite set, from the remainder individuals in the elite set and the new generated $\xi + \rho$ solutions.
7. Stopping criterion: repeat 3–6 until the maximum generation $MaxGen_{RE}$ is reached.

13.3.2 Procedures of ES

The procedures of a $(1+\lambda)$ ES are stated as follows [26]:

1. Initialization of the elite individual: initialize a random individual in the search space as the elite.
2. Mutation of the elite individual: generate λ children from the elite using Gaussian mutation, with mutation variance $MutVar$.
3. Adaptation of the mutation variance: if the proportion of improved children is higher than a given threshold, for example, $1/5$, then increase $MutVar$ with a factor P_{ES}; else decrease $MutVar$ with a factor $1/P_{ES}$, $P_{ES} > 1$.
4. Update of the elite individual: choose the best one as the new elite among the original elite and generated children.
5. Stopping criterion: repeat 2–4, until the maximum generation $MaxGen_{ES}$ reaches.

13.3.3 Complexity of REES

The order of complexity is $O(\mu^2)$ in Step 5 and $O((\mu+\rho)^2)$ in Step 6 of REES. It is not large for small μ and ρ. Main runtime comes from the evaluation of objective function. In each generation, the algorithm evaluates objective function for $\mu + \rho + \xi + \mu \times \lambda \times MaxGen_{ES}$ times. Setting $MaxGen_{ES}$ as a small number, the consumption of function evaluation is limited, especially when μ, i.e. the number of solutions we want to obtain, is not too large.

13.3.4 Modification of REES

REES works well in locating multiple niches. If λ and $MaxGen_{ES}$ of the ES are set to a large number, the accuracy of located optima will be satisfactory while evaluation times of the objective function will increase sharply, as seen from the complexity analysis of the algorithm. Thus we introduce a consequent simplex method as the local search operator which uses identified solutions as starting points. It is expected that with the local search heuristic, each solution converges to the nearest local optimum. Modified REES maintains the ability of locating multiple niches, and improves the accuracy of solutions.

For comparison convenience, niching techniques discussed in Sect. 13.4 are all followed by the simplex method as a consequent local search operator. Therefore, accuracy of the multi-optimum optimization is not the issue.

13.3.5 Parameter Settings of REES

According to the discussion above, we conclude several experimental rules for the parameter settings of REES:

1. The minimum, maximum, and initial value of evolution range $a_{min,j}$, $a_{max,j}$ and $a_{init,j}$ for each design variable a_i affect the formation of subpopulations. Proper settings of these parameters are related to the prior knowledge of fitness landscape. Generally speaking, small minimum evolution range leads to extensive fine search, thus is unnecessary if the consequent local search operator is appended. Maximum evolution range should be large enough to cover the whole search space, but too large maximum evolution range results in too much competition between niches, thus destroy stable evolution of the population. Initial evolution range is not important because of the half-adaptive mechanism of REES. For the same reason, settings of evolution range parameters are not stringent, which is different from the niche radius problem.
2. Size of the elite set μ denotes the expected number of optima. Trade-off should be made between the multimodal optimization performance and the computation overload.
3. Appropriate choice of the number of extra competition individuals ρ will improve the quality of the whole population. But too large ρ may obstacle stable convergence of the population.
4. Setting of λ in the $(1+\lambda)$ ES also needs a trade-off. Large λ improves the quality of solutions, but increases the computation overhead.
5. The mutation variance $MutVar$ denotes the expected search space for each $(1+\lambda)$ ES. We set them as the current evolution range of design variables.
6. The adaptive parameter P_{RE} in RE reflects the expected convergence rate of solutions. Large P_{RE}, i.e. large expected convergence rate, results in unstable population, while small P_{RE} leads to low quality population.

Similarly, the adaptive parameter P_{ES} in ES reflects the expected convergence rate of each ES search process. We set both P_{RE} and P_{ES} as 1.2 through this chapter.

7. Generation numbers $MaxGen_{RE}$ and $MaxGen_{ES}$ are almost proportional to the computation time. But preliminary experiments indicate that large $MaxGen_{ES}$ is unnecessary, because the REES framework may tolerate the relatively low accuracy of each ES search process.

13.4 Numerical Experiments

The modified REES algorithm is compared with other niching techniques in two test functions, according to the quality and quantity of effective optimal solutions.

13.4.1 Performance Criteria

Evaluation of the performance of multi-optimum search depends on the engineering requirements. Here we use two performance criteria: niche number (NN) and niche objective (NO). In this chapter, a local optimum is considered to be identified if at least one solution converges to it after the consequent local search operator.

Niche Number NN

At the end of the optimization process, the niche number is defined as the number of identified local optima which reside in the feasibility region. NN characterizes the quantity of effective solutions.

Niche Objective NO

The niche objective is the sum of the objective value of local optima identified by the niching technique. NO characterizes the quality of effective solutions.

13.4.2 Test Functions

Two two-dimensional test functions are used in this chapter. The first one has a simple fitness landscape [20]:

$$\min_{x_1,x_2} \quad f_1(x_1, x_2) = \sum_{i=1}^{2} -x_i \times \cos(2\pi(x_i - 5))$$
$$\text{s.t.} \quad 0 \leq x_i \leq 9.2, \quad i = 1, 2.$$

The second test function has a relatively rugged fitness landscape, known as the Ripple function [36]:

$$\min_{x_1,x_2} \quad f_2(x_1, x_2) = \sum_{i=1}^{2} \{ (\tfrac{1}{4})^{(\frac{x_i - 0.1}{0.8})^2} \times [\sin(5\pi x_i)^6 + \tfrac{\cos(500\pi x_i)^2}{10}] \}$$
$$\text{s.t.} \quad 0 \leq x_i \leq 1, \quad i = 1, 2$$

13.4.3 Comparison in Test Functions

REES is compared with three niching methods: standard crowding where crowding factor CF equals to 2 (SCA), standard crowding where crowding factor CF equals to population number (SCB), and deterministic crowding (DC). For the convenience of comparison, these methods are all followed by a consequent simplex search. Parameters of each algorithm are listed in Table 13.1.

Table 13.1. Algorithm parameters

Algorithm	Parameters
REES	$MaxGen_{RE} = 15$, $\mu = 40$, $\rho = 10$, $\lambda = 5$, $MaxGen_{ES} = 2$
SCA	maximum generation = 100, population number = 40, $CF = 2$
SCB	maximum generation = 100, population number = 40, $CF = 40$
DC	maximum generation = 100, population number = 40

Execute the four algorithms 100 times for each test function, compute the mean value and standard variance of performance criteria, as shown in Table 13.2 and Table 13.3. For both test functions, REES and SCB can identify nearly 40 local minima, which is close to the expected number of identified optima. In the simple fitness landscape of f_1, REES and SCB have similar search ability. REES is better than SCB in niche objective NO, and slightly worse than SCB in niche number NN. In the complicated fitness landscape of f_2, REES outperforms SCB in both niche number NN and niche objective NO. The main reason is that the standard crowding algorithm has difficulty overcoming genetic drift in rugged fitness landscapes. Therefore individuals converge to a few eminent solutions along with generations. The other two algorithms, SCA and DC, both fail in identifying enough local optima, because of the influence of selection error and genetic drift.

Table 13.2. Mean value (MEAN) and standard variance (STD) of niche number NN and niche objective NO of four algorithms for f_1

Group	MEAN of NN	STD of NN	MEAN of NO	STD of NO
1	38.1500	1.1315	−449.0278	16.4659
2	2.2400	0.4522	−28.7402	6.8227
3	39.3600	0.8229	−394.4150	24.3628
4	11.6100	2.6927	−153.3006	33.8637

Table 13.3. Mean value (MEAN) and standard variance (STD) of niche number NN and niche objective NO of four algorithms for f_2

Group	MEAN of NN	STD of NN	MEAN of NO	STD of NO
1	31.8300	2.0521	−41.8183	2.0168
2	2.0600	0.3712	−3.5912	0.6755
3	24.5100	1.6361	−35.7460	2.1747
4	5.5200	1.3961	−10.5463	2.6811

13.5 Practical Engineering Optimization

In this section, the modified REES algorithm is applied to the practical optimal design of recording optics of the varied-line-spacing holographic grating (VLSHG) in the National Synchrotron Radiation Laboratory (NSRL).

13.5.1 Recording Optics of VLSHG

Holographic gratings are fabricated by recording interference fringes of two coherent sources in the photoresist coated on grating blanks. Given the shape of blanks, their focal properties can be adjusted by altering the properties of two recording light sources. Ling [19] introduced auxiliary uniform-line-spacing gratings to generate aspherical wave-fronts to fabricate varied-line-spacing holographic gratings. VLSHGs are able to correct high order aberrations in diffractive optical systems and are widely used in high resolution spectrometers and monochromators.

The recording optical system is shown in Fig. 13.1. It consists of two coherent point light sources of wavelength λ_0, C and D, and two uniform-line-spacing gratings, G_1 and G_2. Four distance parameters and four angle parameters, p_C, q_C, p_D, q_D, γ, η_C, δ, η_D, are known as recording parameters [19].

The aim of optimal design is to find a set of recording parameters to form the expected groove shape, which is generally represented by the expected groove density $n_E(w)$ in the Y axis:

$$n_E(w) = n_{0E}(1 + b_{2E}w + b_{3E}w^2 + b_{4E}w^3), \quad -w_0 \le w \le w_0. \quad (13.3)$$

Here n_{0E}, b_{2E}, b_{3E}, b_{4E} are expected groove density parameters, w is the coordinate in the Y axis, and w_0 is the half-width of the grating to record.

For a given set of recording parameters, the practical groove density $n_P(w)$ can be calculated from a complicated function g of recording parameters [19]:

$$n_P(w) = g(p_C, q_C, p_D, q_D, \gamma, \eta_C, \delta, \eta_D). \quad (13.4)$$

We formulate the objective function as minimizing the integral of square errors of the expected groove density and the expected groove density in the Y axis [20]:

$$\min \quad f = \int\limits_{-w_0}^{w_0} (n_P(w) - n_E(w))^2 dw. \tag{13.5}$$

We evaluate (7) by numerical integral with proper step size. The objective function for minimization is multidimensional and multimodal, and its evaluation is time-consuming.

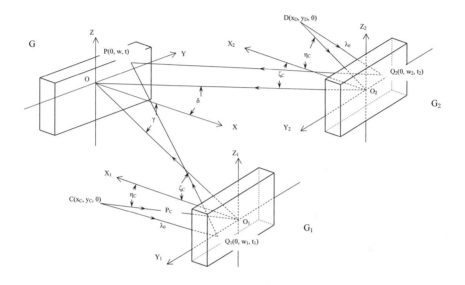

Fig. 13.1. Schematic diagram of the recording system consists of two coherent point sources, C and D, two auxiliary uniform-line-spacing gratings, G_1 and G_2, and a plane grating blank G. Recording parameters are four distance parameters and four angle parameters

13.5.2 Multimodal Optimal Design of VLSHG

The optimal design of VLSHG is a typical multi-optimum search problem with a multimodal function. The main reason for multi-optimum search is the requirement to handle implicit constraints. There are two kinds of constraints, explicit and implicit, in the optimization process. Firstly, recording parameters must satisfy the upper and lower bound constraints. These kinds of explicit constraints come from the restriction of the size of recording table. Secondly, in the practical recording system, there are many other optical elements not shown in the schematic diagram, such as beam splitter, light filter, et cetera. All these optical elements must do not disturb each other. For the ease of maintenance, it is also expected that they are evenly placed in the recording table. Furthermore, improper recording parameters may bring on extra demand of auxiliary optical elements, or extra requirement of fine calibration. These kinds of implicit constraints are difficult or inconvenient to be described in a closed mathematical formulation.

To tackle the implicit constraint problem, we adopt the following optimal design procedures:

1. Construct the objective function and explicit constraints, and search for multiple solutions with the multimodal optimization algorithm.
2. Draw schematic diagrams for solutions, and exclude those possibly violating implicit constraints. Notice that with the consequent local search operator, some solutions may violate explicit constraints, and must be excluded too.
3. Evaluate concrete design schemes for the qualified solutions with the aid of optical manufacturing engineers, select a final solution, and keep several proper solutions as alternatives in case of unexpected difficulties.

The multimodal optimization algorithm plays an important role in our optimal design procedures for both satisfying implicit constraints and providing backup solutions. Considering the complexity of the fitness landscape for the VLSGH design, finding the global optimum is almost an impossible task. Thus the objective of optimization is to find several groups of acceptable solutions, and the main concern of the multi-optimum optimization algorithm is the diversity of solutions. According to numerical experiments in Sect. 13.4, the modified REES algorithm is a proper choice for the VLSHG design because of its capability of multi-optimum search in a complicated fitness landscape.

13.5.3 Modified REES for VLSHG Design

The VLSHG to design in NSRL has desired groove density parameters: $n_{0E} = 1.4000 \times 10^3$ groove/mm, $b_{2E} = 8.2453 \times 10^{-4}$ mm^{-1}, $b_{3E} = 3.0015 \times 10^{-7}$ mm^{-2}, $b_{4E} = 0.0000 \times 10^{-10}$ mm^{-3}. Half width of the grating is $w_0 = 90$ mm and the recording wavelength is $\lambda_0 = 413.1$ nm. Auxiliary grating G_2 has groove density $n_2 = 1.0000 \times 10^3$ groove/mm and the diffraction order equals to $+1$. Auxiliary grating G_1 has groove density $n_1 = 0$ groove/mm, i.e. a plane mirror. Upper bounds and lower bounds of recording parameters are from the physical constraints of recording table. For distance parameters, upper bounds are 2000 mm, lower bounds are 100 mm. For angle parameters, upper bounds are $\pi/2$, lower bounds are $-\pi/2$.

To optimize the objective function with modified REES, the algorithm parameters are: $MaxGen_{RE} = 30$, $\mu = 40$, $\rho = 20$, $\lambda = 20$, $MaxGen_{ES} = 2$. Four sets of feasible recording parameters are shown in Table 13.4. Corresponding groove parameters are shown in Table 13.5, denoted as n_0, b_2, b_3, and b_4. Corresponding objective functions are also provided, denoted as f. Schemes of the solutions are shown in Fig. 13.2.

Table 13.4. Four sets of feasible recording parameters

Group	γ(rad)	η_C(rad)	δ(rad)	η_D(rad)	p_C(mm)	q_C(mm)	p_D(mm)	q_D(mm)
1	-0.0999	0.0999	0.4990	1.1427	574.4	752.0	427.8	989.9
2	-0.5970	0.5970	0.0156	1.2673	510.4	316.4	254.7	808.1
3	0.3865	-0.3865	1.2709	0.6494	782.5	710.6	112.4	1451.4
4	-0.1748	0.1748	0.4164	0.9876	510.0	518.0	520.9	380.8

Table 13.5. Corresponding groove parameters and objective functions of the optimized recording parameters

Grp	n_0(groove/mm)	b_2(mm^{-1})	b_3(mm^{-2})	b_4(mm^{-3})	f
1	1.3999×10^3	8.2457×10^{-4}	3.0028×10^{-7}	-0.0005×10^{-10}	9.9202×10^{-3}
2	1.3999×10^3	8.2472×10^{-4}	3.0047×10^{-7}	-0.0007×10^{-10}	9.8671×10^{-3}
3	1.4000×10^3	8.2440×10^{-4}	3.0017×10^{-7}	-0.0002×10^{-10}	8.5979×10^{-5}
4	1.4001×10^3	8.2453×10^{-4}	3.0017×10^{-7}	0.0002×10^{-10}	1.0050×10^{-2}

13.5.4 Discussion of Solutions

Before discussion of the optimized solutions, we need to clarify several issues regarding the practical recording optics:

1. Though we refer to two light sources C and D in the schematic diagram in Fig. 13.1 and previous discussions, it is obligatory to use a single laser generator to generate the two light sources. Because the coherence of light sources can not be guaranteed by using two laser generators. Beam splitters and light filters are applied to generate two coherent light sources, as shown in the schematic diagram in Fig. 13.3. A plane wave from light source S reaches the beam splitter BS, i.e. a half-transparent-half-reflecting mirror, and is split to two plane waves. The two plane waves are filtered by light filters LF_1 and LF_2, and transformed to two spherical waves, i.e. point light sources C and D. Furthermore, directions of the principle rays from C and D are adjusted by using extra plane mirrors. Therefore, a proper solution of the recording optics needs to leave appropriate room for the placement of beam splitters, light filters, and other auxiliary optical elements. On the other hand, large distance between C and D requires long light distance before the plane waves reach the light filters, thus we need to adjust beam splitters and light filters carefully to guarantee the coherence and parallelism of the two plane waves.

2. Adjustment of the distance and angle parameters are always challenging in the recording optics. Generally we require the largest tolerance is 1 mm for distance parameters and 0.001 rad for angle parameters. Proper recording parameters are able to mitigate the burden of system adjustment. For example, if the lengths of the incident arm and the emergent arm are not

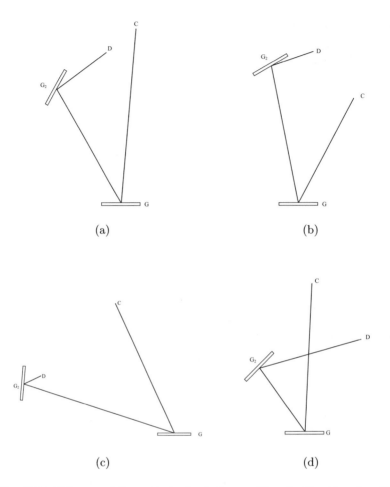

Fig. 13.2. Schemes of the optimized solutions, with only the +1 order diffractive light shown. Here **(a)**–**(d)** are corresponding to Groups 1–4 of Table 13.2

in the same magnitude, adjustment error in the shorter arm will introduce a much larger system error than that of the longer arm. Regarding the angle parameters, grazing incident ray, i.e. incidence angle near to $\pi/2$, should be avoided.

3. The auxiliary plane gratings will generate multiple orders of diffractive rays. The orders of diffractive rays depend on the recording parameters. Generally diffractive rays from order -5 to order $+5$ are observable. Note that we only need order $+1$ for the recording, thus other diffractive rays

may interfere the recording process. One solution to this problem is to use screens to absorb other lights. Another and more preferred solution is to choose a proper set of recording parameters such that the lights of other orders will not reach the grating G.

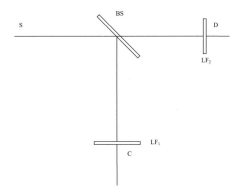

Fig. 13.3. A plane wave from light source S reaches the beam splitter BS, and is split to two plane waves. The two plane waves are filtered by light filters LF_1 and LF_2, and transformed to two spherical waves C and D

The recording parameters Group 1 and Group 2 of Table 13.2 are both proper solutions. Group 2 is slightly inferior to Group 1 because of the small imbalance between the incident arm p_D and emergent arm q_D, with a ratio of 1/3.

In Group 3, the incident arm p_D is much smaller than the other arms. This magnitude of imbalance is intolerable for optical system calibration. We can use a predetermined threshold to penalize this kind of solution, but it will introduce twelve extra inequality constraints for four distance parameters, and result in a more complicated optimization problem. For recording parameters in Group 4, light sources C and D are too far, thus the plane waves have to travel a long distance before reach the light filters. Therefore the laser generator should have a high stability to guarantee the parallelism of light sources. Furthermore, the optics are exposed in the air, thus the coherence is vulnerable to stray lights and vibrations. Large light distance will increase the risk of incoherence of the two plane waves. Hence Group 4 sacrifices the system reliability comparing to Group 1 and Group 2. Though Group 3 and Group 4 are feasible solutions, they are not proper solutions from the viewpoint of practical fabrication. Noting that Group 3 is the best one regarding the objective function value, this fact underlines the necessity of the multimodal optimization algorithm and the usefulness of our optimal design procedures.

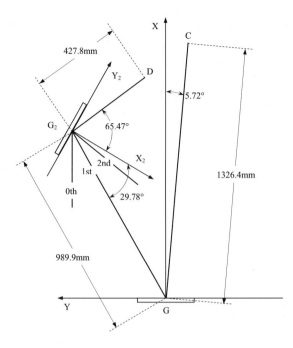

Fig. 13.4. Final recording optics adopts the optimized recording parameters in Group 1 of Table 13.2. Light originated from D is diffracted at auxiliary grating G_2, and the $+1$ order diffractive light reaches G. Light originated from C reaches G directly because the auxiliary plane mirror G_1 can be omitted. The other diffractive lights will not disturb the recording process when recording parameters are properly selected

We finally leave Group 2 as a backup solution, and select Group 1 as the practical recording parameters, as shown in Fig. 13.4. Note that by properly selecting recording parameters, the other diffractive lights will not disturb the recording process while only the $+1$ order diffractive light reaches G.

13.6 Conclusion

In this chapter, we focus on an important kind of optimization problem for practical applications: multimodal function optimization, i.e. a search for multiple optimal solutions. A survey of the main niching techniques for multi-optimum search is given. According to the method of dealing with the niche radius problem, niche techniques are classified as explicit, implicit, and adaptive methods.

A specified adaptive niche radius method, REES, is selected to demonstrate the usefulness of multi-optimum search in a practical engineering design. To improve the local search ability, a consequent local search operator is applied to modify the original REES. Numerical experiments indicate that REES outperforms the standard crowding algorithm and the deterministic crowding algorithm, regarding both quantity and quality of solutions.

The characteristics of REES are fit for our practical multi-optimum design problem, i.e. VLSHG design. In the VLSHG design problem, various implicit constraints, which are difficult or inconvenient to be described in a closed mathematical formulation, emerge from the requirement of practical fabrication. To cope with implicit constraints, multiple solutions are expected for trials by optical engineers. The modified REES algorithm is applied to the practical design of a VLSHG in NSRL, and achieves satisfactory results.

References

1. Alba, E., Dorronsoro, B.: The exploration/exploitation tradeoff in dynamic cellular genetic algorithms. IEEE Transactions on Evolutionary Computation **9**, 126–142 (2005)
2. Beasley, D., Bull, D., Martin, R.: A sequential niche technique for multimodal function optimization. Evolutionary Computation **1**, 101–125 (1993)
3. Cedeno, W., Vemuri, V.: Analysis of speciation and niching in the multi-niche crowding GA. Theoretical Computer Science **299**, 177–197 (1999)
4. Crutchley, D., Zwolinski, M.: Globally convergent algorithm for DC operating point analysis of nonlinear circuits. IEEE Transactions on Evolutionary Computation **7**, 2–10 (2003)
5. Deb, K., Goldberg, D.: An investigation of niche and species formation in genetic function optimization. In: Proceedings of International Conference on Genetic Algorithms, pp. 42–50 (1989)
6. Elo, S.: A parallel genetic algorithm on the CM-2 for multi-modal optimization. In: Proceedings of International Conference on Evolutionary Computation, pp. 818–822 (1994)
7. Francois, O.: An evolutionary strategy for global minimization and its Markov chain analysis. IEEE Transactions on Evolutionary Computation **2**, 77–90 (1998)
8. Gan, J., Warwick, K.: A genetic algorithm with dynamic niche clustering for multimodal function optimization. In: Proceedings of International Conference on Artificial Neural Network and Genetic Algorithms, pp. 248–255 (1999)
9. Gan, J., Warwick, K.: Dynamic niche clustering: a fuzzy variable radius niching technique for multimodal optimisation in GAs. In: Proceedings of IEEE Congress on Evolutionary Computation, pp. 215–222 (2001)
10. Gan, J., Warwick, K.: Modelling niches of arbitrary shape in genetic algorithms using niche linkage in the dynamic niche clustering framework. In: Proceedings of IEEE Congress on Evolutionary Computation, pp. 43–48 (2002)
11. Goldberg, D.: Sizing populations for serial and parallel genetic algorithms. In: Proceedings of International Conference on Genetic Algorithms, pp. 70–79 (1989)
12. Goldberg, D., Richardson, I.: Genetic algorithms with sharing for multimodal function optimization. In: Proceedings of International Conference on Genetic Algorithms, pp. 41–49 (1987)

13. Gomez, J.: Self adaptation of operator rates for multimodal optimization. In: Proceedings of IEEE Congress on Evolutionary Computation, pp. 1720–1726 (2004)
14. Harik, G.: Finding multimodal solutions using restricted tournament selection. In: Proceedings of International Conference on Genetic Algorithms, pp. 24–31 (1995)
15. Hocaoglu, C., Sanderson, A.: Planning multiple paths with evolutionary speciation. IEEE Transactions on Evolutionary Computation **5**, 169–191 (2001)
16. Im, C., Kim, H., Jung, H., Choi, K.: A novel algorithm for multimodal function optimization based on evolution strategy. IEEE Transactions on Magnetics **40**, 1224–1227 (2004)
17. Lee, C., Cho, D., Jun, H., Lee, C.: Niching genetic algorithm with restricted competition selection for multimodal function optimization. IEEE Transactions on Magnetics **34**, 1722–1725 (1999)
18. Li, J., Blazs, M., Parks, G., Clarkson, P.: A species conserving genetic algorithm for multimodal function optimization. Evolutionary Computation **10**, 207–234 (2002)
19. Ling, Q., Wu, G., Liu, B., Wang, Q.: Varied line spacing plane holographic grating recorded by using uniform line spacing plane gratings. Applied Optics **45**, 5059–5065 (2006)
20. Ling, Q., Wu, G., Wang, Q.: Restricted evolution based multimodal function optimization in holographic grating design. In: Proceedings of IEEE Congress on Evolutionary Computation, pp. 789–794 (2005)
21. Liu, X., Lee, B.: Optimal design for ultra-broad-band amplifier. Journal of Lightwave Technology **21**, 3446–3454 (2003)
22. Liu, X., Lee, B.: Optimal design of fiber Raman amplifier based on hybrid genetic algorithm. IEEE Photonics Technology Letters **16**, 428–430 (2004)
23. Logot, G., Walter, M.: Computing robot configurations using a genetic algorithm for multimodal optimization. In: Proceedings of International Conference on Evolutionary Computation, pp. 312–317 (1998)
24. Mahfoud, S.: Crossover interactions among niches. In: Proceedings of International Conference on Evolutionary Computation, pp. 188–193 (1994)
25. Mahfoud, S.: Genetic drift in sharing methods. In: Proceedings of International Conference on Evolutionary Computation, pp. 67–71 (1994)
26. Michalewicz, Z.: Genetic Algorithms + Data Structures = Evolution Programs. Springer, Berlin (1996)
27. Miller, B., Shaw, M.: Genetic algorithms with dynamic niche sharing for multimodal function optimization. In: Proceedings of International Conference on Evolutionary Computation, pp. 786–791 (1996)
28. Nakashima, T., Ariyama, T., Yoshida, T., Ishibuchi, H.: Performance evaluation of combined cellular genetic algorithms for function optimization problems. In: Proceedings of International Symposium on Computational Intelligence in Robotics and Automation, pp. 295–299 (2003)
29. Nijssen, S., Bäck, T.: An analysis of the behavior of simplified evolutionary algorithms on trap functions. IEEE Transactions on Evolutionary Computation **7**, 11–22 (2003)
30. Petrowski, A.: A clearing procedure as a niching method for genetic algorithms. In: Proceedings of International Conference on Evolutionary Computation, pp. 798–803 (1996)

31. Rudolph, G.: Self-adaptive mutations may lead to premature convergence. IEEE Transactions on Evolutionary Computation **5**, 410–414 (2001)
32. Sareni, B., Krahenbuhl, L.: Fitness sharing and niching methods revisited. IEEE Transactions on Evolutionary Computation **2**, 97–106 (1998)
33. Sareni, B., Krahenbuhl, L., Nicolas, A.: Niching genetic algorithms for optimization in electromagnetics I – fundamentals. IEEE Transactions on Magnetics **34**, 2984–2991 (1998)
34. Sareni, B., Krahenbuhl, L., Nicolas, A.: Efficient genetic algorithms for solving hard constrained optimization problems. IEEE Transactions on Magnetics **36**, 1027–1030 (2000)
35. Singh, G., Deb, K.: Comparison of multi-modal optimization algorithms based on evolutionary methodologies. In: Proceedings of the Genetic and Evolutionary Computation Conference (GECCO-2006), pp. 1305–1312. The Association for Computing Machinery (ACM), New York (2006)
36. Thomsen, R.: Multimodal optimization using crowding based differential evolution. In: Proceedings of IEEE Congress on Evolutionary Computation, pp. 1382–1389 (2004)
37. Tsutsui, S., Fujimoto, Y., Hayashi, I.: Extended forking genetic algorithm for order representation (o-fGA). In: Proceedings of International Conference on Computational Intelligence, pp. 639–644 (1994)
38. Ursem, R.: Multinational evolutionary algorithms. In: Proceedings of IEEE congress on Evolutionary Computation, pp. 1633–1640 (1999)
39. Yin, X., Germay, N.: A fast genetic algorithm with sharing scheme using cluster analysis methods in multimodal function optimization. In: Proceedings of International Conference on Artificial Neural Networks and Genetic Algorithms, pp. 450–457 (1993)

GPBG: A Framework for Evolutionary Design of Multi-domain Engineering Systems Using Genetic Programming and Bond Graphs

Jianjun Hu[1], Zhun Fan[2], Jiachuan Wang[3], Shaobo Li[4], Kisung Seo[5], Xiangdong Peng[6], Janis Terpenny[7], Ronald Rosenberg[8], and Erik Goodman[9]

[1] Department of Computer Science and Engineering, University of South Carolina, SC 29208, USA jianjunh@cse.sc.edu
[2] Department of Mechanical Engineering, Technical University of Denmark, Building 404, DK-2800 Lyngby, Denmark zf@mek.dtu.dk
[3] United Technologies Research Center, Systems Department, East Hartford, CT 06108, USA wangj2@utrc.utc.com
[4] CAD/CIMS Institute, Guizhou University, Guiyang, Guizhou 550003, China lishaobo@gzu.edu.cn
[5] Department of Electronics Engineering, Seokyeong University, Seoul 136-704, Korea ksseo@skuniv.ac.kr
[6] Department of Electrical and Computer Engineering, Michigan State University, East Lansing, MI 48824, USA pengxian@egr.msu.edu
[7] Department of Engineering Education, Virginia Tech, Blacksburg, VA 24061, USA terpenny@vt.edu
[8] Department of Mechanical Engineering, Michigan State University, East Lansing, MI 48824, USA rosenber@egr.msu.edu
[9] Department of Electrical and Computer Engineering, Michigan State University, East Lansing, MI 48824, USA goodman@egr.msu.edu

14.1 Introduction

Current engineering design is a multi-step process proceeding from conceptual design to detailed design and to evaluation and testing. It is estimated that 60–70% of design decisions and most innovation occur in the conceptual design stage, which may include conceptual design of function, operating principles, layout, shape, and structure. However, few computational tools are available to help designers to explore the design space and stimulate the product innovation process. As a result, product innovation is strongly constrained by the designer's ingenuity and experience, and a systmatic approach to product innovation is strongly needed.

Many engineering design problems, such as mechatronic systems, can be abstracted as a design space exploration problems in which a set of building blocks or modules need to be assembled/connected together to compose a system satisfying a set of given functional requirements. In many cases, such as analog circuit design, it is relatively easy to simulate a product design model to evaluate its functional performance via simulation software such as P-SPICE, while it is extremely hard to come up with an innovative design solution out of the almost unlimited number of design candidates of the topological design space. An efficient topological search technique is needed to help to improve this process.

In recent decades, evolutionary computation has emerged as an effective and promising search/optimization technique that is suitable for large-scale non-linear multi-modal engineering optimization problems. In particular, genetic programming (GP) has been used as an attractive approach for engineering design innovation in a variety of domains, including design of analog circuits, digital circuits, chemical molecules, control systems, et cetera [12]. Such work employs GP as a topologically open-ended search technique for functional design innovation – achieving given behaviors without pre-specifying the design topology – and has achieved considerable success. While electrical circuits and block diagrams are well suited for the design problems in analog circuit design and controller synthesis, many engineering design problems cover multiple domains, including, for example, mechanical, electrical, and hydraulic subsystems. Since 2001, we have been developing a new framework, called GPBG [19], for automated synthesis of multi-domain systems using genetic programming and bond graphs [8], which are a well-established modeling tool for multi-domain systems.

In this chapter, we will detail how an engineering design problem can be solved under the GPBG framework using several design synthesis problems: a vibration absorber, a MEMS filter design, and a controller design in suspension systems. The rest of the chapter is organized as follows. Section 14.2 presents a survey of applications of evolutionary algorithms in engineering design synthesis that are more than just parameter optimization. Section 14.3 introduces the GPBG framework, which exploits genetic programming and bond graphs for automated synthesis of dynamic systems. In Sect. 14.4, first, the vibration absorber design problem is used to illustrate the basic approach to mapping the engineering design problem into a topology space search problem using genetic programming. We then use a MEMS filter design problem to show how expert domain knowledge can be incorporated into the evolutionary synthesis and greatly improve the efficiency of this approach. The third application uses a different, direct-encoding GPBG approach for synthesizing controllers of suspension systems. A brief summary of evolving robust designs is then presented. Finally, the conclusions and future research are highlighted in Sect. 14.5.

14.2 Related Work

Automated synthesis of dynamic systems has been investigated intensively in the past ten years. Most of that work is related to analog circuit synthesis, as pioneered by Koza and his colleagues [10, 12]. Their work in automated analog circuit synthesis, including low-pass, high-pass, and asymmetric band-pass filters, is described in [10,11]. Lohn and Colombano [14] proposed a linear representation approach to evolve analog circuits. Ando and Iba [1] suggested another simple linear genome method to evolve low-pass and band-pass filters with small numbers (<50) of components. In our previous work, we applied GP to the lowpass analog filter design problem [2], MEMS [3], the printer mechanism design, active-passive dynamic system design [22], all using bond graphs as the modeling and simulation tool. Controllers, or dynamic systems represented as block diagrams have also been synthesized automatically using genetic programming by Koza et al. [13]. This work has led to the invention of a patentable controller having better performance than a standard PID controller.

14.3 The GPBG Framework for Evolutionary Design

In this section, we present a generic methodology for open-ended computational synthesis of multi-domain dynamic systems based on bond graphs [9] and genetic programming – the GPBG approach = Genetic Programming + Bond Graphs.

14.3.1 Genetic Programming

Genetic programming is a derivative of genetic algorithms that is characterized by its capability to evolve programs. In typical GP, an individual is represented as a syntax/GP tree composed of functions and terminals defined by the user according to the problem. Each function has one or more inputs, while terminals have no inputs. Both functions and terminals can be executed to generate some output or do some processing such as inserting a new component into a developing/growing analog circuit. Genetic programming's open-ended topological search capability has been widely applied to computational synthesis of analog circuits, controllers, mechatronic systems, quantum circuits, et cetera.

14.3.2 Bond Graphs

The bond graph is a multi-domain modeling tool for analysis and design of dynamic systems, especially hybrid multi-domain systems, including mechanical, electrical, pneumatic, hydraulic, etc., components. Details of notation and methods of systems analysis related to the bond graph representation can be found in [9]. Figure 14.1 illustrates a small bond graph that represents the accompanying electrical system.

A typical simple bond graph model is composed of inductors (I), resistors (R), capacitors (C), transformers (TF), gyrators (GY), 0-Junctions (J0), 1-junctions (J1), sources of effort (SE), and sources of flow (SF). In this chapter, we are only concerned with linear dynamic systems and do not include transformers and gyrators as components.

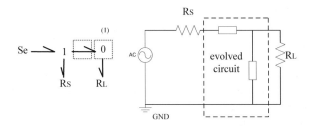

Fig. 14.1. A bond graph and its equivalent circuit. The dotted boxes in the left bond graph indicate modifiable sites at which further topological manipulations can be applied (to be explained in the next section)

In the context of electric circuit design, a bond graph consists of the following types of elements:

- C, I, and R elements, which are passive one-port elements that contain no sources of power, and represent capacitors, inductors, and resistors.
- Power source elements including Se and Sf, which are active one-port elements representing sources of voltage and current, respectively. In addition, when the current of a current source is fixed as zero, it can serve as an ideal voltage gauge. Similarly, when the voltage of a voltage source is fixed as zero, it can serve as an ideal current gauge.
- Transformer (TF) and gyrator (GY), which are two-port elements. Power (i.e., product of voltage and current) is conserved in these elements, but the values of voltage and current may be changed in the elements.
- 0-junctions and 1-junctions, which are multi-port elements for representing series and parallel relationships among elements. They serve to interconnect elements into subsystem or system models.
- Bonds, which are used to connect any two elements in the bond graph.

A unique characteristic of bond graphs is their use of 0- and 1-junctions to represent the series and parallel relationships among components in circuits. In fact, it is this concept that led to the foundation of the bond graph field [15]. Junctions transform common circuits into a very clean structure with few loops, which can otherwise make circuits appear very complicated. Figure 14.1 shows the comparison of a circuit diagram and a corresponding bond graph. The evaluation efficiency of the bond graph model is further improved due to the fact that analysis of causal relationships and power flow between elements and subsystems can reveal certain system properties and inherent characteristics. This makes it possible to discard infeasible design candidates even before numerically evaluating them, thus reducing time of evaluation to a large degree. In addition, as virtually all of the circuit topologies created are valid, our system does not need to check validity conditions of individual circuits to avoid singular situations that could interrupt the running of a program evaluating them.

14.3.3 GPBG = GP + Bond Graphs

By combining the topological search capability of GP and the multi-domain representation feature of bond graphs, GPBG provides an appealing approach for open-ended synthesis of multi-domain systems. To map an engineering design problem into the GPBG framework, the design space of target design solutions must first be identified, including all component types, component interfaces, and connection types. Then, depending on how we encode the bond graphs using the GP tree, two types of approaches have been used within GPBG framework. One is the developmental GPBG approach, similar to Koza's work on evolving analog filters, in which the bond graph phenotypes are grown from an embryo bond graph by executing the GP tree program to manipulate the topology. The other approach is the direct encoding GPBG, in which the GP trees directly encode the bond graph topology. This method shares some similarity to the GP-based controller synthesis approach by Koza [13].

Developmental GPBG

The problem of automated synthesis of bond graphs involves two basic searches – the search for a good topology and the search for good parameters for each topology – in order to be able to evaluate its performance. Based on Koza's work [10] on automated synthesis of electronic circuits, we created a developmental GPBG system for synthesizing mechatronic systems, including:

1. An embryo bond graph with modifiable sites at which further topological operations can be applied to grow the embryo into a functional system.
2. A GP function set, composed of a set of topology manipulation and other primitive instructions which will be assembled into a GP tree by the evolutionary process (execution of this GP program leads to topological and parametric manipulation of the developing embryo bond graph).
3. A fitness function to evaluate the performance of candidate solutions.

We use the analog filter synthesis problem as an example to illustrate the developmental GPBG approach [2]. In this problem, the design space consists of bond graphs composed of capacitors (C), inductors (I), resistors (R), 1-junctions (J1) and 0-junctions (J0) (we omitted transformers and gy-rators for the sake of simplicity). We have two types of elements in a bond graph. One is the elements including C/I/R/J1/J0 which have one or more interface ports. The second type of elements are two-port bonds. The applica-ble topological operations on node elements include replacing the component type (on C/I/R/J1/J0), and adding a new C/I/R component (on J1/J0). The topological operations on bonds include inserting a new J1/J0. This operator set will enable the GPBG to evolve a large number of bond graphs. Then we develop an alternative, "basic" GP function set for this problem, includ-ing {Insert_J0/J1, Add_C/I/R, and Replace_C/I/R}. Figure 14.2 and 14.3 shows how these topological manipulation GP functions work. Note that for Add_C/I/R functions, we can have one or more branches that accept numeric subtrees to set the component parameters. More examples of developmental GP are presented in Sect. 14.4.1 and Sect. 14.4.2.

The developmental GPBG framework enables us to do simultaneous topol-ogy and parameter search. Compared to the direct encoding GPBG below, it can evolve much more diverse topology types including those with loops and has more flexibility during the evolutionary search process, but at the cost of a less intuitive GP tree.

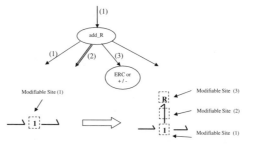

Fig. 14.2. The add_R function adds a resistor to a junction

GPBG with Direct Encoding

One interesting observation of typical bond graphs is that most bond graphs do not have loops and are themselves tree-structured. This makes it natural to use the GP trees themselves to represent/encode the bond graph structure. We thus proposed the direct-encoding GPBG approach for evolving tree-type bond graph models. In this approach, 1-junction and 0-junction are used as GP functions with two input variables (ports). The capacitors/resistors/inductors

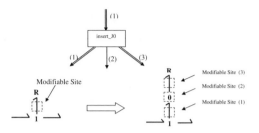

Fig. 14.3. The Insert_J0 inserts a new 0-junction into a bond

are all GP functions with one input that connects to a numeric subtree to establish the component size (parameter value). We also have plus/minus arithmetic operators in addition to ERC random terminals. One such function set is shown in Table 14.1, which is used to synthesize controllers for a suspension system in Sect. 14.4.3. Since there is a one-to-one correspondence of the GP tree and bond graph topology, one can easily build a bond graph by following the topologies in the GP trees.

Table 14.1. GP function set for suspension controller synthesis

Name and description	Function arity
J0 – Junction (0)	2
J1 – Junction (1)	2
R Element (R)	1
C Element (C)	1
I Element (I)	1
Arithmetic + (+): add two ERCs	2
Arithmetic – (–): Subtract two ERCs	2
Ephemeral Random Constant (ERC)	0

With this encoding approach, each GP tree is a bond graph. Actually, here, all GP trees are binary trees. Clearly, this can only represent a subset of bond graphs compared to the developmental GPBG approach, but enjoys simplicity in implementation. Since the standard GP needs to specify the arity of the GP functions, this limits the number of ports of the junctions to be three, which is a shortcoming of this approach. However, this disadvantage can be ameliorated by defining a larger arity (e.g., 8) and then defining a null-element terminal. In this way, we can evolve 6-arity GP trees in which many of the ports are simply empty. Most real-world bond graph models have fewer than 8 ports. One possible disadvantage maybe that this may greatly increase the search space.

14.4 Case Studies of Evolutionary Synthesis Using GPBG

In this section, we applied GPBG to three real-world design problems, including synthesis of a passive vibration absorber, a MEMS filter, a robust analog filter, and a controller for a suspension system. These examples are used to illustrate the following unique advantages of GPBG based evolutionary design compared to conventional design approaches:

- topologically open-ended exploration of design space (the vibration absorber design problem);
- easy and natural incorporation of domain knowledge into the design process (the MEMS filter design problem);
- the capability to evolve novel and unconventional design solutions (the suspension system design problem).

Each problem will start with the description of the design space and the configuration of GPBG, including design embryo, GP function set, and fitness function. We also discuss the strategies to evolve robust designs using the GPBG framework.

14.4.1 Synthesis of Mechanical Vibration Absorber

Problem Description

Vibration absorbers are a class of dynamic systems, and can be modeled as analog circuits, block diagrams, bond graphs, et cetera. One special characteristic of these particular dynamic systems is that the building blocks usually have a fixed number interface ports and may not be connected arbitrarily.

In this section, we are mainly interested in synthesizing passive vibration absorbers to reduce the vibration response of primary systems of various configurations. Figure 14.4 shows a primary system and its corresponding bond graph model. The design task is to attach some new components to the primary system such that the frequency response at the excitation frequency ω be minimized. Figure 14.5 shows the first vibration absorber, invented by H. Frahm in 1911, and its bond graph model. The frequency response of the stand-alone primary system and the primary system with vibration absorber is shown in Fig. 14.6. It can be seen that the vibration absorber can significant quench the response of the primary systems at the excitation frequency.

In this design problem, the objective is to synthesize a vibration absorber such that the frequency response

$$f_{raw} = |TF(j\omega)||_{\omega=\omega_0} \tag{14.1}$$

of the primary system mass (displacement) at the frequency ω of excitation force $f = f_0 \times \sin\omega t$ is minimized. This problem is extracted from [7]. We want

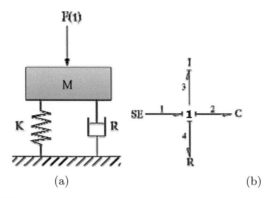

Fig. 14.4. The bond graph structure of a primary system and its bond graph model (**a**) The primary system under perturbation of excitation force $F(t)$. (**b**) The bond graph model of the embryo system

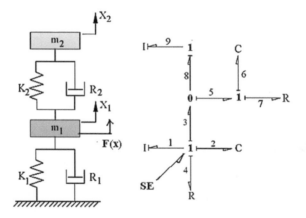

Fig. 14.5. The bond graph structure of the first patented vibration absorber and its bond graph model

to see if the GPBG system can reinvent the first patented vibration absorber, shown in Fig. 14.5. The parameters of the primary system are as follows: m_p = 5.77 kg; k_p=251.132 × 1e6 N/m; c_p= 192.92 kg/s. The parameters of the standard passive absorber solution are as follows: m_a = 0.227 kg; k_a=9.81e6 N/m; c_a= 355.6 kg/s.

GPBG Configuration

The design space of passive vibration absorbers is composed of masses (R), springs (C), and dampers, corresponding to Resistor(R), Capacitor (C) and

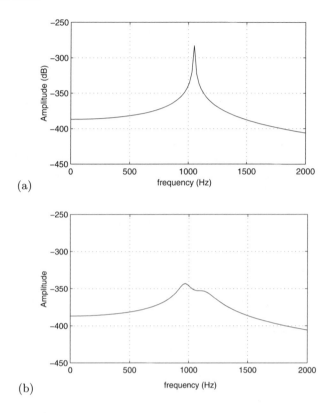

Fig. 14.6. Frequency responses of the primary system under perturbation of excitation force $F(t)$, without and with a vibration absorber: **(a)** without vibration absorber; **(b)** with vibration absorber

Inductor (I), respectively. Following the GPBG framework outlined before, we used the bond graph embryo in Fig. 14.1 for this design problem. The modifiable site is the 1-junction. Since it is not physically realistic to have many masses attached to the primary structures, we limit the maximum number of masses to two in all the experiments.

In our earliest work [2], a "basic" GP function set was used for evolutionary synthesis of analog filters. In that approach, the GP functions for topological operation included {Insert_J0/J1, Add_C/I/R, and Replace_C/I/R}, which allowed evolution of a large variety of bond graph topologies. The shortcoming of this approach is that it tended to evolve redundant and sometimes causally ill-posed bond graphs [18]. Later, we used a causally well-posed modular GP function set to evolve more concise bond graphs with much less redundancy [6]. However, that encoding had a strong bias toward a chain-type topology and thus may have limited the scope of topology search [5]. Here we have improved the basic function set in [2] and developed the following hybrid function set approach to reduce redundancy while enjoying the flexibility of topological exploration:

F={Insert_J0E, Insert_J1E, Add_C/I/R, EndNode, EndBond, ERC}

where the Insert_J0E, Insert_J1E functions insert a new 0/1-junction into a bond while attaching at least one and at most three elements (from among C/I/R). Figure 14.7 illustrates the operation of the Insert_J0E function. EndNode and EndBond terminate the development (further topology manipulation) at junction modifiable sites and bond modifiable sites, respectively; ERC represents a real number (Ephemeral Random Constant) that can be changed by Gaussian mutation. In addition, the number and type of elements attached to the inserted junctions are controlled by three "flag" bits. A flag mutation operator is used to evolve these flag bits, each representing the presence or absence of the corresponding C/I/R component. Compared with the basic function set approach, this hybrid approach can effectively avoid adding many bare (and redundant) junctions. At the same time, Add_C/I/R (illustrated in Fig. 14.8) still provides the flexibility needed for broad topology search. For any of the three C/I/R components attached to each junction, there is a corresponding parameter to represent the component's value, which is evolved by a Gaussian mutation operator in the modified genetic programming system used here. This is different from our previous work in which the "classical" numeric subtree approach was used to evolve parameters of components. Figure 14.9 shows a GP tree that develops an embryo bond graph into a complete bond graph solution. Our comparison experiments [5] showed that this function set was more effective on both an eigenvalue and an analog filter test problem.

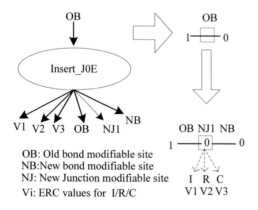

OB: Old bond modifiable site
NB: New bond modifiable site
NJ: New Junction modifiable site
Vi: ERC values for I/R/C

Fig. 14.7. The Insert_J0E GP function inserts a new junction into a bond along with a certain number of attached components

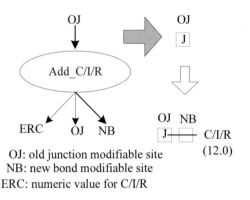

OJ: old junction modifiable site
NB: new bond modifiable site
ERC: numeric value for C/I/R

Fig. 14.8. The Add_C/I/R GP function adds a C/I/R component to a junction

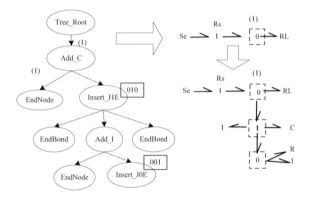

Fig. 14.9. An example of a GP tree, composed of topology operators applied to an embryo, generating a bond graph after depth-first execution (numeric ERC nodes are omitted). Note that the 010 and 001 are the flag bit sets showing the presence or absence of attached C/I/R components

The fitness function for candidate design evaluation is defined as:

$$f_{norm} = \frac{\text{NORM}}{\text{NORM} + f_{raw}} \tag{14.2}$$

where f_{raw} is the frequency response as defined in (14.1). NORM is a normalization term aimed at adjusting the f_{norm} into the range of [0,1]. This process transforms the minimization of deviation from target frequency response into a maximization of fitness process as used in our GP system. Since tournament selection is used as the selection operator, the normalization term can be an arbitrary positive number. Here NORM is set to 10, and the fitness range is [0,1].

According to (14.1), we calculate the frequency response $X_1(s)/F(s)$ where X_1 is the displacement of the primary mass. However, we can only extract from a bond graph the source effort signal $\dot{X}(s)$. We use the following procedure to get the f_{raw}:

1. calculate A, B, C, D matrices from a given bond graph;
2. convert A, B, C, D into transfer function TF_{raw};
3. $TF_{norm} = TF_{raw} \times 1/s$ is equal to $X_1(s)/F(s)$;
4. convert TF_{norm} back to A', B', C', D' matrices and simulate its frequency response with MATLAB.

Design Experiments

Compared to the evolutionary synthesis of electrical circuits, a mechanical vibration absorber usually has a much smaller number of components. So the topological and parameter search can be greatly decreased. We used a bond graph simulation engine and developed the GPBG platform based on the Open beagle GP framework by Christian Gagne [4]. Most of the experiments are finished in less than an hour. Some of them take only a few minutes. Here we set the maximum number of components to be 7. Other standard GP parameters are summarized in Table 14.2.

Table 14.2. Experimental parameters for vibration absorber synthesis

Parameter	Value	Parameter	Value
No. of subpopulations	5	Tournament Selection Size	7
Sub population size	400	pCrossover	0.4
Maximum evaluation	100000	pMutationStandard	0.05
Migration Interval	5 gen	MutateMaxDepth	3
Migration Size	40	pMutationParameter	0.3
Init.MaxDepth	3	pSwitchBit	0.2
Init.MinDepth	2	pSwapSubtree	0.05
StronglyTyped	True	TreeMaxDepth	7

Figure 14.10 shows an evolved single frequency vibration absorber and its frequency response compared to the responses of the primary structure without any absorber and with a standard passive absorber invented in 1912. It is very interesting that the frequency response of the evolved vibration absorber has a very deep spike at the excitation frequency to minimize the frequency response at that single frequency. If the excitation frequency is relatively constant with little shifting, our evolved absorber will achieve better performance at that specific frequency. Our evolved vibration absorber utilizes

one damper (I) and several springs (C), sharing similarity to the original absorber invention of 1912. We found the GPBG framework is very flexible for vibration absorber synthesis. In addition to this single-frequency vibration absorber, we have also synthesized novel dual frequency and band-pass passive vibrators, which will be reported elsewhere.

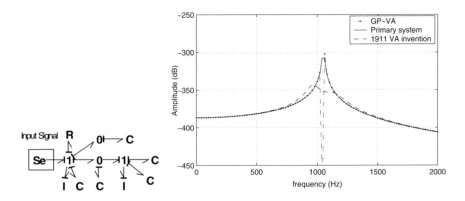

Fig. 14.10. The evolved single-frequency vibration absorber and its performance compared to a standard vibration absorber

14.4.2 Synthesis of MEMS Filters: Knowledge Incorporation in GPBG

Due to the complexity of real-world engineering design problems, hands-free automated synthesis system can rarely provide entirely satisfactory solutions. It may be that the computational demand is too high for current inexpensive computing hardware or the design solutions are hard to implement using physical components or they violate some design constraints. It is thus strongly desirable to incorporate expert/human knowledge into the evolutionary synthesis process to create some kind of interactive evolutionary synthesis tools that help human designers make better decisions and explore under-explored design spaces.

Problem Description

In this section, we try to synthesize MEMS (micro-electro-mechanical systems) band-pass filters to examine how domain knowledge can be conveniently included into the evolutionary synthesis process. Due to its multi-domain and intrinsically three-dimensional nature, MEMS design and analysis is very complicated. However, the multi-domain property of MEMS models makes them suitable for representation as bond graphs. In this MEMS filter synthesis problem, the goal is to automatically generate bond graph models of MEMS filters to meet particular design specifications.

One distinct characteristic of MEMS filter design with vibration absorber synthesis and analog filter design is that, due to manufacturing constraints, MEMS filters are usually composed of a restricted and specialized set of components. Two popular topologies for micromechanical band-pass filters, built using surface micromachining, are topologically composed of a series or concatenation of Resonant Units (RUs) and Bridging Units (BUs), or RUs and Coupling Units (CUs). Figure 14.11 illustrates the layouts and corresponding bond graph representations of two such filter topologies, labeled I and II.

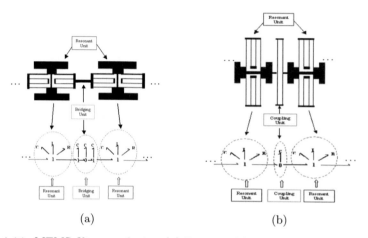

(a)   (b)

Fig. 14.11. MEMS filter topologies. **(a)** Layout of filter topology I **(b)** Layout of filter topology II

From this figure, it is clear that here the building blocks of MEMS filter design are high-level modules that are tailored to a specific fabrication process. Design solutions composed of arbitrary topologies of basic primitive components will be difficult to manufacture.

GPBG Configuration

For this band-pass filter design problem, we use the bond graph model shown in Fig. 14.12 as the design embryo of the GPBG framework. The accompanying block diagram indicates that this implementation will accept an electrical voltage signal as input and produce a voltage signal as output, but the interior components will be implemented as micromechanical elements.

To incorporate the domain knowledge of MEMS filters, we propose the realizable GP function set concept, which manipulates topologies composed

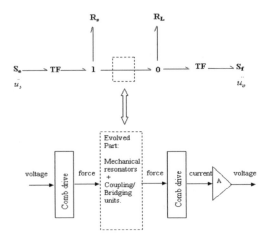

Fig. 14.12. MEMS filter design embryo in bond graph and block diagram forms

Table 14.3. Realizable function set for MEMS filter synthesis

Function	Description
insert_RU	Insert a Resonator Unit
insert_CU	Insert a Coupling Unit
insert_BU	Insert a Bridging Unit
add_RU	Add a Resonator Unit
insert_J01	Insert a 0-1-junction compound with elements
insert_CIR	Insert a special CIR compound
insert_CR	Insert a special CR compound
Add_J	Add a junction compound
+	Sum two ERCs
−	Subtract two ERCs
endn	End terminal for add functions
endb	End terminal for insert functions
endr	End terminal for replace functions
erc	Ephemeral Random Constant (ERC)

of manufacturable modules by adding, removing, or replacing these modules as units. We use the following GP function set in Table 14.3 as our modular GP function set, which can impose domain knowledge of design constraints on the final synthesis results for guaranteed manufacturability of the design under current or anticipated manufacturing technology. By using only operators in a realizable function set, we seek to guarantee that the evolved design is physically realizable and has the potential to be manufactured. This concept of realizability may include stringent fabrication constraints to be fulfilled in some specific application domain.

Examples of operators, namely insert_CU and insert_RU, are illustrated in
Fig. 14.13. Examples of basic operators are available in our earlier work [2].
Figure 14.13(a) explains how the insert_BU function works. A Bridging Unit
(BU) is a subsystem composed of three capacitors with the same parameters,
attached together with a 0-junction in the center and 1-junctions at the left
and right ends. After execution of the insert_BU function, an additional modifiable site (2) appears at the rightmost newly created bond. As illustrated
in Fig. 14.13(b), a resonator unit (RU), composed of one I, R, and C component all attached to a 1-junction, is inserted in an original bond with a
modifiable site through the insert_RU function. After the insert_RU function
is executed, a new RU is created and one additional modifiable site, namely
bond (3), appears in the resulting phenotype bond graph, along with the
original modifiable site bond (1). The newly-added 1-junction also has an additional modifiable site (2). As components C, I, and R all have parameters to
be evolved, the insert_RU function has three corresponding ERC-typed sites,
(4), (5), and (6), for numerical evolution of parameters.

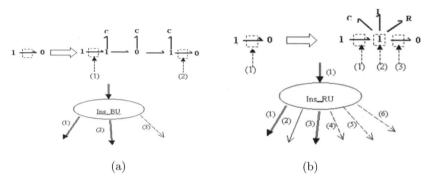

(a)   (b)

Fig. 14.13. Two realizable GP functions for MEMS filter design. **(a)** Insert_BU
(b) Insert_RU

Filter performance is measured by the magnitude ratio of the frequency response for the voltage across RL to the input voltage u_s. The desired frequency
response has unity magnitude ratio in the pass band (316–1000 Hz), and zero
magnitude ratio outside the pass band. The frequency range of interest is
0.1Hz–100KHz. To evaluate fitness within the frequency range of interest, 100
points are sampled at equal intervals on a log scale. The magnitudes of the
frequency response at the sample points are compared with their desired magnitudes. The differences are computed and the sum of all squared differences is
taken as raw fitness. The normalized fitness is calculated according to (14.2).

Design Experiments

We used a strongly-typed version of lilgp [17] to generate bond graph models. The major GP parameters were as shown in Table 14.4.

Table 14.4. Experimental parameters for vibration absorber synthesis

Parameter	Value
Population size	500 in each of 13 subpopulations
Initial population	half_and_half
Initial depth	4–6
Max depth	50
Max_nodes	5000
Selection	Tournament (size=7)
Crossover	0.9
Mutation	0.3

Results of the experiments show the capability of the GPBG approach for finding realizable designs for micro-electro-mechanical filters. Figure 14.14(a) shows the fitness improvement curve of a typical genetic programming run, in which K is defined as the number of resonator units used in the MEM filter design. It is shown that as evolution progresses, the fitness value undergoes continual improvement. It is also observed that as fitness improves, the value of K also becomes larger. This observation is supported by the reasoning that a higher-order system with more resonator units has the potential of having better system performance than its lower-order counterpart. The system frequency responses at generations 27, 52, 117 and 183 are shown in Fig. 14.14(b), with increased K value and performance evaluation.

The use of realizable function sets can be made less rigid to assist the designer in exploring more novel topologies for MEMS filter design. The designer may use a function set in which not all elements are guaranteed to be strictly realizable. Instead, a different set of design knowledge is incorporated in the evolutionary process – i.e., a semi-realizable function set may be used to relax the topological constraints with the purpose of finding new topologies not discovered before but still usually realizable after careful interpretation. Figure 14.15 gives an example of a novel topology evolved for a MEM filter design by incorporating a special CIR component into the semi-realizable function set.

The work presented in this section analyzes the promise of MEMS design synthesis at the system level using the GPBG approach. The basic GP function set imposes very few constraints on design, while the realizable function set used for MEMS design features relatively few but structurally more complex devices in the component library. The use of a realizable function set guarantees that the phenotypes generated can be built using existing or anticipated manufacturing technology. Large-scale component reuse and assembly

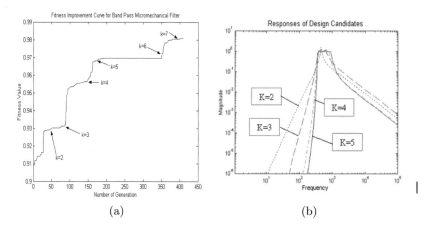

(a)   (b)

Fig. 14.14. (a) Fitness progress over generations (b) Frequency responses of design candidates at different generations

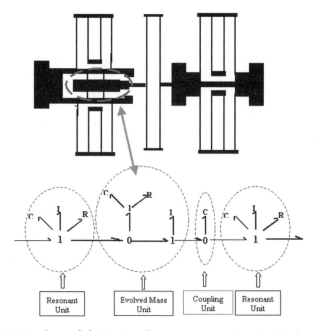

Fig. 14.15. A novel design topology using a semi-realizable function set

of MEMS is expected to show more applicability and promise of this method for MEMS design.

14.4.3 Synthesis of Suspension System Controllers

Problem Description

Suspension systems are important subsystems of most wheeled vehicles. From a system design point of view, there are two main types of disturbances acting on a vehicle, namely road and load disturbances. Road disturbances have the characteristics of large magnitude in low-frequency disturbances (such as hills) and small magnitude in high-frequency disturbances (such as road roughness). Load disturbances include the variations of loads induced by accelerating, braking and cornering. A good suspension design is concerned with disturbance rejection from these disturbances to the outputs (e.g. vertical position of vehicle mass), the basis for evaluating performance. In general, a suspension system needs to be "soft" to insulate against road disturbances and "hard" to insulate against load disturbances.

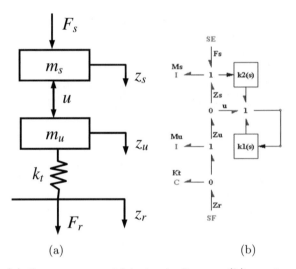

(a)   (b)

Fig. 14.16. (a) Quarter-car model in iconic diagram **(b)** quarter-car suspension control with both road and load disturbances

A quarter-car iconic model is illustrated in Fig. 14.16(a). The sprung mass m_s (kg), consists of the main vehicle body supported by the suspension. The unsprung mass m_u (kg), consists of hub, wheel and tire. The tire is modeled as a spring with stiffness k_t (N/m). z_s, z_u, and z_r are the vertical positions of the sprung mass, the unsprung mass and the road disturbance input, respectively. Force Fs is the load force disturbance input. Force u represents any possible suspension force.

From the point of view of a multi-port mechatronics network, the quarter-car suspension system can be viewed externally as a two-port network [21], with its corresponding mixed immittance matrix G defined as:

$$\begin{bmatrix} F_r \\ \dot{z}_s \end{bmatrix} = \begin{bmatrix} G_{11}(s) & G_{12}(s) \\ G_{21}(s) & G_{22}(s) \end{bmatrix} \begin{bmatrix} \dot{z}_r \\ F_s \end{bmatrix} \tag{14.3}$$

where F_r represents the applied force from the tire to the road. The matrix G can be obtained from the following equations of system motion, together with specified suspension force u:

$$m_s \ddot{z}_s = -u + F_s \tag{14.4}$$

$$m_u \ddot{z}_u = u + k_t(z_r - z_u). \tag{14.5}$$

When load disturbance is also considered, the suspension system needs to be stiff to loads acting on the sprung mass. This requires in (14.3), $G_{11}(s)$ and $G_{21}(s)$ be set "soft" for road disturbance rejection while $G_{12}(s)$ and $G_{22}(s)$ be set "hard" for load disturbance rejection. For such design requirements, the matrix G fails to be positive-real, which implies active energy input is necessary for such suspension implementation [21].

There is one degree-of-freedom available for the response to each of the road and load disturbances. They can be determined independently if two suitable measurements are available for feedback (e.g. suspension deflection and sprung mass velocity). The suspension design with two measurements is shown in Fig. 14.16(b), with the control law taken to be: $u = \begin{bmatrix} k_1(s) & k_2(s) \end{bmatrix} \begin{bmatrix} z_s - z_u \\ s z_s \end{bmatrix}$, where $k_1(s)$ is collocated control, while $k_2(s)$ is non-collocated control.

In order to synthesize controller $k_1(s)$ and $k_2(s)$, desired performance requirements for road and load disturbance rejection are specified. The desired frequency response for road disturbance $H_1(s)$ is specified in (14.6). The desired load disturbance frequency response $H_2(s)$ is the frequency response specification for $G_{22}(s)$ in the immittance matrix in (14.7). It is obtained by choosing certain suitable parameters in another double skyhook configuration: $u = k_s(z_s - z_u) + c_1 \dot{z}_s - c_2 \dot{z}_u$, with a hard damper and spring configuration with $k_s = 150000$ N/m, $c_1 = 12000$ Ns/m, $c_2 = 6000$ Ns/m. The desired $H_2(s)$ is calculated as:

$$H_1(s) = \frac{\dot{z}_s}{\dot{z}_r}$$
$$= \frac{c_2 k_t s + k_s k_t}{m_s m_u s^4 + (c_1 m_u + c_2 m_s)s^3 + (k_s m_u + k_s m_s + k_t m_s)s^2 + c_1 k_t s + k_s k_t} \tag{14.6}$$

$$H_2(s) = \frac{\dot{z}_s}{F_s}$$
$$= \frac{(m_u s^2 + c_2 s + k_t + k_s)s}{m_s m_u s^4 + (c_1 m_u + c_2 m_s)s^3 + (k_s m_u + k_s m_s + k_t m_s)s^2 + c_1 k_t s + k_s k_t}. \tag{14.7}$$

GPBG Configuration

The design space of the controllers is bond graphs composed of C/I/R components and 1/0 junctions. We use the direct encoding formulation of the GPBG framework as specified in Sect. 14.3.3. There need be no embryo or modifiable site. The bond graphs are directly encoded by the GP trees. We use the following GP function set in Table 14.1. In this function set, the J0 and J1 functions have two inputs, meaning that in this encoding, the 1/0 junctions in the represented bond graphs can only have three ports: two input ports and one output port.

The fitness of a GP individual is evaluated by how accurately it approximates the desired frequency domain specification, minimizing the value of the expression $\|dTF(j\omega) - tTF(j\omega)\|_2$, where $dTF(j\omega)$ is the desired frequency response as specified by $H_1(s)$ and $H_2(s)$, and $tTF(j\omega)$ is the theoretical frequency response of an evolved individual bond graph structure to be evaluated.

Design Experiments

Taking the desired road and load disturbance rejection responses $H_1(s)$ and $H_2(s)$ as evaluation criteria, we used the settings listed in Table 14.5 for the experiments.

The best run of genetic programming using the basic function set in Table 14.1 produced the results shown in Fig. 14.17 for $k_1(s)$, and Fig. 14.18 for $k_2(s)$.

$$k_1(s) = \frac{2128s^3 + 46680s^2 + 1137000s + 4792000}{s^2 + 16.08s + 32.45}$$
$$= \frac{2128(s + 5.011)(s^2 + 16.93s + 449.4)}{(s + 2.366)(s + 13.71)}$$

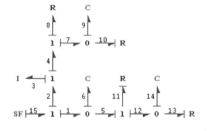

Fig. 14.17. Controller structure in bond graph form for $k_1(s)$

Table 14.5. Experiment settings

Objective:	Design a suspension system composed of two controllers.
Test fixture and embryo:	Two-input, two-output initial suspension system with a sprung mass, an unsprung mass, and a spring.
Program architecture:	Two result-producing GP species, k1 and k2, sharing the following attributes.
Function set:	For construction-continuing subtrees: $F_{ccs-rpb-initial}$ = {f0, f1, R, C, I}. For arithmetic-performing subtrees: F_{aps} = {ADD, SUB}.
Terminal set:	For arithmetic-performing subtrees: T_{aps} = {E}.
Fitness Cases:	41 frequency values in an interval of four decades of frequency values between 0.1 Hz and 1000 Hz.
Raw Fitness:	Taking the desired road and load disturbance rejection responses as evaluation criteria, the raw fitness of a combined solution including individuals from both species is calculated as: $Fitness_{raw} = \sqrt{\dfrac{\sum\limits_{i=1}^{n}(err_1+err_2)^2}{n}}$ n is the number of logarithmically sampled frequency points; err_1 and err_2 are the absolute differences of magnitude between the evolved and the desired road and load disturbance rejection frequency responses, respectively. $err_1 = \left\| G_{12}(j\omega) - G_{12}^s(j\omega) \right\|_2;$ $err_2 = \left\| G_{11}(j\omega) - G_{11}^h(j\omega) \right\|_2$
Normalized Fitness:	$Fitness_{norm} = \dfrac{1.0}{Fitness_{raw}+1.0}$
Parameters:	Each species: 10 subpopulations of 100 individuals; Migration interval: 10 generations; Migration size: 2 individuals Crossover rate: 0.85; Mutation rate: 0.15; initializing tree depth: 2–4; maximum tree depth: 10–17
Result designation:	Best-so-far individual from max fitness species and matching individual from another species.
Termination:	When either species reaches max fitness value 0.99.

The degree of a system can be determined by counting independent storage elements present in the bond graph. The controllers obtained here are of lower order than the controllers obtained using conventional approaches based on the standard controller design theory [20]. The limitation of this approach is that the design space are usually constrained by the limited number of component configurations used in the suspension systems. On the other hand, the GPBG approach can exploit the open-ended topology search capability to evolve novel structure of the suspension systems as well as its controller.

$$k_2(s) = \frac{10320s^3 + 453300s^2 + 40260000s + 437000000}{s^3 + 172s^2 + 5799s + 15890}$$
$$= \frac{10320(s + 12.04)(s^2 + 31.89s + 3517)}{(s + 3)(s + 41.5)(s + 127.5)}$$

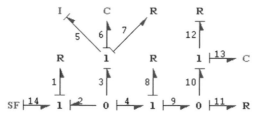

Fig. 14.18. Controller structure in bond graph form for $k_2(s)$

The simultaneous design of structures and controllers distinguishes the GPBG approach from conventional controller design methods.

Figure 14.19 shows the simulation results as MATLAB Bode diagrams comparing desired responses (solid lines) with actual responses (dashed lines) realized by active suspension control evolved from evolutionary computation. The left-hand side shows the road disturbance rejection responses, and the right-hand side shows the load disturbance rejection responses. It demonstrates that the actual responses approximate the desired responses very well.

In summary, using the GPBG framework, we have evolved an active suspension system that has the ability to store, dissipate and to introduce energy to the system, with extra flexibility to achieve improved design performance. It should be noted that in this work, we have assumed that the sensor and the actuator have perfect dynamics. The suspension design will be considerably modified if such assumptions do not hold well.

14.4.4 Automatic Generation of Robust Designs

Although the topic of design for robustness cannot be addressed in detail in this chapter, application of GPBG for robust design has already been demonstrated. In [16], three strategies for using GPBG to synthesize robust passive analog filters were explored. The broadest conclusion was that filters of high robustness to variation in the values of their parameters could be evolved under GPBG, by introducing appropriate stochasticity during evolution of the topology of the filters. It did not require many more filter evaluations to evolve robust structures than to evolve those of similar nominal performance without stochasticity. It was also shown that robustness of designs with component values chosen from small, discrete sets could be improved by using only

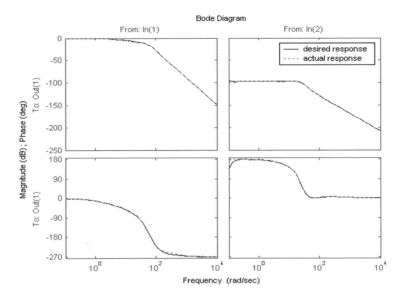

Fig. 14.19. Desired and actual responses of evolved suspension design. Left (road disturbance response). Right (load disturbance response)

the "catalog" values during the evolutionary process, but adding stochastic variation about their nominal values.

14.5 Conclusions and Future Work

This chapter has applied genetic programming and bond-graph simulation – the GPBG approach – to the design synthesis problem in engineering. Three real-world design problems have been examined in detail, including mechanical vibration absorbers, MEMS filters, and suspension system controllers. Experimental results illustrate that the GPBG framework is an effective tool for exploring design space and evolving innovative designs that differ from those produced by human designers.

References

1. Ando, S., Iba, H.: Linear genome methodology for analog circuit design. Tech. rep., Information and Communication Department, School of Engineering, University of Tokyo (2000)
2. Fan, Z., Hu, J., Seo, K., Goodman, E., Rosenberg, R., Zhang, B.: Bond graph representation and GP for automated analog filter design. In: E. Goodman (ed.)

2001 Genetic and Evolutionary Computation Conference Late Breaking Papers, pp. 81–86. San Francisco, California, USA (2001)

3. Fan, Z., Seo, K., Hu, J., Rosenberg, R., Goodman, E.: System-level synthesis of MEMS via genetic programming and bond graphs. In: E. Cantú-Paz et al. (ed.) Genetic and Evolutionary Computation (GECCO-2003), *LNCS*, vol. 2724, pp. 2058–2071. Springer-Verlag, Chicago (2003)

4. Gagné, C., Parizeau, M.: Open BEAGLE: a new versatile C++ framework for evolutionary computation. In: E. Cantú-Paz (ed.) Late Breaking Papers at the Genetic and Evolutionary Computation Conference (GECCO-2002), pp. 161–168. AAAI, New York, NY (2002)

5. Hu, J., Goodman, E.: Robust and efficient genetic algorithms with hierarchical niching and sustainable evolutionary computation model. In: Proceedings of the 2004 Genetic and Evolutionary Computing Conference. Springer, Chicago (2004)

6. Hu, J., Goodman, E., Rosenberg, R.: Topological search in automated mechatronic system synthesis using bond graphs and genetic programming. In: Proc. of American Control Conference ACC 2004. Boston (2004)

7. Jalili, N.: A comparative study and analysis of semi-active vibration-control systems. Journal of Vibration and Acoustics **124**, 593 (2002)

8. Karnopp, D., Margolis, D., Rosenberg, R.: System Dynamics: Modeling and Simulation of Mechatronic Systems, 3rd edn. John Wiley & Sons, Inc., New York (2000)

9. Karnopp, D., Margolis, D., Rosenberg, R.: System Dynamics: Modeling and Simulation of Mechatronic Systems. John Wiley & Sons, Inc., New York (2000)

10. Koza, J., Andre, D., Bennett III, F., Keane, M.: Genetic Programming 3: Darwinian Invention and Problem Solving. Morgan Kaufmann (1999)

11. Koza, J., Bennett III, F., Andre, D., Keane, M., Dunlap, F.: Automated synthesis of analog electrical circuits by means of genetic programming. IEEE Transactions on Evolutionary Computation **1**(2), 109–128 (1997)

12. Koza, J., Keane, M., Streeter, M., Mydlowec, W., Yu, J., Lanza, G.: Genetic Programming IV: Routine Human-Competitive Machine Intelligence. Kluwer Academic Publishers (2003)

13. Koza, J., Keane, M., Yu, J., Bennett III, F., Mydlowec, W.: Automatic creation of human-competitive programs and controllers by means of genetic programming. Genetic Programming and Evolvable Machines **1**(1/2), 121–164 (2000)

14. Lohn, J., Colombano, S.: A circuit representation technique for automated circuit design. IEEE Transactions on Evolutionary Computation **3**(3), 205–219 (1999)

15. Paynter, H.: An epistemic prehistory of bond graphs. In: P. Breedveld, G. Dauphin-Tanguy (eds.) Bond Graphs for Engineers. Elsevier Science Publishers, Amsterdam (1991)

16. Peng, X., Goodman, E., Rosenberg, R.: Comparison of robustness of three filter design strategies using genetic programming and bond graphs. In: R. Riolo, T. Soule, B. Worzel (eds.) Genetic Programming Theory and Practice IV. Springer (2006)

17. Punch, W., Zongker, D.: lilgp – a C system for genetic programming (1995). URL: http://garage.cse.msu.edu/software/lil-gp

18. Seo, K., Fan, Z., Hu, J., Goodman, E., Rosenberg, R.: Dense and switched modular primitives for bond graph model design. In: E. Cantú-Paz et al. (ed.)

Genetic and Evolutionary Computation (GECCO-2003), *LNCS*, vol. 2724, pp. 1764–1775. Springer-Verlag, Chicago (2003)

19. Seo, K., Fan, Z., Hu, J., Goodman, E., Rosenberg, R.: Toward an automated design method for multi-domain dynamic systems using bond graph and genetic programming. Mechatronics **13**(8-9), 851–885 (2003)

20. Smith, M.: Achievable dynamic response for automotive active suspension. Vehicle System Dynamics **24**, 1–33 (1995)

21. Smith, M., Walker, G.: Performance limitations and constraints for active and passive suspensions: a mechanical multi-port approach. Vehicle System Dynamics **33**, 137–168 (2000)

22. Wang, J., Terpenny, J.: Integrated active and passive mechatronic system design using bond graphs and genetic programming. In: B. Rylander (ed.) Genetic and Evolutionary Computation Conference Late Breaking Papers, pp. 322–329. Chicago, USA (2003)

Index

Natural Computing Series

A.A. Freitas: **Data Mining and Knowledge Discovery with Evolutionary Algorithms.**
XIV, 264 pages, 74 figs., 10 tables. 2002

H.-P. Schwefel, I. Wegener, K. Weinert (Eds.): **Advances in Computational Intelligence.**
Theory and Practice. VIII, 325 pages. 2003

A. Ghosh, S. Tsutsui (Eds.): **Advances in Evolutionary Computing. Theory and**
Applications. XVI, 1006 pages. 2003

L.F. Landweber, E. Winfree (Eds.): **Evolution as Computation.** DIMACS Workshop,
Princeton, January 1999. XV, 332 pages. 2002

M. Hirvensalo: **Quantum Computing.** 2nd ed., XI, 214 pages. 2004 (first edition
published in the series)

A.E. Eiben, J.E. Smith: **Introduction to Evolutionary Computing.** XV, 299 pages. 2003

A. Ehrenfeucht, T. Harju, I. Petre, D.M. Prescott, G. Rozenberg: **Computation in Living**
Cells. Gene Assembly in Ciliates. XIV, 202 pages. 2004

L. Sekanina: **Evolvable Components. From Theory to Hardware Implementations.**
XVI, 194 pages. 2004

G. Ciobanu, G. Rozenberg (Eds.): **Modelling in Molecular Biology.** X, 310 pages. 2004

R.W. Morrison: **Designing Evolutionary Algorithms for Dynamic Environments.**
XII, 148 pages, 78 figs. 2004

R. Paton[†], H. Bolouri, M. Holcombe, J.H. Parish, R. Tateson (Eds.): **Computation in Cells**
and Tissues. Perspectives and Tools of Thought. XIV, 358 pages, 134 figs. 2004

M. Amos: **Theoretical and Experimental DNA Computation.** XIV, 170 pages, 78 figs. 2005

M. Tomassini: **Spatially Structured Evolutionary Algorithms.** XIV, 192 pages, 91 figs.,
21 tables. 2005

G. Ciobanu, G. Păun, M.J. Pérez-Jiménez (Eds.): **Applications of Membrane Computing.**
X, 441 pages, 99 figs., 24 tables. 2006

K.V. Price, R.M. Storn, J.A. Lampinen: **Differential Evolution.** XX, 538 pages,
292 figs., 48 tables and CD-ROM. 2006

J. Chen, N. Jonoska, G. Rozenberg: **Nanotechnology: Science and Computation.**
XII, 385 pages, 126 figs., 10 tables. 2006

A. Brabazon, M. O'Neill: **Biologically Inspired Algorithms for Financial Modelling.**
XVI, 275 pages, 92 figs., 39 tables. 2006

T. Bartz-Beielstein: **Experimental Research in Evolutionary Computation.**
XIV, 214 pages, 66 figs., 36 tables. 2006

S. Bandyopadhyay, S.K. Pal: **Classification and Learning Using Genetic Algorithms.**
XVI, 314 pages, 87 figs., 43 tables. 2007

H.-J. Böckenhauer, D. Bongartz: **Algorithmic Aspects of Bioinformatics.**
X, 396 pages, 118 figs., 9 tables. 2007

P. Siarry, Z. Michalewicz (Eds.): **Advances in Metaheuristics for Hard Optimization.**
XVI, 481 pages, 66 figs., 83 tables. 2008

J. Knowles, D. Corne, K. Deb (Eds.): **Multiobjective Problem Solving from Nature.**
From Concepts to Applications. XVI, 412 pages, 178 figs., 53 tables. 2008

P.F. Hingston, L.C. Barone, Z. Michalewicz (Eds.). **Design by Evolution.**
XII, 362 pages, 143 figs., 20 tables. 2008